Digital Forensics and Cyber Investigation

Digital Forensics and Cyber Investigation

FIRST EDITION

Kyung-Shick Choi, Sinchul Back, and
Marlon Mike Toro-Alvarez

cognella®

SAN DIEGO

Bassim Hamadeh, CEO and Publisher
John Remington, Managing Executive Editor
Gem Rabanera, Senior Project Editor
Abbey Hastings, Production Editor
Asfa Arshi, Graphic Design Assistant
Kylie Bartolome, Licensing Associate
Natalie Piccotti, Director of Marketing
Kassie Graves, Vice President of Editorial
Jamie Giganti, Director of Academic Publishing

3970 Sorrento Valley Blvd., Ste. 500, San Diego, CA 92121

Brief Contents

Contents

Kyung-Shick Choi

Computer Ethics in Cybercrime Investigation and Cybersecurity Practices

Claire Seungeun Lee and Kyung-Shick Choi

Introduction

Nowadays, almost everyone utilizes a computer and/or a smartphone or the internet on a daily basis. The internet and cyberspace are very much connected to our lives and, in fact, are a big part of our lives. Within their digitalized life, one might use Alexa or Google Home, which have speech recognition functions, and a smartwatch that has various features for tracking one's health behaviors.

In early January of 2018, a fitness app known as Strava was hacked by a college student named Nathan Ruser. Ruser developed an interest in the Syrian conflict and had a fascination with maps. This combination ultimately led him to determine the potential security risks posed by the fitness app Strava. He noticed lit up in areas in regions of Syria and the Sahara, often not occupied by civilians, that indicated the presence of security forces working out near military bases and other sensitive locations. His findings went viral over the weekend, exposing remote areas and conflict zones where military personnel and government officials are seemingly present. This sensitive information could be used to track the running routes and identify locations where personnel is deployed (O'Brien, 2018, paras. 1–3).

Because of his technical abilities, Ruser was able to get ahold of sensitive information about remote areas and conflict zones where military personnel and government officials are seemingly present. Likewise, our information and location are also susceptible to being leaked. This epitomizes the importance of the user's ethics as well as professionals' ethics. Once individuals attain the necessary skillsets to gather sensitive information, they may begin to feel the urge to tap into their skills and do further things. However, ethically speaking, such actions ought to be reconsidered.

Specifically, ethics concerning computer science and computers, or computer ethics, are important in a highly digitized society. This chapter offers pertinent hands-on and recent case studies that explain key concepts and issues within computer ethics and digital forensics. A case involving Silk Road and serious computer ethics violations will be discussed through the lens of computer ethics vis-à-vis digital forensics.

This chapter is organized as follows: The first section, which aims to understand various cybercrime environment issues in the course of cybercrime investigation, will use a case study to reach our basic understanding of why these issues are important and focus on the definitions and key elements of computer ethics. In the second section, in an effort to understand the ability to apply digital forensic knowledge to cybercrime cases, we will be focusing on computer ethics concerning other stakeholders—namely, individuals, governments, and investigators. We will use some of the latest case studies, Silk Road in particular, and will summarize and provide implications.

Computer Ethics
Cybercriminology and Cybercrime

Criminology, the study of deviance, offenses, and crime, has developed to encompass the themes of crime characteristics, causes, consequences, the mechanisms of victims and offenders, offenses, and the criminal justice system. Cybercriminology, branching from the field of criminology, examines new types of crime in cyberspace, such as "computer crime (affecting the physical computer) and cybercrime (operating through a cyber-platform)," with new technological and industrial changes including the fourth industrial revolution. In examining these new types of crimes, we need new definitions, new ways to investigate such criminal and deviant activities, and new ways to create, understand, and interpret criminal profiles (Choi et al., 2020).

Cybercriminology was founded upon a method of interdisciplinary examination that includes recognizing the reasons for cybercrime—citing studies in criminology, psychology, sociology, computer science, and cybersecurity—to comprehend the role of cybercrime within the criminal justice system. In particular, cybercriminology investigates the causes and outcomes of crime and deviance on the internet, as well as their legitimate issues, ethics, countermeasures, and control systems. This is a dynamic, applied field that lands resources directly into legal practice, particularly in the way of developing laws to affect criminal justice–related policy and its implementation (Choi & Lee, 2018, p. 1).

Various versions of cybercrime exist. One convention of classifying a cybercrime is differentiating between cyber-dependent crime and cyber-enabled crime. On the one hand, cyber-dependent crimes (or 'pure' cybercrimes) are offenses that can only be committed using a computer, computer networks, or other forms of information communications technology (ICT) (McGuire & Dowling, 2013, p. 75). These acts include the spread of viruses or other malware, hacking, and distributed denial-of-service (DDoS) attacks (McGuire & Dowling, 2013, p. 75).

On the other hand, cyber-enabled crimes fall broadly into two main categories: illicit intrusions into computer networks (for example, hacking) and the disruption or downgrading of computer functionality and network space (for example, viruses and DDoS attacks) (McGuire & Dowling, 2013, p. 75). Existing survey evidence reveals that approximately 25 percent of respondents in the 2012 Commercial Victimisation Survey (CVS) suspected that their most recent online 'crime' incidents were committed by an organized group of criminals rather than someone working alone (McGuire & Dowling, 2013, p. 26). Cyber-dependent and cyber-enabled crimes are not, however, just about technical skills; they rely heavily on the behavior of the intended victim. Social engineering tactics are key to deceiving computer users about the purpose of a file or an email they have received (McGuire & Dowling, 2013, p. 26). Case-study research has attempted to categorize particular types of cyber-offenders, producing seven archetypes enumerated here: newbies (who have limited skills and experience and are reliant on tools developed by others, also known as 'script-kiddies'); cyberpunks (who deliberately attack and vandalize); internals (who are insiders with privileged access and who are often disgruntled employees); coders (who have high skill levels); old guard hackers (who have no criminal intent and high skill levels, so would most likely equate to white-hat hackers); professional criminals; and cyberterrorists (Choi et al., 2020; McGuire & Dowling, 2013, p. 24).

Definitions and Key Elements of Computer Ethics

Definitions

The increasing number of cybercrimes highlights the importance of rethinking the ethical dimensions of cybercrimes and cybercrime investigations. While computer ethics has grown due to recent attention—in particular, within the fields of computer science, cybersecurity, and cybercrime—computer ethics, as a field of study, has long existed. Computer ethics, in terms of a concept and a practice, is increasingly gaining attention for its application to the fields of cybercrime and cybersecurity. There is an urgent need to understand this and incorporate it into digital investigation training and education.

Prior to offering a definition and key elements of computer ethics, it is worthwhile to establish a base in order to discuss ethics because computer ethics is ultimately a branch of ethics. Ethics is a systematic study of morality and a branch of philosophy. It involves freedom, privacy, equality, duty, obligations, choice, justification of judgments, rights, claims, and a need to understand (1) the code or set of principles, standards, or rules that guide the moral actions of an individual within a particular social framework and (2) moral judgment and decision.

Computer ethics, a field of study that considers ethical approaches to computers, is an interdisciplinary field that connects computer science with philosophy. A definition of computer ethics provided by James Moor, one of the pioneers in the field of computer ethics, is still very useful to note. Moor (1985) defines computer ethics as "The analysis of the nature and social impact of computer technology and the corresponding formulation and justification of policies for the ethical use of such technology." Essentially, computer ethics is the systematic study of the ethical and social impact of computers in the information society, involving the acquisition, distribution, storage, processing, and dissemination of digital data and how individuals and groups interact with the systems and data.

The subject of computer ethics is not only important for users to be aware of in their daily lives, but it is also important for investigations. We are currently seeing a gap between the knowledge of computer ethics within the field and the social problems and phenomena. More importantly, from an educational perspective, computer ethics is, as yet, not commonly taught to college students studying criminal justice and criminology. However, the study of criminal justice ethics, as part of the study of ethics in law enforcement, is well-established in criminology and criminal justice, which makes the study of computer ethics at the college level a beneficial addition to the field. Nevertheless, computer ethics is a relatively new subject to not only criminology scholars but also to law enforcement. The advent and development of technology and the increasing use of technology and computer techniques offer us the perfect rationale to understand why ethics regarding computers and technology are so important.

Principles of Computer Ethics

Principles of computer ethics can be explained by the Ten Commandments of Computer Ethics. The commandments include (1) Thou shalt not use a computer to harm other people; (2) Thou shalt not interfere with other people's computer work; (3) Thou shalt not snoop around in other people's computer files; (4) Thou shalt not use a computer to steal; (5) Thou shalt not use a computer to bear false witness; (6) Thou shalt not copy or use proprietary software for which you have not paid (without permission); (7) Thou shalt not use other people's computer resources without authorization or proper compensation; (8) Thou shalt not appropriate other people's intellectual output; (9) Thou shalt think about the social consequences of the program you are writing or the system you are designing; and (10) Thou shalt always use a computer in ways that ensure consideration and respect for other humans (Computer Professionals for Social Responsibility, 2011).

These ten commandments of computer ethics speak to issues such as "No harmful use," "No interference," "No illegal use," "No intellectual property violation," and "No social harm," which are ultimately very important for cybercrime victimization/offending and investigation.

Digital Forensic Investigation

Professional Digital Forensics Ethics

The objective behind the formation of a code of ethics is to defend members of the forensic community against various pressures and ramifications that the profession may bring. According to Barnett (2001), Bowen (2009), and Davis (1991) (as cited in Sharevski, 2015), a code of ethics serves as a collaborative agreement in regard to members' duties and it provides a coherent disciplinary reference if an accepted professional norm were to have been infringed upon. In spite of the need for standardized procedures, the field of forensics continues to lack the necessary structure that is required to enforce more efficient practices, effective operating guidelines, and certification and accreditation programs (Sharevski, 2015).

In order to understand the importance of digital forensic ethics, one must understand the field more in depth. According to Bassett et al. (2006, as cited in Sharevski, 2015), computer forensics requires "a well-balanced combination of technical skills, legal acumen, and ethical conduct" (p. 40). Recognizing that the forensics field will often experience ethical issues can be crucial during the development of a code of ethics. Unfortunately, due to the lack of unification among the governing bodies that oversee the operations of the digital forensic field, the ability to enforce as well as manage the various aspects of the digital forensic code of ethics continues to be disregarded, according to Harrington (as cited in Sharevski, 2015).

In efforts to address the need for professional regulations and procedures within the forensic science field, Barnett (as cited in Sharevski, 2015) suggested that there clearly stands a strong correlation between the criminal justice system and the providers of forensic expertise. According to the Federal Judicial Center (as cited in Sharevski, 2015), while the criminal justice system has a demand for newfound services, these services are provided by forensic practitioners, ultimately enabling the system to utilize the available range of tools relative to the proceeding, while the evaluation process of the received forensic products are determined admissible by referring to the rules of evidence.

In order to strengthen the foundation for professionals in the field of forensics, Barnett (as cited in Sharevski, 2015) continues to deliberate on the process of developing a code of ethics and he goes into detail about distinct and extensive introductions that include "the codes of ethics established by the American Academy of Forensic Sciences (2013), American Board of Criminalistics (2013) and the California Association of Criminalists (2010) and the complementary policies and procedures that deal with the issue of enforceability and sanctioning" (p. 40). Barnett (as cited in Sharevski, 2015) also highlights the ethical requirements based on factors such as competency, diligence, significance, reviewability, and disclosure, all while identifying potential ethical disputes relative to both technical capabilities and professional practice.

In efforts to further illustrate, Bowen (as cited in Sharevski, 2015) meticulously explains the drive as well as the reasoning behind the act of ethical misconduct and reinforces it by providing actual case studies that involve issues such as selective bias, deceit, contingency, disputes over dedication and morality, as well as partiality. Upon this foundation, Bowen (as cited in Sharevski, 2015) continues by putting an emphasis on the structural process in regard to the forensic code of ethics. His arguments reflect Barnett's approach and then reinforce his major points by providing data from a research study that focused on ethical conflicts such as dishonorable conduct and the lack of information about ethics.

Code of Ethics and Professional Responsibility

The International Society of Forensic Computer Examiners (ISFCE) strives to support professionals within the field of digital forensics through the distribution of an internationally recognized Certified Computer Examiner certification. This society provides professionals with expert knowledge as well as training on digital matters that are significant at both a micro and macro level. While the ISFCE consistently evaluates their testing standards and principal demands, they expect their members to uphold a high level of integrity in efforts to protect the authenticity of this certification (The International Society of Forensic Computer Examiners, 2020). According to the International Society of Forensic Computer Examiners,

> All CCEs and CCE Candidates must agree to abide by the International Society of Forensic Computer Examiners (ISFCE) Code of Ethics and Professional Responsibility in order to obtain and maintain the CCE certification. If an individual does not abide by this code, their conduct will be subject to examination by the ISFCE Ethics and Professional Responsibility Committee and their Certification could be

suspended or revoked. If this code is changed, all CCEs will be required to re-submit the amended code in order to maintain their certification. The ISFCE feels very strongly about the importance of a sound, enforced Code of Ethics. Maintaining the integrity of the CCE is in the best interest of all CCE certificants. The ISFCE will thoroughly and impartially investigate all suspected or apparent violations of the Code of Ethics and Professional Responsibility by applicants and CCE certificants. (The International Society of Forensic Computer Examiners, 2020, paras. 1-4)

The International Society of Forensic Computer Examiners also states

A Code of Ethics is necessary to protect the integrity of the certification process. The ISFCE feels very strongly about the importance of a sound, enforced Code of Ethics and Professional Responsibility (Code). Maintaining the integrity of the CCE is in the best interest of all CCE certificants. The ISFCE will thoroughly and impartially investigate all suspected violations of the Code by CCE candidates and CCE certificants. (The International Society of Forensic Computer Examiners, 2020, para. 5)

The already complicated digital forensics field becomes even more convoluted when technological innovations are created because it essentially sets up a platform for novel crimes to develop and a place for criminal activity to thrive (Irons & Konstadopoulou, 2007). While complex, there has fortunately been an inclination within the number of investigations, in both civil and criminal cases, to take advantage of digital forensics—as evidenced by the increase of computer forensic services as well as digital forensic practitioners (Irons & Konstadopoulou, 2007). Consequently, with this increase comes the need to recognize potential professional issues that may arise. Therefore, in order to ensure that these forensic practitioners have the necessary expertise and skillsets, it is essential for a professional framework to be developed (Irons & Konstadopoulou, 2007).

Although there is significant overlap in the issues that concern professionality within the computing and digital forensic disciplines, there is a whole other compartment of concern that the digital forensic field faces due to the inherent nature of this domain (Irons & Konstadopoulou, 2007). For a digital forensic investigation to be solved both legally and ethically, those who are working the case must abide by the rules and regulations outlined within the codes they follow. If this professional approach is not complied with, the case may become jeopardized (Irons & Konstadopoulou, 2007). According to Irons and Konstadopoulou (2007), the digital forensics field faces distinct professional issues such as evidential integrity, evidential continuity, as well as legal issues that cases endure among various jurisdictions.

As previously indicated, digital forensics professionals are often provided with the opportunity to decide whether or not they want to handle the evidence discovered ethically. With this advantage, these practitioners may feel enticed to commit technology-facilitated crimes either by handling a case unethically or by using the knowledge they acquire while working the criminal investigation (Irons & Konstadopoulou, 2007). Therefore, in order to preserve the integrity of these forensic services, professionals must have a lengthy background check and obtain a security clearance to work in this field. In addition, a deliberation in regard to legal and ethical issues should also be considered when addressing the future of this field (Irons & Konstadopoulou, 2007).

Digital Forensic Investigation and Computer Ethics

This section will apply the body of knowledge to digital and cybercrime investigations. First, we should understand what digital forensic science is. The field is often defined as "the use of scientifically derived and proven methods toward the preservation, collection, validation, identification, analysis, interpretation, documentation and presentation of digital evidence derived from digital sources to facilitate or further the reconstruction of events found to be criminal, or helping to anticipate unauthorized actions shown to be disruptive to planned operations (Digital Forensic Research Workshop (DFRWS), 2001).

As Figure 1.1 illustrates, three areas where digital forensic science can be applied are law enforcement, military operations, and critical infrastructure protection.

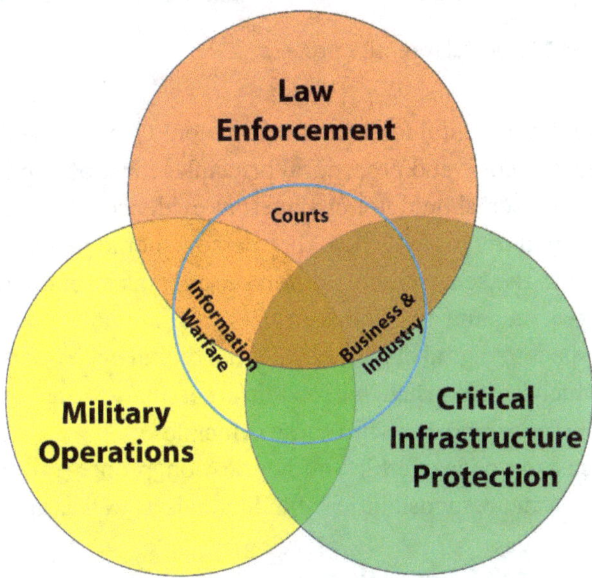

FIGURE 1.1 Three areas of digital forensic science

Digital forensic activities include the secure collection of computer data, the identification of suspect data, the examination of suspect data to determine details such as origin and content, the presentation of computer-based information to courts of law, and the application of a country's laws to computer practice.

The basic methodology of digital forensic activities can be referred to as the three As: (1) Acquire the evidence without altering or damaging the original; (2) Authenticate the image, and (3) Analyze the data without modifying it.

The process includes the following four parts:

- **Prepare for the investigation.** You should identify the purpose of the investigation and the resources required.
- **Acquire evidence.** You, as an investigator, identify the source of digital evidence and capture the evidence.
- **Analyze evidence.** You can report findings.
- **Present and disseminate results.** You can identify tools and techniques, process data, and interpret and analyze results.

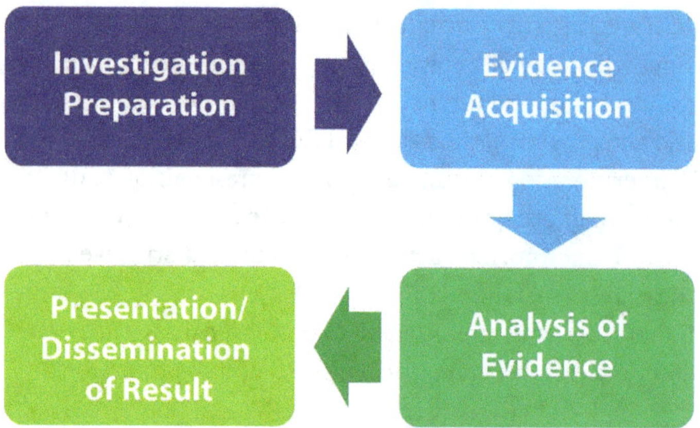

FIGURE 1.2 Digital forensics process

There are six principles of digital forensic investigation: (1) when dealing with digital evidence, all of the general forensic and procedural principles must be applied; (2) upon seizing digital evidence actions taken should not change that evidence; (3) when it is necessary for a person to access original digital evidence, that person should be trained for the purpose; (4) all activity relating to the seizure, access, storage, or transfer of digital evidence must be fully documented, preserved, and available for review; (5) an individual is responsible for all actions taken with respect to digital evidence while the digital evidence is in their possession; and (6) any agency which is responsible for seizing, accessing, storing, or transferring digital evidence is responsible for compliance with these principles.

Each principle discusses a code of ethics, authentication, proper training, methods for preserving data and results, and responsibilities of individuals as well as agencies.

Applied Case: Story of the Ross Ulbricht Arrest

Agents with the U.S. Federal Bureau of Investigation (FBI) investigating Ross Ulbricht, founder of the online black market known as Silk Road, learned that he often ran the site from his laptop, utilizing the wireless networks available at branches of the San Francisco Public Library. When they had enough evidence to arrest him, they planned to catch him in the act of running Silk Road, with his computer on and logged in. They needed to make sure he was not able to trigger any encryption or delete any evidence.

In October 2013, a male and female agent pretended to have a lovers' quarrel near where he was working at the Glen Park branch. According to *Business Insider*, Ulbricht was distracted and got up to see what the problem was, whereupon the female agent grabbed his laptop while the male agent restrained Ulbricht (Bertrand, 2015). The female agent was then able to insert a flash drive in one of the laptop's USB ports, with software that copied key files. According to Joshua Bearman of *Wired*, a third agent grabbed the laptop while Ulbricht was distracted by the apparent lovers' fight and handed it to agent Tom Kiernan (Bearman, 2015).

This story shows the line investigators could cross, violating basic principles of computer ethics and digital forensic investigation. Agent Carl Force, who is also known as French Mai, allegedly created fake IDs and sold Ross Ulbricht information about the government's investigation. Interestingly, he used Drug Enforcement Administration (DEA) PCs to move Bitcoin to personal accounts (more than $800,000) and offer to kill a Silk Road employee. He also blackmailed Ulbricht and asked for some payments in Bitcoin, which can be traded and paid anonymously. As a result of his actions, he was charged with money laundering and wire fraud, among other charges.

FIGURE 1.3 Silk Road

Agents who investigated and were involved in the Ulbrecht case did not follow the six principles of digital forensics introduced earlier in this chapter. When they dealt with digital evidence, they did not apply general forensic and procedural principles. Actions on obtained digital evidence were not ethical. While it is advisable to document all activity relating to the seizure, access, storage, or transfer of digital evidence, agents with criminal intentions did not follow the principles. While it is notable that agents behaved unprofessionally, unethically, and illegally in the Ross Ulbricht Silk Road case, it illustrates the importance of establishing and adhering to principles and procedures of digital forensics in relation to computer ethics.

Summary

Over the years, computer ethics has grown increasingly important. This not only links to issues involving cybercrime and cybersecurity, but it also links to digital forensic investigations. While we think critically about piracy, intellectual property, cybercrime vis-à-vis computer ethics, and the future, we should also think further about ethics for data, artificial intelligence vis-à-vis cybercrime, and cybersecurity on other occasions.

This chapter calls specific attention to computer ethics, which plays a significant role in shaping and creating cybercrime research and investigations. The case of Silk Road is a good example of how computer ethics is intertwined with the principles and process of digital forensics. This illustrates how offenders, investigators, and victims all take part within a digital forensic investigation, and computer ethics must be emphasized due to rapid technological advancements. Challenges for cybercrime research and investigations lie within the changing landscapes of the internet, virtual technologies, and social media. With new forms of crime on the rise, ordinary people are less likely to have the discretion to determine whether or not a particular issue/technology hosts a cybercrime in disguise. This lack of awareness increases the probability of becoming a victim of the newest forms of cybercrime. The absence of computer ethics potentially jeopardizes investigations, investigators, victims, and the law enforcement system. In this regard, it is important for local, regional, federal, and global investigation agencies to provide guidance as well as guidelines in order to adhere to basic yet critical computer ethics during their investigation. Future research should be more preemptive in investigating, predicting, and analyzing the underlying causes and consequences of lawful and unlawful practices of digital forensics, which may benefit from a further understanding and implementation of computer ethics within forensic investigations.

Lab 1: Cybercrime Investigation Scenario

On X month, X day, X year, a Dell notebook computer, serial number VLQLW, was found abandoned along with a wireless PCMCIA card and an external homemade 802.11b antennae. It is suspected that this computer was used for hacking purposes; however, it cannot be tied to a hacking suspect.

The online pseudonym of "Mr. Evil Noodle" and some of his associates have said that he would park his vehicle within range of wireless access points (like Starbucks, airports, and other mobile hotspots) where he would then intercept internet traffic, attempting to get credit card numbers, usernames, and passwords.

Mr. Evil Noodle offers online users free software, illegal images, free music, and free movie files that contain Trojan horses on P2P sites. Mr. Evil Noodle has attempted to steal personal information through Trojan horses. Mr. Evil Noodle will frequently engage in spoofing and phishing based attacks via email scams to defraud victims. Mr. Evil Noodle is also responsible for a recent DDoS attack that led to the shutdown of multiple major corporation servers.

Mr. Evil Noodle's darknet site has had about a million registered users from all over the world with many illegal items and services available to users. Mr. Evil Noodle's darknet site, an illicit online market functioning within the "deep web," represents a contemporary drug market operation. The illegal nature of this site is readily discernible to any user searching for illegal drugs, such as cocaine and heroin. These drugs are visibly advertised on the site's homepage and, as of today, there are approximately 13,000 listings for controlled substances. The marketplace made the sale of everything from narcotics to weaponry possible.

Mr. Evil Noodle sought to anonymize the transactions on the darknet in two prominent ways. The darknet site operates on a special network of computers spread throughout the world, designed to conceal the actual IP addresses of the computers on that network and therefore conceal the identities of their users. This network is currently known as "the onion router," or more commonly as the "Tor" network. Every communication made in this Tor network is bounced through several relay points in the network and wrapped in a layer of encryption at each of those points. This makes the recipient of a message unable to trace the communication back to its originating IP address. Similarly, the Tor network enables websites to operate on the network in such a way that the true IP address of the server hosting the site is hidden.

Mr. CIC, who graduated from the Cybercrime Investigation and Cybersecurity program, is working on the Mr. Evil Noodle case as a cybercrime investigator. Mr. CIC is an agent of the FBI. Being an agent for an undisclosed amount of time, he is currently assigned to the computer intrusion squad in the FBI's New York Field Office (as that is where the investigation of this case has taken place) and has received extensive training in the use of computer equipment to conduct criminal activity. As an agent of the federal government, Mr. CIC is completely authorized to investigate violations made against the laws of the United States of America and to act on warrants issued under the authority of the United States.

Lab Assignment 1.1: Google Takeout

On December 1, 2019, a man named Curry went to the library where he utilized the public computers to complete his assignments. After a hard day's work, Curry packed up his things and decided to head home. Unfortunately, he forgot to sign out of his Google account. The infamous Mr. Evil Noodle happened to stumble upon this and decided to acquire all the information that was contained within the Google account. The stolen data comprised personal credit card information, contact information, calendars, and Google Map information. Not even a few days later, a text message was sent to Curry's cell phone stating, "A payment was made with an AA bank card." Curry was enraged and scurried to the bank as fast as he could in order to cease this fraudulent payment. When he arrived, the bank explained that Curry was able to receive a refund for the money because the payment was not authorized by the owner of the account.

Below are useful tools to acquire Google account information:

Google Dashboard is an internet program that provides users with an easy-to-read summary of all the data that Google collects on them. The main purpose of this program is to amalgamate that data so the user can manage it along with all of their Google service accounts in one location.

Google Takeout is a project by the Google Data Liberation Front that allows users of Google products, such as YouTube and Gmail, to export their data to a downloadable archive file.

Please answer the following questions (2–3 pages):

1. How did Mr. Evil Noodle acquire all of the information linked to Curry's Google account?
2. How could Mr. Evil Noodle share the acquired information with other criminals or cybercriminals?
3. If Mr. CIC were to investigate this case, how would he go about doing so?
4. Prevention: How would one stop Mr. Evil Noodle from acquiring data within a Google account?
5. What are your thoughts on this type of case? What would your reaction be as a victim?

Lab Instruction

In this lab, you will have to check all of the data on your Google account and download it.

1. Go to Google Takeout
2. Select data you want to download
3. Click 'Next step'
4. Choose 'Send download link via email' for Delivery method

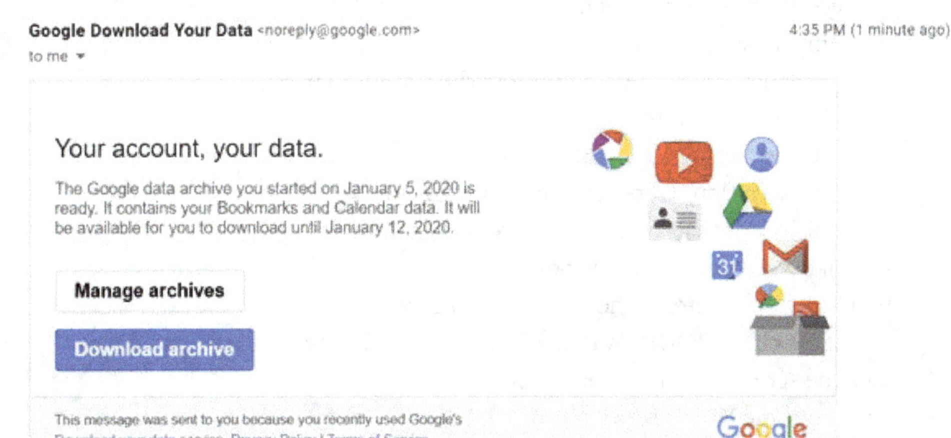

Image 1.1

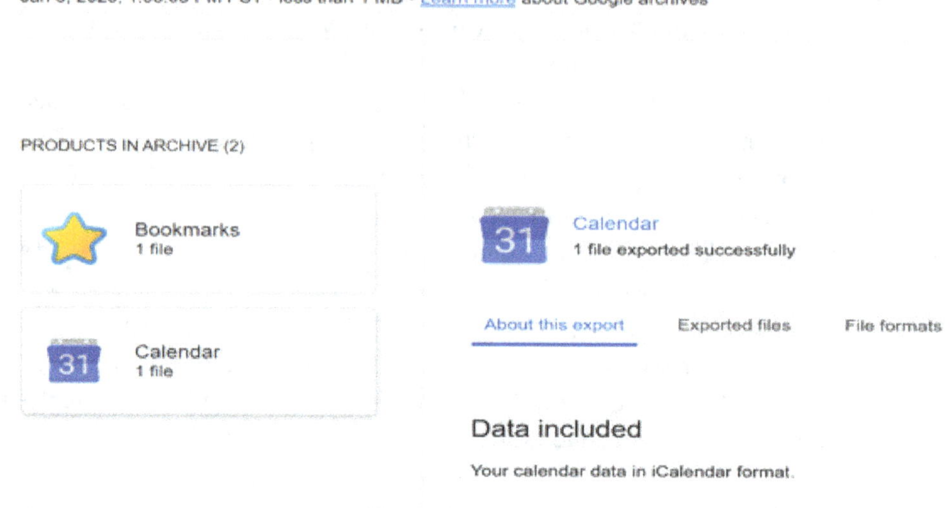

Image 1.2

5. Click 'Create export'
6. It takes some time if you have a large amount of information, but you will receive an email. When the archive is received, open the file and see what you get.

Lab Assignment 1.2: Creating a Case with EnCase

Once you have successfully logged into the lab, you will see the EnCase icon on your screen. To begin, you will create a case and add the hard disks on your remote workstation as a live preview.

1. Double-click the EnCase icon on your desktop to launch the EnCase software
2. On the Home tab, single-click '**New Case**' to begin creating a new case
3. In the Options dialog window, type or select the following:
   a. Name: **Lab1**
   b. Base case folder: **C:\EnCases\Cases**
   c. Check the option for **Use base case folder for primary cache** (**Your EVIDENCE DRIVE**)
   d. Secondary evidence cache will remain blank
   e. Templates: **#5 Flexible**
   f. Double-click the **Value** column in the Case info pane:
      • Case number: **001**
      • Examiner Name: **Your Name**
      • Description: **Cybercrime Investigation**
      • Select option to **"Backup Every 30 Minutes"**
   g. Take a screenshot at this point
   h. Click 'OK' to create the new preview case
   i. Please check YOUR EVIDENCE DRIVE for your case folder titled Lab Assignment
4. Place the screenshot in a Word document
5. Name your file YOUR LAST NAME_LAB1
6. Submit your lab report

References

Barnett, P. D. (2001). *Ethics in forensic science: Professional standards for the practice of criminalistics.* CRC Press.

Bassett, R., Bass, L., & Brien, P. O. (2006). Computer forensics: An essential ingredient for cyber security. *Journal of Information Science and Technology, 3*(1), 22–32.

Bearman, J. (2015). The rise and fall of Silk Road: The untold story of Silk Road, Part I. *Wired.* https://www.wired.com/2015/04/silk-road-1/.

Bertrand, N. (2015, January 22). The FBI staged a lovers' fight to catch the kingpin of the web's biggest illegal drug marketplace. *Business Insider.* https://www.businessinsider.com/the-arrest-of-silk-road-mastermind-ross-ulbricht-2015-1.

Bowen, R. (2009). *Ethics and the practice of forensic science (International forensic science and investigation).* Boca Raton: CRC Press.

Choi, K. S., & Lee, C. S. (2018). The present and future of cybercrime, cyberterrorism, and cybersecurity. *International Journal of Cybersecurity Intelligence & Cybercrime, 1*(1), 1–4.

Choi, K. S., Lee, C. S., & Louderback, E. R. (2020). Historical evolutions of cybercrime: From computer crime to cybercrime. In T. J. Holt & A. Bossler (Eds.), *The Palgrave handbook of international cybercrime and cyberdeviance* (pp. 27–43). Palgrave Macmillan.

Computer Professionals for Social Responsibility (2011). The ten commandments of computer ethics. http://cpsr.org/issues/ethics/cei/.

Digital Forensic Research Conference. (2001, August). *A road map for digital forensic research.* Digital Forensic Research Workshop (DFRWS).

Irons, A. D., & Konstadopoulou, A. (2007). Professionalism in digital forensics. *Digital Evidence and Electronic Signature Law Review, 4,* 45–50.

McGuire, M., & Dowling, S. (2013). *Cybercrime: A review of the evidence. Summary of key findings and implications. Home Office Research Report 75.*

Moor, J. (1985). What is computer ethics? *Metaphilosophy, 16*(4), 266–275.

O'Brien, S. A. (2018, January 29). *How a 20-year-old Australian student discovered U.S. military's secret sites.* CNN. money.cnn.com/2018/01/29/technology/strava-nathan-ruser/index.html

Sharevski, F. (2015). Rules of professional responsibility in digital forensics: A comparative analysis. *Journal of Digital Forensics, Security and Law, 10*(2), Article 3.

The International Society of Forensic Computer Examiners. (2020). *Code of ethics and professional responsibility.* https://www.isfce.com/ethics2.htm

Credits

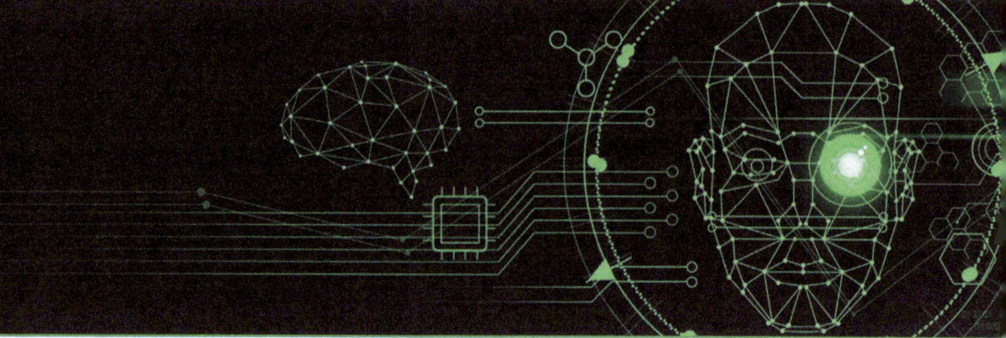

Overview of Digital Evidence and Digital Forensics

Kyung-Shick Choi

Introduction

Currently, the way law enforcement officers do their job is rapidly changing as the business world transitions from manual records to computer records; this evolutionary transition is having a major impact. During the execution of a search warrant, more and more computer devices are being discovered and seized; however, most law enforcement officers do not possess knowledge concerning the processing of computer data and related evidence.

Of course, the obvious use of computers for cybercrime includes internet fraud, storing criminal data, infiltrating private computers belonging to others, and the illegal use of various services. To successfully obtain a warrant for access to computers and the information they contain, there must be legitimate need and law enforcement must develop an organized approach to examine investigative and intelligence information while preserving the information for subsequent admission in court.

It is critical that experienced law enforcement officers become experts in the rules of evidence and preserve and process the evidence appropriately, because proper documentation and expert testimony is necessary to effectively deal with the seizure and processing of computer-related evidence. In an effort to develop a greater understanding and expertise, law enforcement officers become trained in this new forensic science. Frequently, a specialist must be present during the first moments of the raid. It is important to note that a criminal could have installed destructive programs or taken other defensive measures, which only a trained person can recognize, defeat, and document before evidence is destroyed.

Principles of Digital Evidence

Digital evidence is a term coined by the Scientific Working Group on Digital Evidence (SWGDE) and the International Organization on Computer Evidence (IOCE) in October 1999. Digital evidence is defined as information and data of investigative value which is stored on or transmitted by an electronic device in binary form. Digital evidence is obtained when information and/or physical objects are collected or stored for vetting purposes. The term evidence is used to imply that the collector of evidence is recognized by the courts. Moreover, the process of collecting evidence is also assumed to be a legal process and should adhere to rules of evidence in that locality.

According to Casey (2011), the term "digital evidence" includes "any and all digital data that can establish that a crime has been committed or can provide a link between a crime and its victim or a crime and its perpetrator" (p. 1). Essentially, digital data is a pattern of numbers that represents information of various kinds, encompassing text, images, audio, and video. Digital evidence is becoming more substantial and important to investigative efforts with the increasing use of technology and the internet.

Digital evidence is digital information that can substantiate the relationship between the crime itself and the suspect/victim. Digital evidence consists of data in the storage media, including hard drive or USB media, temporarily maintained information in the computer memory, data transferred through a network, and so on.

Characteristics of Digital Evidence

Digital evidence is unseen to the naked eye. Digital evidence can be easily modified and fragile; the original and duplicates cannot be differentiated. Digital evidence has massive material and various formats. Wired or wireless networks transfer the digital evidence. Digital evidence is:

- Invisible to the naked eye
- Easily modified and fragile
- Indistinguishable between an original and a duplicate
- A massive amount of information
- Exists in various formats
- Transferred through wired/wireless networks

Digital Evidence

Many kinds of digital evidence can be found at the crime scene:

- Computer systems: Personal computers, web servers, messenger servers, game servers, file servers, database servers, email servers, etc.
- Data storage: HDD, FD, CD, DVD, Zip, tape, USB storage, memory stick, cloud, etc.
- Network devices: Router, switch, firewall, iOS, IPS, etc.
- Other digital devices: Cellular phones, tablets, global positioning systems (GPS), media players, closed-circuit television (CCTV), IoT devices, digital cameras/camcorders, etc.

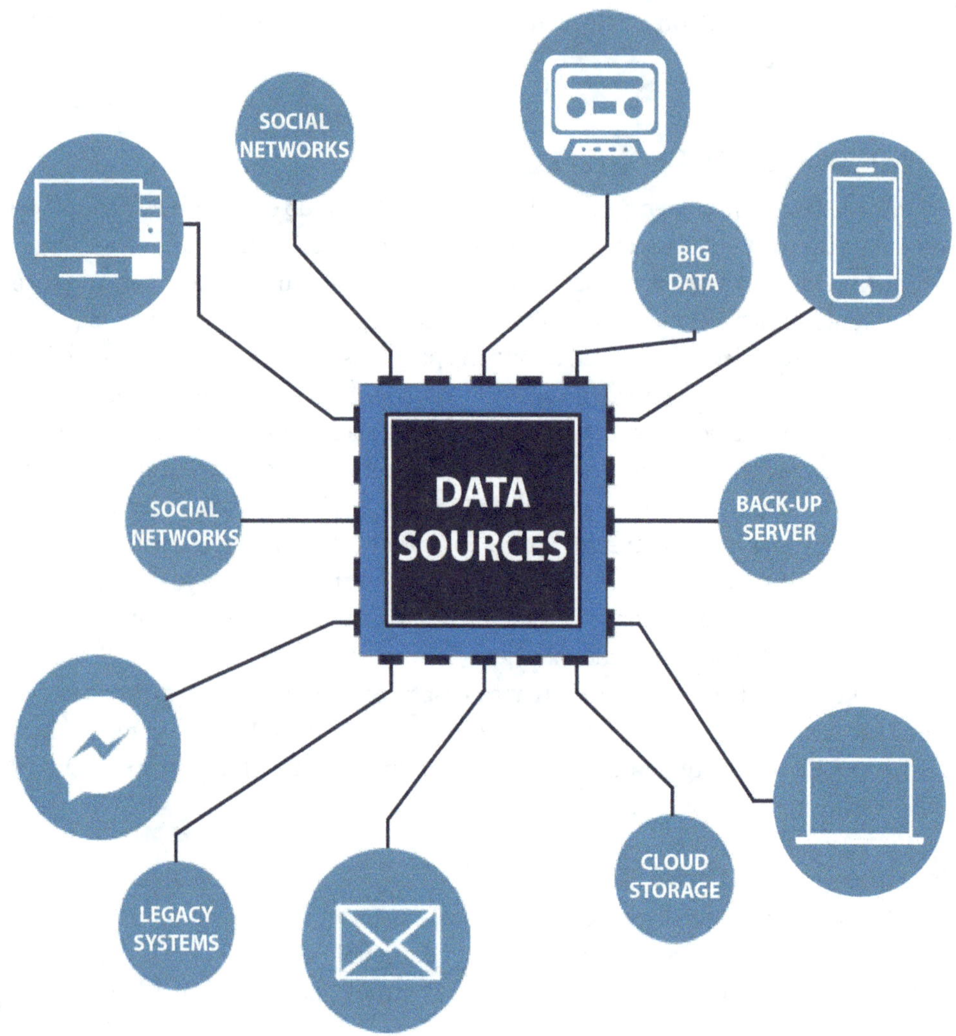

A computer system is the most conventional digital evidence. Computer systems include personal computers, web servers, messenger servers, game servers, file servers, database servers, and email servers. A server is the computer system that provides a service to the clients responding to the request from the client. The service includes email, web, database, games, and so on.

Data storage includes the popular storage devices of the early 20th century such as floppy disks, compact discs, digital versatile disks, zip drives, tape drives, down to the devices in use today such as hard disk drives, USB storage (commonly referred to as flash drives or thumb drives), memory sticks, and cloud storage devices. These forms of data storage could be found all around the crime scene and are very reliable sources of evidence.

The main network devices include routers, switches, firewalls, intrusion-detection systems, and intrusion protection systems. Each network device may maintain login records or the network traffic—for cybercrime investigations, these are valuable resources. With the advancement of technology, new digital devices have been manufactured including smartphones, tablets, global positioning systems, portable media players, closed-circuit television, and digital cameras that are compatible with network devices. These advanced devices each use digital information in

Image 2.2

the data process to provide a full understanding of what the digital media is like, how the data processes, and where we can find the exact information in them. Additionally, closed-circuit television, also known as CCTV, has the digital media to record and play the scene.

Carter (1995) proposed the following criminal justice cybercrime categories:

- Computer as the target (e.g., computer intrusion, data theft, techno-vandalism, techno-trespass)
- Computer as the instrumentality of the crime (e.g., credit card fraud, telecommunications fraud, theft, or fraud)
- Computer as incidental to other crimes (e.g., drug trafficking, money laundering, child pornography)
- Crimes affiliated with the prevalence of computers (e.g., copyright violation, software piracy, component theft)

The United States Federal Guidelines for Searching and Seizing Computers (Jarrett et al., 2009) also outlines a solid set of guidelines and cybercrime categories:

- Hardware as contraband or fruits of crime
- Hardware as an instrumentality
- Hardware as evidence
- Information as contraband or fruits of crime
- Information as an instrumentality
- Information as evidence

In relation to the Federal Guidelines context, hardware refers to all of the physical components of a computer, and information refers to the data and programs that are stored or transmitted while using a computer. The remaining three categories refer to information which falls under the forms of digital evidence.

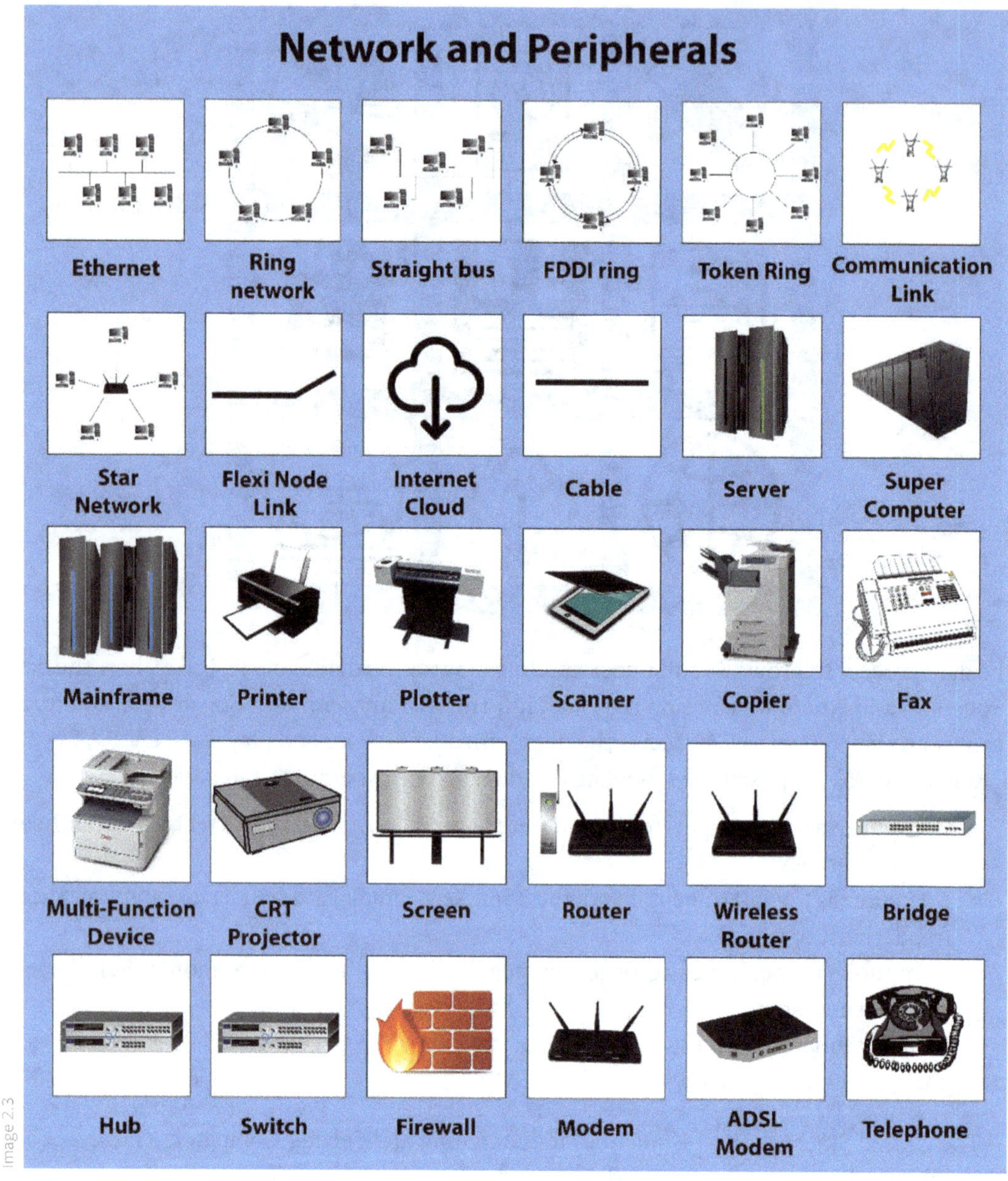

Image 2.3

Notably, each of the categories contains individualized legal procedures that must be followed. Most often, a warrant is required to search and seize evidence, yet some exceptions include consent, exigency, and evidence in plain view. However, in some situations, privacy laws may protect the evidence.

Hardware

Hardware as Contraband or Fruits of Crime

The property that was obtained by criminal activity, such as computer equipment that was stolen or purchased using stolen credit card numbers is the fruit of the crime. The objective for seizing contraband or fruits of crime is to prevent and deter future crimes.

In the event law enforcement officers decide to seize evidence in the category of items, a court will examine if the circumstances would have led a reasonably cautious agent to believe that the object was contraband or a fruit of crime.

Hardware as an Instrumentality

Hardware as an instrumentality applies in the event computer hardware has contributed to a significant role in a crime. For example, computer crackers tend to use hardware type sniffers to compile passwords that can then be used to obtain unauthorized access to computers. Since the cracker utilizes computer hardware that is particularly manufactured or configured to commit a specific crime, it is evident that it is an instrumentality.

When determining whether or not a piece of hardware can be seized as an instrumentality of crime, it is important to note that "significant" is the operative word in the definition of instrumentality (Casey, 2011).

At the end of the day, it is ultimately up to the courts to determine whether or not an item played a significant role in the crime at hand. To date, the courts have been quite liberal on this issue. According to the Eastern District Court of Virginia, a computer with related accessories was an instrumentality because it contained a file that specified the growing characteristics of marijuana plants (Casey, 2011).

Hardware as Evidence

According to the Federal Rule of Criminal Procedure 41(b), "search for and seize any property that constitutes evidence of the commission of a criminal offense."

The category of "hardware as evidence" includes computer hardware that is neither contraband nor the instrumentality of a crime. For example, if a scanner used to digitize child pornography has unique scanning abilities that connect the hardware to the digitized images, it could be seized as evidence.

Information

Information as Contraband or Fruits of Crime

A private citizen is not permitted to possess contraband information. A typical type of information as contraband is encryption software. In the event a criminal is caught, but all of the incriminating digital evidence is encrypted, decoding the evidence to prosecute the criminal may not be an option.

Information as fruits of crime includes illegal copies of computer programs, stolen trade secrets and passwords, and any other information that was obtained by criminal means.

Information as an Instrumentality

Programs that computer crackers use to break into computer systems are the instrumentality of a crime. Computer programs that record passwords when a person logs into a computer can be an instrumentality, and computer programs that crack passwords often play a significant role in a crime.

Information as Evidence

Information as evidence is the richest category of all. Several of our daily actions produce a trail of digits. All service providers—such as telephone companies, ISPs, banks, and credit card companies—keep some information about their customers.

At this time, the U.S. Computer Assistance Law Enforcement Act, which took effect in 2000, compels telephone companies to store detailed records of their customers' calls for an indefinite period of time.

If a suspect utilized an encrypted email to correspond with another individual—an individual suspected of a crime—around the time a crime was committed, this could be treated as sufficient probable cause to obtain a warrant to examine the email or even search the second person's computer or residence.

In other words, our daily activities that are linked to any digital form (information) can be considered as evidence, when applicable.

Understanding Digital Forensic Investigation

The following lesson briefly discusses the basic phases of the forensic process and operating principles and requirements for forensic investigations.

A criminal investigation is defined as "the process of legally gathering evidence of a crime that has been or is being committed" (Brown, 2001, p. 3). The way evidence is collected and processed in cybercrime cases is relatively similar to other criminal cases; however, the investigators need to confront the bulk of the actual evidence in electronic or digital format.

A cybercrime investigation can still be assimilated into a thorough report for prosecution. The manner in which a case is investigated is heavily influenced by the composition of the investigation team. In other words, cybercrime investigation uses the same principle as other traditional criminal investigations but applies those principles to digital evidence.

In order to properly discuss the forensic process, it is important to understand the terminology of digital forensics. Generally, forensic science is defined as the application of science to the law. Digital forensics, otherwise referred to as computer and network forensics, has many definitions. Generally, it is considered the application of science to the identification, collection, examination, and analysis of data while preserving the integrity of the information and continuing a strict chain of custody for the data. Distinct pieces of digital information or data are formatted in a specific way. The amount of data from many sources is ever increasing among organizations. For example, data can be stored or transferred by standard computer systems, networking equipment, computing peripherals, mobile devices including smartphones and tablets, consumer electronic devices, and various types of media, among other sources. Due to the variety of data sources, digital forensic techniques can be used for many purposes, such as investigating crimes and internal policy violations, reconstructing computer security incidents, troubleshooting operational problems, and recovering from accidental system damage (Kent et al., 2006).

What Is Digital Forensics?

According to the Technical Working Group for Education and Training in Digital Forensics (2007), "Digital forensics is the science of identifying, collecting, preserving, documenting, examining, analyzing, and presenting evidence from computers, networks, and other electronic devices" (p. 10).

Computer forensics or digital forensics is a skillset that is acquired to investigate and analyze electronic data in the interest of determining potential legal evidence. Computer forensics enables the systematic and precise identification of evidence in computer-related crime and abuse cases.

Forensic Staffing

Today, almost every investigation team should have some capability to perform computer and network forensics. If an investigation team lacks such a capability, the investigation will encounter great difficulty in determining what events have occurred within the computer systems and networks.

Let's imagine investigators are working a case involving sourcing and distribution of illegal drugs via the internet. Upon seizing the computer of the suspect, the investigator passes it on to the analyst for processing and tells the analyst that it is believed that the suspect has been in correspondence with other small-scale dealers using email and email chat. However, the investigator fails to tell the analyst that there is no information on how payments have been exchanged. The criminal has been exchanging funds via PayPal using the cover of a fake email account. But since this analyst is unaware that the method of payment is not known to investigators, matters related to email or eBay are not investigated. Thus, it is reasonable to state that the majority of the fault here lies with the investigator who lacked the full details, not the analyst, despite the lack of investigation.

Even though the example appears as a simple case, one investigator would find it almost impossible to resolve the case. Commonly, the primary users of forensic tools and techniques within an investigation team can be divided into three groups: forensic investigators, IT professionals, and incident handlers.

Forensic Investigators

Within law enforcement agencies, forensic investigators are typically responsible for investigating allegations of criminal activities. The forensic investigator serves as a skilled digital forensic analyst with in-depth knowledge of the circumstances surrounding the case and the ability to apply comprehensive knowledge directly within the forensic examination. Commonly, larger police agencies tend to have dedicated cybercrime units. It is important to note that the objectives must be clearly stated during the investigation, and the forensic investigator must have as much background information on the case as possible. The forensic investigator typically utilizes several forensic techniques and tools. Investigators within private organizations may include legal advisors and personnel of the human resources department.

IT Professionals

This group comprises technical support staff and system, network, and security administrators. IT professionals do not heavily use forensic techniques, yet they do tend to use tools specific to their area of expertise during their routine work (e.g., monitoring, troubleshooting, data recovery).

Incident Handlers

This part of the investigation team responds to a multitude of computer security incidents, including unauthorized data access, inappropriate system usage, malicious code infections, and denial-of-service attacks. Similar to forensic investigators, incident handlers often use a wide variety of forensic techniques and tools to complete their investigations.

DISCUSSION

Please list devices that are important to digital evidence collection and explain why the devices you selected are important.

Before we start discussing the process itself, it is useful to spend a little extra time considering a philosophical principle of digital forensics—Locard's principle.

Every action performed on a digital device will leave a "digital footprint." It may be fragile and transient, but it will exist.

The founding father of modern forensic science, Professor Edmond Locard, indicated that

> wherever he steps, whatever he touches, whatever he leaves, even unconsciously, will serve as a witness against him. Not only his fingerprints or his footprints, but his hair, the fiber from his clothes, the glass he breaks, the tool mark he leaves, the paint he scratches, the blood or semen he deposits. All of these, and more, can be a mute witness against him. This is evidence that does not forget, it's not confused by the excitement of the moment, it's not absent because human witnesses are, it's factual evidence that cannot be wrong, cannot perjure itself, it cannot be wholly absent. Only human failure to find it, study and understand it can diminish its value (Kirk, 1953, p. 2).

In summary, every contact leaves a trace. This principle holds true in relation to digital evidence. Although electronic or digital evidence is not easy to read or understand, each and every action performed on a digital device will leave a digital footprint. Even though the footprint may be fragile and transient, it will exist.

Locard's principle in action:

- The objective of the digital investigator is to locate and preserve the digital footprint without contaminating it.
- In pre-internet days, data found on a computer system was sufficient evidence.
- Today, outside influences such as hacking, Trojan horses, and viruses can easily manipulate digital evidence.

All factors must be taken into consideration by the examiner today, including the possibility that evidence may be tainted by unknown users. Digital forensic investigators have to locate and preserve the digital footprint without any contaminations.

In pre-internet days, the mere presence of data on the computer was sufficient evidence; however, those days have passed. For example, perpetrators may be located on the other side of world when they commit a crime. IP tracing techniques are commonly used to locate the origin of criminal activities, and are now needed to prove the history of files found on a computer system to indicate whether the crime scene is located where the perpetrator performed the act or is at the victim's location.

Prior to the internet era, there was a remote possibility that the system had been accessed online by an unauthorized individual. During that time, nobody had heard of a Trojan horse (a malicious piece of software) nor any virus activity that would be contained in the stand-alone system. The digital forensic investigator must now take all of these factors into consideration.

Operating Principles and Requirements for Forensic Investigation

Operating Principles for Forensic Investigations

Please note, the following information is based on the *2012 ACPO Good Practice Guide for Digital Evidence* (Williams, 2012).

Although computer forensics as a discipline has only existed for around 20 years, some universal operating principles have developed. The wording of these principles is that of the Association of Chief Police Officers (ACPO). The operating principles are widely utilized in other nations and among professional associations.

- **Principle 1.** No action taken by law enforcement agencies, persons employed within those agencies or their agents should change data that may subsequently be relied upon in court.
- **Principle 2.** In circumstances where a person finds it necessary to access original data, that person must be competent to do so and be able to give evidence explaining the relevance and the implications of their actions.
- **Principle 3.** An audit trail or other record of all processes applied to digital evidence should be created and preserved. An independent third party should be able to examine those processes and achieve the same result.
- **Principle 4.** The person in charge of the investigation has overall responsibility for ensuring that the law and these principles are adhered to.

Forensic examiners rarely work on the original evidence item. There are, however, circumstances when this cannot be avoided. The operating principles emphasize that a competent person, in this case, a qualified examiner, must perform any work on the original evidence; thus, they should be able to clearly understand and express whether any changes have been made and what those changes are. The cases in which an examiner will work with original evidence are very rare.

With the advent of mobile telephones, however, this has changed to some degree; in many cases, the only way to extract data from a mobile telephone is to access the device itself. In these cases, the examiner must be skilled and show that they have taken every reasonable precaution to prevent the device from accessing the mobile phone network while it was in their possession.

The principle of digital forensics is that no action should change data in the computer or storage media. In case of unavoidable access to the data, all evidence should be created in order to explain the reason for the access and the implication of the action to the data. An audit trail of all processes applied to digital evidence should be created and preserved. An officer responsible for the investigation has overall responsibility for the whole process.

DISCUSSION

Please search the operating principles for forensic investigations and consider the local, state, and federal levels. Do these basic principles differ between the federal and state levels? If so, how?

Media Type	Reader	Typical Capacity[16]	Comments
Primarily Used in Personal Computers			
Floppy disk	Floppy disk drive	1.44 megabytes (MB)	3.5-inch disks; decreasing in popularity
CD-ROM	CD-ROM drive	650 MB–800 MB	Includes write-once (CD-R) and rewritable (CD-RW) disks; most commonly used media
DVD-ROM	DVD-ROM drive	1.67 gigabytes (GB)–15.9 GB	Includes write-once (DVD±R) and rewritable (DVD±RW) single and dual layer disks
Hard drive	N/A	20 GB–400 GB	Higher capacity drives used in many file servers
Zip disk	Zip drive	100 MB–750 MB	Larger than a floppy disk
Jaz disk	Jaz drive	1 GB–2 GB	Similar to Zip disks; no longer manufactured
Backup tape	Compatible tape drive	80 MB–320 GB	Many resemble audio cassette tapes; fairly susceptible to corruption from environmental conditions
Magneto optical (MO) disk	Compatible MO drive	600 MB–9.1 GB	5.25-inch disks; less susceptible to environmental conditions than backup tapes
Advanced Technology Attachment (ATA) flash card	PCMCIA slot	8 MB–2 GB	PCMCIA flash memory card; measures 85.6 x 54 x 5 mm
Used by Many Types of Digital Devices			
Flash/Jump drive	USB interface	16 MB–2 GB	Also known as thumb drives because of their size
CompactFlash card	PCMCIA adapter or memory card reader	16 MB–6 GB	Type I cards measure 43 x 36 x 3.3 mm; Type II cards measure 43 x 36 x 5 mm
Microdrive	PCMCIA adapter or memory card reader	340 MB–4 GB	Same interface and form factor as CompactFlash Type II cards
MultiMediaCard (MMC)	PCMCIA adapter or memory card reader	16 MB–512 MB	Measures 24 x 32 x 1.4 mm
Secure Digital (SD) Card	PCMCIA adapter or memory card reader	32 MB–1 GB	Compliant with Secure Digital Music Initiative (SDMI) requirements; provides built-in data encryption of file contents; similar in form factor to MMCs
Memory Stick	PCMCIA adapter or memory card reader	16 MB–2 GB	Includes Memory Stick (50 x 21.5 x 2.8 mm), Memory Stick Duo (31 x 20 x 1.6 mm), Memory Stick PRO, Memory Stick PRO Duo; some are compliant with SDMI requirements and provide built-in encryption of file contents
SmartMedia Card	PCMCIA adapter or memory card reader	8 MB–128 MB	Measures 37 x 45 x 0.76 mm
xD-Picture Card	PCMCIA adapter or xD-Picture card reader	16 MB–512 MB	Currently used only in Fujifilm and Olympus digital cameras; measures 20 x 25 x 1.7 mm

Image 2.4

Source: National Institute of Standards and Technology (2006), Guide to Integrating Forensic Techniques into Incident Response.

Requirements for Understanding Forensic Examination
Digital Devices and Media

Let's explore important materials that will be subject to a forensic examination before we focus on cybercrime investigations.

As previously mentioned, digital forensic investigators possess a comprehensive understanding of technology. Although a deep knowledge of computers and other digital devices is not required for composing forensic report, it is critical to digest the general knowledge of digital devices and media for effective forensic examinations.

Prior to attempting to collect or examine files, digital forensic investigators should have a reasonably comprehensive understanding of files and file systems. Initially, the investigator should be mindful of the variety of media that may contain files.

File Storage Media

The widespread use of computers and other digital devices has led to a significant increase in the number of different media types that are used to store files. In addition to traditional media types such as hard drives, files are often stored on consumer devices such as tablets and smartphones, as well as on newer media types, such as flash memory cards, which were made popular by digital cameras. The types of commonly available media are listed below.

Two Types of Digital Devices: Programmable vs. Non-Programmable Devices

Programmable Devices

Programmable devices utilize an operating system—software that allows third-party programs to run—for example, a personal computer with Microsoft Windows operating system and a word processor.

Non-Programmable Devices

Most other digital devices are non-programmable; however, this does not mean that the user cannot direct the device to take certain actions or that the data may not be stored upon them—the device would be pretty useless if this were the case. Rather it means, in effect, that the manufacturer controls the options available to the user. To clarify this, a personal computer may be loaded with Microsoft Windows or another operating system. The user may then choose to load and execute word processing, graphics spreadsheets, communications, or any other type of program at will. Therefore, the device is programmable. In contrast, a digital camera may allow the setting of shutter speeds or aperture or even the use of preset image manipulation, like red-eye removal; however, it will not allow the user to load a program that turns it into a music player or a voice recorder, in the context in which we use the term—it is non-programmable.

Personal Computers (PCs)

PC networks provide many benefits in relation to sharing information and resources; however, from an investigative point of view, having many locations to store or hide data is the downside.

Computer Hardware Components

CPU: Part of the computer which performs data processing.

POST: A diagnostic test of the computer's hardware for presence and operability during the boot sequence prior to running the operating system.

ROM (read-only memory): This is a form of memory that can hold data permanently. ROM is nonvolatile memory, meaning the data remains when the system is powered off. Having these properties (read-only and nonvolatile) makes ROM ideal for files containing start-up configuration settings and code needed to boot the computer (ROM BIOS).

RAM (random access memory): A computer's main memory is its temporary workspace for storing data, code, settings, and so forth. RAM is usually volatile memory, meaning that upon losing power, the data stored in memory is lost.

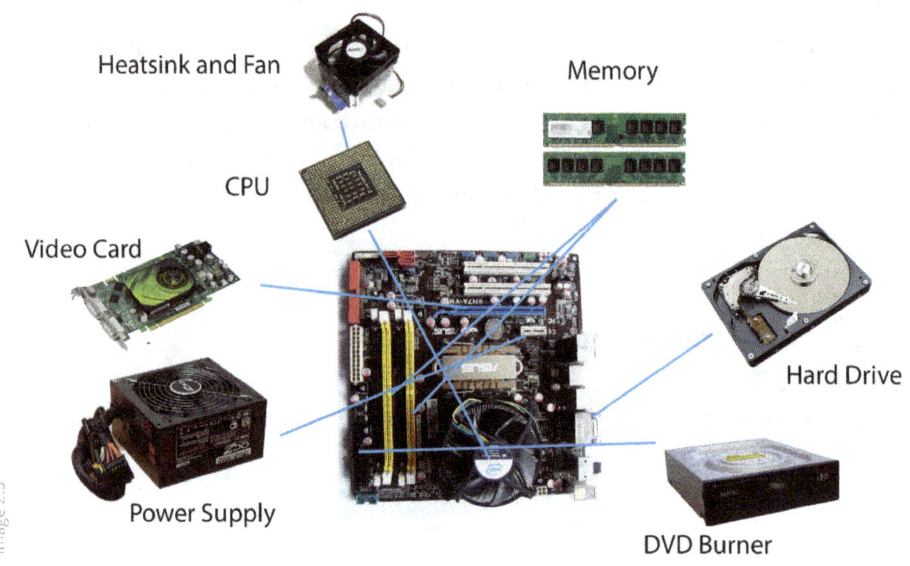

Two Types of Memory

There are two types of memory: RAM and a hard disk drive.

- **RAM** (random access memory) is the short-term working memory of the computer. RAM is volatile flash memory, meaning that when it loses power, it loses data. Everything that a user performs on the system goes through RAM. Let's say we work on a text document. As we type the content of the document, the CPU is processing the input and storing it in RAM, then displaying output on the monitor. Nothing has occurred over at the hard drive since the creation of the file.The contents in RAM will be lost when the power is down.
- Contents of RAM may be captured to the hard disk drive as a swap file.
- The swap file may be useful to the examiner because it may contain password information, network connections, user accounts, emails, or even complete documents.
- The contents of RAM are always changing (volatile memory).

In certain circumstances, the contents of RAM may be captured to a hard disk—otherwise known as a swap file. Virtual memory is used to ensure the system runs efficiently. Often, data will be swapped from RAM to a file called the pagefile.sys, and then swapped from the pagefile back to RAM. The Pagefile.sys is a file on the hard drive and could contain any data that was once present in RAM. The contents of this file may be of great use to the examiner. There are times when RAM cannot be acquired for various reasons. If RAM cannot be acquired, it is important to acquire the virtual memory files on the hard drive.

Image 2.6

The **hard disk drive** is the magnetic storage provided by the hard disk drive that can be used as digital evidence.

- Equate to the long-term memory of a computer.
- Stores documents, images, programs, etc.
- In the early days, hard disks could be as small as five megabytes, but now it is easy to find drives with a capacity of 500 gigabytes or more.

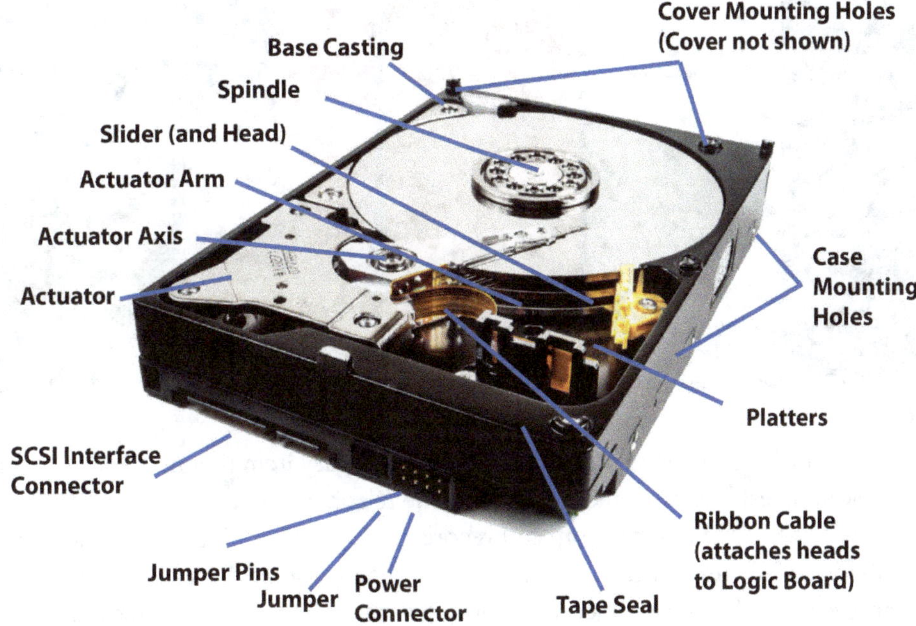

Image 2.7

The sheer volume of data dictates that the search for and extraction of digital evidence must now be subject to a much tighter focus than it once was. When one moves on to file servers, the volume of data to be examined gives real cause for concern. File servers control many terabytes of data.

Generally, the hard disk drive is of more interest to us in the context of digital evidence.

Hard disk drives are a huge filing systems. In addition to the more obvious contents within the hard disk drive system—documents, images, programs, etc.—there are many different items of data that enable a comprehensive understanding of the previous use of the computer.

The sheer volume of data dictates that the search for and extraction of digital evidence is now subject to a much tighter focus than it once was. Modern versions of data storage are far more powerful and have an ability to store a considerable amount of data. For example, today's SSDs (Solid State Drives) are complex pieces of equipment with varying technologies and capacities. We can think of an SSD as its own independent computer, which consists of flash memory modules and stores 4 TB. SSDs are becoming more affordable and reliable.

There is always a ray of hope when it comes to recovering data. For example, users often synchronize the content of their iPad with a host PC. Therefore, if an iPad loses its data, the synchronized data may be recoverable on the host PC.

Mobile/Smart Phone

Basic functions of mobile phones include storing address books, scheduling information, etc.; call records and SMS (short message system) text messages. A smartphone is a hybrid between a mobile telephone and a portable tablet.

Image 2.8

In only 10 years, mobile phones transformed from a luxury item to a necessity consumer item. Even the simplest and most inexpensive mobile telephones have the capability to store address books, scheduling information, call records, etc. Mobile device browsers store the browsing history of the user, which can be found in the mobile device cache folder for that browsing application. The cache folder of the browser history indicates the timestamps. Various mobile applications can be installed, and videos, as well as images, are being uploaded on social media. Excellent evidence may be found in call records and apps; however, SMS is often of even more value. The nature of the information users are willing to transmit using tech services is quite astounding.

When carrying a smartphone, a person is equipped with several gigabytes of internal storage. It is often necessary to obtain billing records to gain a complete picture of the use of a mobile telephone. Since the primary function of the mobile telephone network provider is to make a profit, the service company certainly retains important billing information that can prove useful in cybercrime investigations.

Digital Camera/Video Camera

First-class evidence is provided by digital cameras and video cameras. They have high-capacity memory cards, or sticks, which can store any type of file. This capacity makes digital cameras and video cameras ideal mediums for transporting concealed data. Surprisingly, high-capacity memory cards or storage sticks often used in digital cameras/video cameras can be used to store files of any type. This capability makes an ideal medium for transporting concealed data.

Satellite Navigation Units have become extremely common and relatively inexpensive in a very short period of time, and they can contain crucial evidence relating to the movements of a suspect. Search and Seizure of Digital Evidence.

For most cybercrime cases, investigators need search warrants to search and seize evidence. To obtain a search warrant, an officer must prepare an affidavit describing the basis for probable cause. This lesson focuses on the search and seizure process for cybercrime investigations.

A probable cause, in short, is defined as a reasonable belief that a person has committed a crime. The affidavit must specify the area to be searched and the evidence being sought. If granted, the search warrant allows the officer only a limited right to violate a citizen's privacy. For example, if there is probable cause of digital evidence on a USB, this would not justify seizing every computer on the premises (Brenner & Frederiksen, 2002). If the investigator wants to seize a computer and analyze it later, the probable cause statement should demonstrate the impracticality or danger of examining the computer on the premises and the need to confiscate and analyze it off site (Volonino et al., 2007, p. 56).

The Search Warrant Requirement

Generally, the Fourth Amendment has been interpreted to require that a search warrant contains a complete analysis and description of the place to be searched by law enforcement officers.

DISCUSSION

What information should a law enforcement officer include during the search for digital evidence? Why?

Consider the Fourth Amendment below:

Amendment IV

The right of the people to be secure in their persons, houses, papers, and effects, against unreasonable searches and seizures, shall not be violated, and no Warrants shall issue, but upon probable cause, supported by Oath or affirmation, and particularly describing the place to be searched, and the persons or things to be seized.

Do individuals have a reasonable expectation of privacy in the contents of their laptop computers, USB drives, or cell phones? If the answer is yes, then the government ordinarily must obtain a warrant, or fall within an exception to the warrant requirement, before it accesses the information stored inside (Jarrett et al., 2009, p. 3).

Many cybercrime investigation training materials relating to the handling of digital evidence recommend including the following phrase: "including but not limited to." This wording will allow officers to show that they are looking for specific evidence, but there is the potential for the suspect to have additional evidence that may not have been accounted for in the search warrant.

Thus, it is important that law enforcement officers provide a complete list of every item that could potentially contain digital evidence, and then include a statement concerning how digital evidence may take many forms and so there is the potential that the suspect may have additional pieces of technology that were not included. It is essential to check the scope of the search warrant and the assigned jurisdiction's specific search and seizure restrictions/ rules. For example, courts differentially treat the evidence located in Cloud, an App, or Browser.

Additionally, it is also important that law enforcement officers who request a search warrant ensure that the signing judge understands what is being seized. Many cybercrime investigators recommend that all officers who request search warrants involving digital evidence and computers carry a pocket dictionary of computer terms. The judge may then be provided with these easy-to-understand definitions, should there be any doubt as to what the search warrant is calling for in terms of seizures.

Preplanning Associated with the Search Warrant

Please note that, pursuant to the search warrant, personnel must ensure they only seize computers that are listed by the search warrant.

ECPA

Investigators need to consider whether the actions of their search warrant will affect any federal legislation. There is the possibility that the Electronic Communications Privacy Act of 1986 (ECPA) may govern the actions of government agents if the computer-related evidence to be seized is used in the transmission of emails or other electronic communications. ECPA regulates the amount of information that law enforcement officers may obtain with certain levels of service. Currently, an officer needs the following level of service to obtain information about a potential suspect under the ECPA:

A **subpoena** can provide basic subscriber information which includes names, addresses, local and long-distance telephone connection records, session times and duration, length of service, types of service used, telephone numbers or IP addresses, sources of payment, and the content of emails that are older than 180 days and have been previously opened by the owner.

In the event a suspect has a screen name that indicates the user is an MSN customer, by obtaining a subpoena drafted with this online identity, the investigator could get the name, address, and billing information for the person who registered that screen name with MSN.

A **court order** can provide transactional information. The addresses of past email correspondents are beneficial to investigators, particularly investigators who are handling an investigation involving child pornography or identity fraud. A list of past email correspondents will allow investigators to estimate how many individuals in the public could have been exposed or victim to the individual's behavior.

A **search warrant** can provide the actual content of email messages. The use of a search warrant will allow investigators to examine email contents as well as audit logs, addresses, and billing information. With a search warrant, **investigators can obtain everything associated with an account.**

The search warrant, while allowing for the most collection of evidence, is also the most difficult level of service to obtain as it requires the investigator to complete a statement of underlying facts supporting the basis for justifiable issuance of a search warrant. Thus, the investigator must demonstrate that a crime has been committed, that the individual who possesses the account is linked to the crime, and that the electronic account contains information relevant to the investigation of the case. **In the event that investigators fail to consider the provisions of the ECPA, they may find the evidence is inadmissible.**

PPA

The Privacy Protection Act (PPA) is another article of federal legislation that investigators should consider. The objective is to protect unpublished materials—books, magazines, etc.—from having their materials released by law enforcement officers prior to making the materials accessible to the public.

When agents have reason to believe that a search may result in a seizure of materials relating to First Amendment activities such as publishing or posting materials on the internet, they must consider the effect of the Privacy Protection Act (PPA), 42 U.S.C. § 2000aa. Every federal computer search that implicates the PPA must be approved by the Justice Department, coordinated through CCIPS at (202) 514-1026 (Jarrett et al., 2009).

If the materials are legal, then law enforcement officers are under a duty to return any publishing materials back to the owner as soon as possible.

Warrantless Search Doctrines and Technological Evidence

If law enforcement personnel determine that the option of using a search warrant is not available, what should happen? If this is the case, the government holds the responsibility to prove that the circumstances justified not using a search warrant.

Warrantless Consent Searches

Several factors should be considered when attempting to seize a computer or search a hard drive on the basis of consent. Does the individual granting the consent have a legitimate right to consent to the search?

In *United States v. Smith* (1998) the court found that a search of a suspect's computer was valid despite the fact that the suspect's girlfriend, not the suspect, provided consent. In the opinion of the court there was evidence of the two individuals sharing a common living space; therefore, an expectation of privacy in the computer could be overruled by consent from either party. Additionally, the court considered the fact that the passwords did not protect any computer files that were obtained, a fact that led to the belief that the files were accessible by anyone within the house.

It is important to note that the absence of a search warrant allows an individual to revoke consent or, even worse, wait until trial and then claim that they attempted to revoke consent in the middle of an examination. A variety of court decisions to discuss revoking consent have occurred, but the following court decisions from the State of Florida may in fact illustrate this concept best. Consent may be withdrawn by any of the following:

- The individual may revoke consent verbally (*State v. Hammonds*, 1990).

- The individual may revoke consent through an intentional act, such as grabbing the investigator's hand in order to stop the search of a particular area (*Jimenez v. State*, 1994).
- The individual may revoke consent to a search by fleeing from the search (*Davis v. State*, 1986).

Having and executing a search warrant to search the contents of the computer will prevent a suspect from claiming any of the above actions were attempted during the search of the computer.

Searches Based on Exigent Circumstances

The use of exigent circumstances to justify the warrantless seizure of evidence has long been granted to law enforcement personnel. However, exigent circumstances surrounding a seizure must be verified by indicating the following:

- That a reasonable person would believe that entry was necessary to prevent physical harm to law enforcement personnel or others in the surrounding area;
- That the warrantless seizure was necessary to prevent the destruction of valuable evidence;
- That the warrantless seizure was necessary to prevent the escape of a suspect; or
- That the warrantless search was necessary to prevent any further consequences that could delay legitimate law enforcement efforts. (Jarret et al., 2009)[1]

In the future, warrantless seizures of digital evidence under exigent circumstances could play a significant role in high-technology crime investigations. Particularly, think of the example of the *United States v. David* (1991). In David's case, without a search warrant, law enforcement officers obtained a suspect's electronic date book. In agreeance with the actions of the officer, the court found that in that situation there was sufficient proof that the suspect knew law enforcement personnel were seeking evidence within the device. In an effort to prevent the information from being found, the suspect was deleting the evidence. However, the court also agreed that the exigency ended the moment the date book was seized and there was no further danger to the integrity of the evidence. The seizure of a technological device without a warrant should be considered as a possibility by law enforcement officers, but the search of the same device should be conducted under the provisions of a valid search warrant.

The Search and Seizure Guidelines

The issue of what constitutes legal authority to conduct a search depends upon whether the investigator is investigating a civil or criminal case. In the event that the investigator is investigating a criminal case, the legal authority to conduct a search is under the local jurisdiction. The investigator needs a search warrant that specifies the scope of the search within the legal

---

1   See *United States v. Reed*, 935 F.2d 641, 642 (4th Cir. 1991); see also *United States v. Plavcak*, 411 F.3d 655, 664-65 (6th Cir. 2005) (agents appropriately seized computer without warrant when targets were caught burning relevant documentary evidence and then ran from residence carrying computer); *United States v. Trowbridge*, 2007 WL 4226385, at *4-5 (N.D. Tex. Nov. 29, 2007) (agents appropriately seized computers without a warrant based on exigent circumstances where agents were concerned for their safety during a fast-moving investigation and it was likely that computer evidence would be destroyed).

framework of a criminal investigation. The investigator must keep in mind the guarantees provided by the Fourth Amendment in the U.S. Constitution.

There are many controversial issues on evidence admissibility to the court due to the characteristics of digital evidence. To prohibit these issues, it is imperative to follow an adequate digital forensic procedure.

> **DISCUSSION**
> Consider the way in which you would undertake a computer forensic search.
> List the points that you consider most essential and explain why.

The digital forensic seizure process has three stages: Stage 1, collect preliminary data at the site; Stage 2, determine the environment for the investigation; and Stage 3, secure and transport evidence.

Stage 1. Collecting Preliminary Data at the Site

Let's first outline what should be done in the seizure process. Upon arrival to the site, where a search and seizure warrant will be executed, we should look for and take note of the digital evidence at the site. Prior to conducting a seizure process, we should acknowledge what types of digital evidence should be examined and try to find out all the important aspects of the investigation. Overall, collecting preliminary data at the site is crucial.

Once the investigator has ascertained the legal authority and scope of the investigation, the next steps include fully documenting the scene by taking pictures and then gathering initial information.

The investigator should be prepared to consider a number of important questions prior to the evidence collection at the site.

- What types of digital evidence am I looking for? (Photographs, documents, databases, spreadsheets, financial records, email, etc.)
- What is the skill level of the suspect in question? (The suspect's capability to alter or destroy evidence)
- What kind of hardware is involved? (PC or Mac computer)
- What kind of software is involved? (Hacking tools, MS Office, accounting software, etc.)
- Do I need to preserve other types of evidence? (Fingerprints, DNA, or trace evidence)
- What is the computer environment like? (Network, ISP, OS, usernames and passwords, etc.)

A high-tech response team should have a digital forensic go-bag with the appropriate tools and equipment for preparing the evidence collection at the site.

- Mobilized forensic workstation
- Forensic tools and evidence storage media
- Hardware/Software write blocker
- Screwdriver tool kit
- Common chargers and adapters
- Anti-static bags
- Documentation equipment

Stage 2. Determining the Environment for the Investigation

The next step is to determine whether to do the forensic work on-site or transfer the equipment to a trusted lab environment after acquiring initial information. Various considerations should be taken into account. However, the primary consideration is regularly the integrity of the evidence. Best practice is to always do your examination in a trusted environment, such as a forensics lab.

Factors to consider when deciding where to conduct the examination should include:

- **Integrity of the evidence collection process.** Is the evidence volatile enough that by waiting or transporting the evidence back to the lab I am going to risk losing it?
- **Estimation of the time required to do an examination.** A short examination is usually worth doing on site. If the investigation is extremely critical or appears that it may take some time, the best practice is to do this type of exam in the lab.
- **Equipment resources.** Will the equipment you bring on-site be sufficient to handle the examination thoroughly and professionally? Can you bring the equipment from the lab or will that be impractical?

The most essential factor is the integrity of the evidence collection process. Examples of considerations for an operating system (OS) include:

- Booting a computer will cause the attributes of hundreds of files to be altered.
- Booting does not necessarily mean that the content of the files has been changed; however, it could potentially affect the overall evidential integrity of the drive. The OS maintains records of the attributes of each file stored on the system, which includes the file name and the time and date that the file was created, modified, and accessed.
- Booting a computer equipped with Microsoft Windows will cause the attributes of literally hundreds of files to be altered—while this does not necessarily mean that the content of the file has been changed, it potentially affects the overall evidential integrity of the drive.

The digital information in a computer that is turned on must be collected by the investigator. A great amount of valuable information can be found from the computer memory, so we have to gather as much information as possible from the powered-up computers. Since the contents of the memory can easily disappear when the computer is turned off to seize it, the information in the memory should be secured. Usually, adequate equipment for securing specific volatile information is needed to get information from the computer memory.

On-scene, we typically have two ways to conduct the digital evidence acquisition process: (1) using the agency mobilized forensic workstation and (2) using the suspect's system. Regardless of where or how the acquisition proceeds, there are a few concerns to be aware of.

- If you encounter a device with removable or fixed disks, how will you proceed?
- Is the suspect computer system being powered-on (live box) or powered-off (dead box)?
- What form factor or type of technology is used on the suspect's computer device?

If the target system is powered-on, it is already booted to the suspect's OS. Thus, we need to proceed with caution. The nature of a live system, changes are occurring, no matter what. Anything that is performed on the live system is making changes. Documentation of anything and everything witnessed and or performed is crucial during the acquisition process. During the examination process, you should record contents and current states of digital evidence in all processes. Because other people may get involved in the data process, detailed information of the examination process should be recorded to help other people. It is not such an easy job to find the lead to solve the case from digital evidence because the suspect has a tendency to hide the critical information related to wrongdoing.

Because the actual investigation will be following the acquisition of the critical information incriminating the suspect, the investigator should find out the information. It is quite a difficult job to process investigations against digital media with huge amounts of data, and the amount of digital evidence is always increasing.

There are two distinctive seizure patterns: personal computers and computer servers. The personal computer is a stand-alone computer generally used by a user. When seizing a personal computer, you should collect digital evidence promptly because the data in the memory can be easily changed and removed. Also collect other digital media like flash drives, digital camcorders, phones, tablets, as well as hard disk drives.

When you seize computer servers used by more than one user, consider the amount of data seized, the number of connected users, and the intricacy of computer manipulations. If there is more than one connected user, you should adopt a different strategy for search and seizure than when examining a normal personal computer. When you seize the server, for the most part, you should collect data from the live system on site through the network after consulting with the computer system administrator.

Stage 3. Securing and Transporting Evidence

Many of the basic procedures for seizing cybercrime evidence are exactly the same as in the seizure of any physical piece of evidence. The sequence of common steps is as follows:

1. Document the evidence.
2. Tag it.
3. Bag it.
4. Transport it to the forensics lab.

It is essential that the investigator should be able to demonstrate the location of the evidence throughout its time in the possession of the agency and who has had access to it. In many jurisdictions, it is acceptable to group together certain items of a similar nature and from the same location. It would be appropriate to have one exhibit comprising, for example, five flash drives from a desk drawer. It would not be appropriate to present an exhibit of 10 flash drives found in different locations at the same premises. The location at which the item was found can be highly relevant to the investigation, as is common with other physical evidence. Making detailed notes of what was found and where it was found is very important. If possible, establish who has had and who has not had access to the computer.

Upon beginning your initial documentation, you should do the following:

- Locate all evidence to be seized.
- Record a general description of the room, including

  - Type of media found;

- All peripheral devices attached to the computer(s);
- Make, model, and serial numbers of all devices (computers or otherwise) to be seized; and
- What types of media devices are located in, near, or on the computer.

- Note all wireless devices.
- Make use of chain of custody forms.

Each cable should be labeled and numbered, with the corresponding number being marked on a label attached to the computer. This precise labeling will make it much easier to reassemble the machine in the laboratory. In the case of computers, the machine should be photographed in place prior to disconnecting any cables.

Hard Drive/Computer Details

Description:			
Manufacturer:	Model #:		Serial #:

Chain of Custody

Date/Time	From:	To:	Reasoe:
Date:	Name/Organization:	Name/Organization:	
Time:	Signature:	Signature:	
Date:	Name/Organization:	Name/Organization:	
Time:	Signature:	Signature:	
Date:	Name/Organization:	Name/Organization:	
Time:	Signature:	Signature:	
Date:	Name/Organization:	Name/Organization:	
Time:	Signature:	Signature:	
Date:	Name/Organization:	Name/Organization:	
Time:	Signature:	Signature:	
Date:	Name/Organization:	Name/Organization:	

Image 2.9

Taking photographs of the scene would be the best way to preserve this type of documentation, in addition to writing down this information. The layout of the room should be photographed by the investigator, and if it is a fairly large area, anything within 20 feet of the computers.

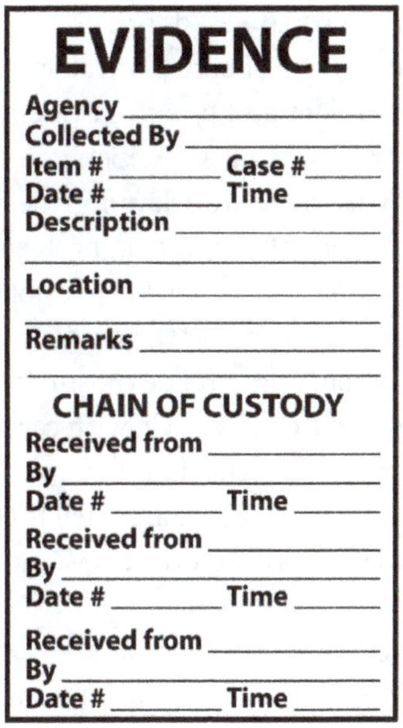
EVIDENCE
Agency
Collected By
Item # Case #
Date # Time
Description
Location
Remarks
CHAIN OF CUSTODY
Received from
By
Date # Time
Received from
By
Date # Time
Received from
By
Date # Time

Image 2.10

Tag the Evidence

The next step is to prepare all items that are going to be transported back to the forensic lab by tagging them. The tags serve as backup documentation and show a physical documentation audit trail, since you are photographing everything. Specifically, the tags should note the time, date, location, and general condition of the evidence.

Bag the Evidence

Upon finishing the tagging of all evidence items, each evidential item should be securely packed using paper packaging. Damage to the system, in some circumstances, can be caused by plastic packaging as it can create static electricity. Small media, sticky notes, and USB drives can be placed into anti-static bags that are quite small. Items such as external hard drives or computers, that tend to be larger, can be transported via anti-static boxes.

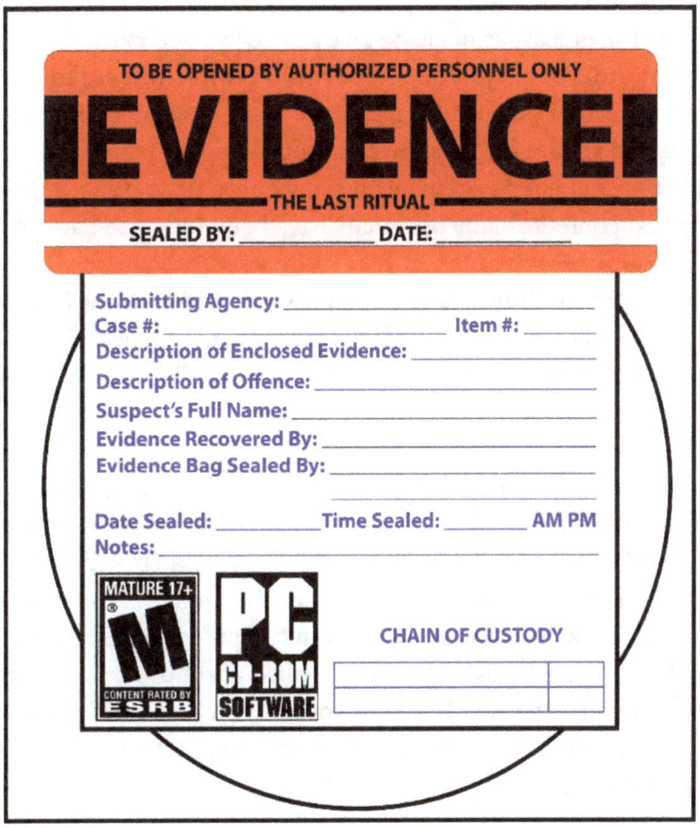
TO BE OPENED BY AUTHORIZED PERSONNEL ONLY
EVIDENCE
THE LAST RITUAL
SEALED BY: DATE:
Submitting Agency:
Case #: Item #:
Description of Enclosed Evidence:
Description of Offence:
Suspect's Full Name:
Evidence Recovered By:
Evidence Bag Sealed By:
Date Sealed: Time Sealed: AM PM
Notes:
MATURE 17+
M
CONTENT RATED BY ESRB
PC CD-ROM SOFTWARE
CHAIN OF CUSTODY

Image 2.11

Transport the Evidence

The process of actually transporting the evidence back to your lab should be considered. As the amount and types of evidence to be transported vary, the tools and equipment you need for transporting will depend on these factors.

When there are delays and the evidence is exposed to environmental factors that could destroy or degrade the evidence, problems tend to arise. During transport, the equipment should be kept as far as reasonably possible from electromagnetic sources such as a police radio. Other negative effects on the evidence include factors such as high humidity, high heat, excessive vibration, and direct exposure to sunlight. The safest vehicle for transportation would be an air-conditioned vehicle with cargo tie-downs on anti-static mats. Upon arrival to the forensics lab, document the time and date and complete the chain of custody forms. The evidence should be stored in a secure area that is controlled with limited access. Ideally, a locked vault providing labeled shelves or drawers would house and organize your evidence.

Investigative Techniques on Seizure of Evidence

The original evidence must remain unchanged from the moment it comes into the possession of law enforcement. There are a number of investigative guidelines that can greatly assist the investigator in achieving this objective:

- **Seek the advice of a trained forensic practitioner.** More information may be divulged by a suspect when they are unsettled by the initial presence of the investigator. The advice of an expert should be sought at the planning stage if the seizure of computer equipment is to be part of a planned operation. Ideally, at the time of seizure, the forensic practitioner should be present or at least available to give advice by telephone.
- **The suspect should be physically separated from the evidence being seized.** The suspect must be physically separated from the evidence to be seized under all circumstances and should not be allowed to touch any evidential item. It is extremely easy to obstruct a computer system by pressing only a few keys to cause total destruction of evidential content. Many criminals have been found to have such routines on their computers.
- **Question the suspect with regard to any passwords and as to the location of possible evidence.** In order to have a successful investigation, the suspect should be questioned with regard to any passwords used and the location of possible evidence. It is a basic investigative principle that the suspect may divulge more information when they are unsettled by the initial presence of the investigator. By the time the investigator is ready for a formal interview, the suspect will have had time to think and may be a great deal more reticent.
- **Make a thorough and detailed search.** A tiny flash drive can house a great deal of data. The search should be complete as soon as possible; although, the severity of the case may be a deciding factor in the time that can be allotted. If chargers or cables for mobile devices are present, seize them. Switching on a modern computer causes hundreds of files to be changed. Even worse, evidential data may have been destroyed. If a computer system is switched on, it should be powered down as soon as practicable (after photographing the screen) by removing the power cable from the machine. In the case of a laptop computer, the battery should be removed.

Routine shutdown procedures can result in the erasure of temporary files that may be relevant to the investigation. In a similar way to the user-initiated processes mentioned above, routines can be constructed that will destroy data if certain keypresses are not input at the time of shutdown.

Common Hiding Places for Physical Evidence

People often make a written note of passwords and other relevant information. Even if a password on a sticky note is not seized by officers, the note may be visible in the photographs of the scene and the password could be read. A special search in the area of the suspect's desk and the area close to it should be conducted.

- Look for notes stuck on the underside of the desk and for small portable memory devices held with gum to the underside of the desk or the back of a drawer.
- Criminals often hide data on CDs in a commercial music CD case, behind the original disk.
- Look for any cables that cannot be identified—they may connect to a remote storage device.
- If a suspect has a network installed, follow any cables to establish whether there are more computers or storage devices.
- Check every electrical outlet on the premises. What is plugged in? Wireless devices can be secreted well away from the computer.

Rules of Mobile Phones

- Mobile phones should be treated differently than computer systems.
- If the phone is switched on, it is better to leave it in that state.
- Switching it off may invoke a security code (PIN number).
- Special bags are available to prevent the phone from connecting to the network.
- If a phone does connect, the calls or messages that may be received could result in the erasure of earlier items.
- A mobile phone should be taken immediately to the forensic lab, as its battery life is limited.

'Airplane Mode PLUS' for Mobile Phones

- Ensuring that cell signal PLUS Bluetooth, and Wi-Fi are all disabled.
- Airplane mode will not always toggle off both Bluetooth and Wi-Fi.

Rules of Tablets and Satnav Devices

- Tablets devices suffer from limited battery life.
- Best practice is to leave them in whatever state they have been seized in.
- Ensure that they are delivered to a competent examiner as soon as is practicable.
- Satnav should be enclosed in an anti-static bag and then placed within a sturdy metal box, thus avoiding as far as possible the reception of satellite signal.

Summary

Addressed in this module were digital evidence principles and search and seizure processes. A thorough description of the various technological devices expected to be encountered at the scene of the search should be expressed in the search warrant. The courts have allowed some jurisdictions to utilize the phrase "including but not limited to" in an effort to recognize that there may be scenarios in which it is impossible to think of every possible type of technology that could be encountered. If the use of a warrant using such terminology is challenged, then the investigator who drafted the warrant could argue that their attempt to list every known technological device commonly encountered during an investigation of such a crime should prevent the appearance of a fishing expedition.

There are certain situations when an investigator may be unable to take the necessary time to draft a valid search warrant. In such a situation, there has been precedent established which would allow for the warrantless search and/or seizure of technological and digital evidence. Currently, the best rule to follow in situations that involve a warrantless seizure is to use one of the established warrantless search doctrines to conduct a seizure of the evidence and immediately work to obtain a well-drafted search warrant to search the internal portions of the evidence.

It is vital that the original evidence remains in initial format from the very moment it comes into the possession of law enforcement. Setting good policies and procedures that serve as guidelines is a key component of methodical investigation. These policies and procedures are constructed not only to delineate the process of an investigation, but also to assist in the management of a computer forensics lab.

Lab 2: VR Project

The purpose of this VR project is to facilitate developing one of the major cybercrime investigation objectives: search and seizure of digital evidence. For the most part, within cybercrime cases, investigators need search warrants to search and seize evidence. To attain a search warrant, an officer must prepare an affidavit that describes the basis for probable cause. This VR project focuses on the learning assessment on the search and seizure process for cybercrime investigations as a virtual hands-on lab.

Objectives

- Learn how to properly search for and seize digital evidence at an electronic crime scene.
- Learn various cybercrime environment issues within the course of the search and seizure of digital evidence.

Fun Law Enforcement Training!

Visit: https://centercicboston.org/STAGESOFTRAINING

Players, are you ready to put your investigative skills to the test?

Join Catch the Digital Evidence to learn the process of collecting digital evidence at a crime scene. Within this VR simulation, players will have the opportunity to experience what it is

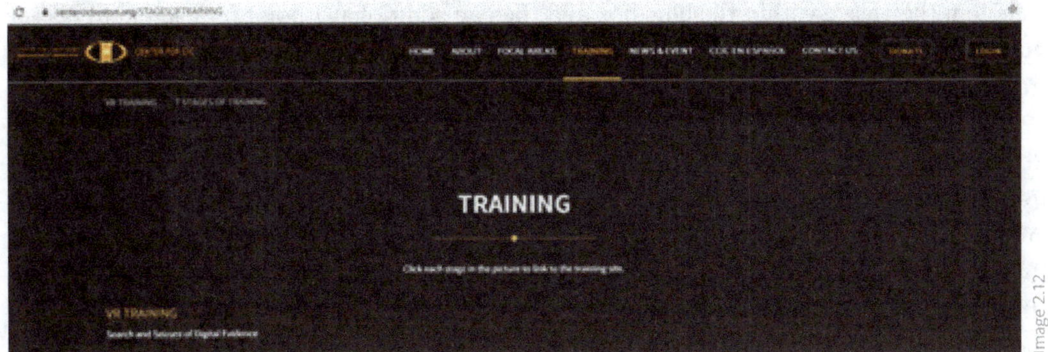

Image 2.12

like to be an investigator including obtaining a search warrant, collecting evidence at a crime scene, and properly transporting the evidence to a lab.

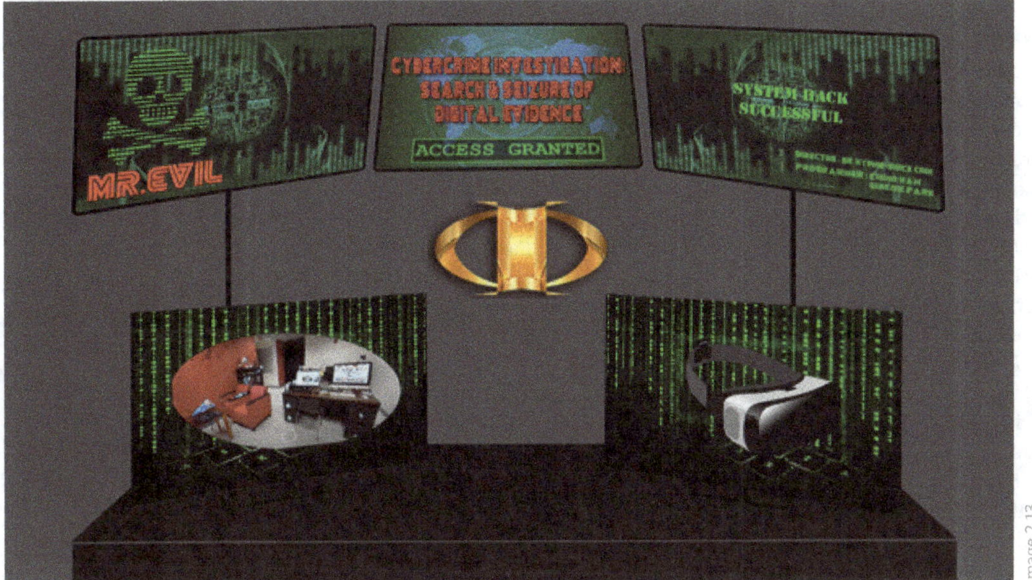

Image 2.13

How to Play

Collect the Evidence!

When players first enter the VR environment, they instantly step foot into the shoes of an investigator. As the lead investigator, they begin by utilizing the battering ram to enter the crime scene. Throughout the collection process, investigators acquire each piece of evidence described within their search warrant. Doing good so far?

Bag and Tag!

After investigators collect each item, they will proceed to the evidence collection station. Here, they will place the item within an anti-static bag and tag the evidence with essential information from the blue evidence box. Investigators will then continue by placing each bagged and tagged evidence item into the brown evidence box. If all goes well, investigators will earn 1 point for each item they place within the blue box and then an additional 1 point for each item placed within the brown box. Congratulations on your success thus far. Keep moving forward. You are so close to solving this case!

How to End

Transportation of Evidence!

Investigators complete their crime scene investigation by taking the two boxes of evidence (blue and brown) and loading them into their choice of transportation to ship the evidence off to a forensic lab. During the final steps of the simulation, it is crucial for investigators to choose wisely in regard to which vehicle they select to transport the two boxes of evidence. The designated vehicle is key to keep the evidence safe and secure. When the simulation has been successfully completed, investigators will have obtained a maximum of 110 points. Sounds to me like you are an expert investigator after all!

References

Brenner, S. W., & Frederiksen, B. A. (2002). Computer searches and seizures: Some unresolved issues. *Michigan Telecommunications and Technology Law Review, 8*(1), 39–114. http://repository.law.umich.edu/cgi/viewcontent.cgi?article=1186&context=mttlr

Brown, M. F. (2001). *Criminal investigation: Law and practice* (2nd ed.). Butterworth-Heinemann.

Carter, D. L. (1995). Computer crime categories: How techno-criminals operate. *FBI Law Enforcement Bulletin, 64*, 21.

Casey, E. (2011). *Digital evidence and computer crime: Forensic science, computers, and the internet.* Academic Press.

Jarrett, H. M., Bailie, M. W., Hagen, E., & Judish, N. (2009). Searching and seizing computers and obtaining electronic evidence in criminal investigations. *US Department of Justice, Computer Crime and Intellectual Property Section Criminal Division.*

Kent, K., Chevalier, S., Grance, T., & Dang, H. (2006). Guide to integrating forensic techniques into incident response. *NIST Special Publication, 10*(14), 800–886. https://csrc.nist.gov/publications/detail/sp/800-86/finalKirk, P. L. (1953). *Crime investigation; physical evidence and the police laboratory.* Interscience.

Rule 41. Search and Seizure. (n.d.). https://www.law.cornell.edu/rules/frcrmp/rule_41

Scientific Working Group on Digital Evidence (SWGDE), & United States of America. (2000). Digital evidence: Standards and principles.

Volonino, L., Anzaldua, R., & Godwin, J. (2007). *Computer forensics: Principles and practices.* Pearson/Prentice Hall.

West Virginia University Forensic Science Initiative. (2007). Technical working group for education and training in digital forensics. *US Department of justice.*

Williams, J. (2012). ACPO good practice guide for digital evidence. *Association of Chief Police Officers.*

Credits

IMG 2.11b: Source: https://brandslogo.net/pc-cd-rom-44969.

IMG 2.12: Source: https://centercicboston.org/STAGESOFTRAINING.

IMG 2.13a1: Source: https://pixabay.com/id/illustrations/keamanan-internet-kejahatan-maya-4700815/.

IMG 2.13a2: Source: https://commons.wikimedia.org/wiki/File:2017_Petya_cyberattack_screenshot.jpg.

IMG 2.13b: Source: https://pixabay.com/id/illustrations/teknologi-globalisasi-komunikasi-3369659/.

IMG 2.13c: Source: https://pixabay.com/id/illustrations/keamanan-internet-kejahatan-maya-4700815/.

IMG 2.13d1: Source: https://www.pxfuel.com/en/free-photo-xzjid.

IMG 2.13e1: Source: https://pixabay.com/illustrations/virtual-reality-play-glasses-2055227/.

IMG 2.13e2: Source: https://www.pxfuel.com/en/free-photo-xzjid.

The Process of Digital Forensics

Kyung-Shick Choi

Introduction

Thus far, we've examined factors to be taken into consideration at the time of seizure, including the types of devices or media that may be seized. During this lesson, we'll expand to review how different types of evidence can be extracted and the process that takes place for this to occur. Additionally, during this lesson, we will discuss securing data integrity.

In terms of cybercrime investigation, what are we looking for? Examples include:

- Word processing documents
- Unlawful images
- Accounting data
- Email messages
- Contact information
- Call records
- Text messages
- Internet usage records
- Copied intellectual properties (music, video, etc.)

These factors are just the tip of the iceberg. Digital forensic investigation involves far more than *just* data recovery. In order for a piece of data to be present on storage media, it must have been processed by a digital device.

Processing digital evidence stored on media includes:

- Processing involving creation, copying, or downloading;
- Searching for digital evidence will focus on what is stored on internal or external media;

- Investigator then needs to trace the processing of all the evidence in order to establish its provenance;
- Investigator must recognize the evidential significance of what they have found;
- Investigator should be able to create a cohesive evidence collection; and
- Investigator must then go back and apply scientific methodology to demonstrate the scientific validity of that evidence.

The focus of the search for digital evidence is looking for evidence stored on either internal media (for example, a hard drive) or external media (for example, a flash drive or a hard drive). Upon finding the evidence, in order to establish its provenance, the investigator will then look for various traces of the processing of all the digital evidence. The evidential significance of what they have found must be recognized by the investigator who can then use it to create a strong case based upon that evidence.

Often, a skilled investigator will locate evidence very quickly and then return and apply scientific methodology to determine the scientific validity of that evidence.

What, Why, When, How, Where, and Who

During the investigation, it is important to have six friends: their names are What, Why, When, How, Where, and Who.

> I keep six honest serving men
> They taught me all I knew
> Their names are What and Why and When
> And How and Where and Who
> —Kipling, R. J. (1902). The elephant's child. *Just so stories*. Macmillan

Rudyard Kipling—a writer in the 19th century—could never have dreamed of the capabilities of the modern computer. However, his words are highly relevant during the search for and extraction of digital evidence.

What

What files are on the computer or other device, and how can they assist us in the pursuit of justice? Files include word processing documents, graphic images, spreadsheets, email or any other type of data produced by a program.

In the forensic examination of a digital device, fragments of previously existing files may be the most crucial factor. Examples include programs like hacking tools, virus production programs, and steganography programs (steganography is a method of concealing data within an innocent looking file) or a "smoking gun" such as stolen intellectual property and collections of child pornography.

At the simplest level, crucial red flags can be found in the presence or occasionally absence of files. Other times, the presence of certain programs on a system may assist in proving the capability to perform certain actions.

Why

The presence of certain digital files is usually the starting point for the analysis. "Why" is the context of motivation. The retrieval of such diverse data as bank statements, email, or threads of text messages may greatly assist in establishing motivation.

The context of motivation may not be thought to be in the province of digital evidence; however, the retrieval of such diverse data as bank statements, email threads, or text messages may greatly assist in establishing motivation.

When

It may be crucial to the prosecution to establish when the suspect created, modified, or accessed files on a computer system.

The digital investigator should create a reliable timeline that can clearly demonstrate the history of the evidential data. A computer literate suspect may go to considerable lengths to attempt to cover their tracks, thus this may not be a simple task.

How

Most examiners will spend much time to find the data. How data came to be on a computer or other digital device may be a very subjective question. In many cases, the important question is not whether the evidence exists, but how it came to exist. Excluding allegations of impropriety, challenges generally take the form of casting doubt as to the creator or user of the data.

Where

The location of data within the system can be vital evidence. If unlawful images are found in the temporary internet files folder, it may be possible to assert that there was a lack of intent. If a "My Documents/Child Images/Boys/Six to eight" folder is found, the origin of the evidential matter can be extremely important.

Questions the investigator will ask include, Where is the evidence located? and Where did it come from? Microsoft Windows utilizes a system-created folder called temporary internet files. Each image that's viewed on the internet will be saved into this temporary folder.

No version of Windows, at this time, automatically creates a folder entitled, "My Documents/Child Images/Boys/Six to eight." Importantly, establishing the origin of the evidential matter can be essential for cybercrime investigation.

Who

This is often a challenging issue in many cybercrime cases. It is often possible to prove the existence of data on a system and the history of the evidence, while it can be extremely difficult to prove beyond all reasonable doubt who really placed the evidence on the computer. Sometimes it is necessary to resort to traditional investigative methods to establish who could, or could not, have access to the computer.

Consider this case:

A considerable volume of child pornography was found on a computer. Who downloaded it?

- The owner of the computer
- Another man living in the house of the computer owner

Evidence: The images were downloaded when the suspect was alone in the house.

The suspect was the owner of the premises from which the computer was seized. Also living in the house was another man and, perhaps unsurprisingly, the suspect had alleged that the other man was the culprit. The investigator then prepared a detailed timeline of the unlawful files. When the investigator obtained the routers of the two individuals, it became very clear that the images had been downloaded when the suspect was alone in the house, or at least when the other man was at work. Without this essential evidence, it would have been virtually impossible to prove the identity of the offender.

Digital Forensic Process: Securing Data Integrity

The data related to a specific event is identified, labeled, recorded, and collected, and its integrity is preserved during the digital evidence collection process. In the second phase—examination—forensic tools and techniques appropriate to the types of data that were collected are executed. This effort is taken to identify and extract the relevant information from the collected data while protecting its integrity.

A combination of automated tools and manual processes may be used for the purpose of examination. The analysis phase then follows, which involves analyzing the results of the examination to derive useful information that addresses the questions that were the momentum for performing the collection and examination. The final phase involves reporting the results of the analysis. Detailing the analysis may include describing the actions performed, determining what other actions should be performed next, and recommending improvements to policies, guidelines, procedures, tools, and other aspects of the forensic process.

The process for performing digital forensics comprises the following basic phases (Kent, 2006):

- **Collection:** *Identifying, labeling, recording, and acquiring data from the possible sources of relevant data, while following procedures that preserve the integrity of the data.*
- **Examination:** *Forensically processing collected data using a combination of automated and manual methods, and assessing and extracting data of particular interest, while preserving the integrity of the data.*
- **Analysis:** *Analyzing the results of the examination, using legally justifiable methods and techniques, to derive useful information that addresses the questions that were the impetus for performing the collection and examination.*
- **Reporting:** *Reporting the results of the analysis, which may include describing the actions used, explaining how tools and procedures were selected, determining what other actions need to be performed (e.g., forensic examination of additional data sources, securing identified vulnerabilities, improving existing security controls), and providing recommendations for improvement to policies, procedures, tools, and other aspects of the forensic process.*

Evidence Collection: Integrity of Data

Critically, in dealing with the digital evidence, the integrity of the digital media must be maintained. The first thing you should do when you seize the digital evidence at the crime scene is the forensic duplication of the digital evidence. The digital evidence can be easily modified or skewed by external impact; thus, you should never be working on the seized digital media. As the investigator, you must work on the duplicated one, while keeping the original one preserved.

Sometimes roadblocks occur. If trouble arises with seizing the hard disk drive in cases of seizing computer servers, then investigators should obtain data through the network by utilizing a dd command[1] or using other imaging collectors.

After forensic duplication, you should "image" the hard disk drive using forensic tools (e.g., Autopsy, AXIOM, EnCase, FTK, X-Way). A disk image is defined as a single file or storage device containing the complete content and structure representing a data storage medium or device, like a hard drive. Usually, a disk image is created by creating a complete sector-by-sector copy of the source medium, resulting in perfectly replicating the structure and contents of a storage device.

Upon creation of the disk image, the investigator is now able to perform extra analyses against the disk image utilizing forensic tools.

After the investigator finds the evidence, the evidence should then be categorized and documented for later processing. During the length of the process, the investigator should use caution when handling digital evidence and document all processes, including details of the steps and processes they worked on.

> **THE NORMAL DIGITAL FORENSIC PROCESS**
> 1. Create a forensic image of the hard disk drive by using specialized forensic hardware or software. The most crucial issue is not to change a single bit or byte on the subject media.
> 2. Clone copies to a similar hard drive.
> 3. Acquire information into special files to emulate the original disk. The common factor is that the acquisition will employ write-blocking hardware.
> 4. Image the RAM prior to it being powered down—very detailed notes must be taken. Successful use of this facility is dependent upon the circumstances of the case and the skills of the examiner.

There are several versions that a forensic image might take, including making exact clone copies to a similar hard drive or receiving the information into special files that can then mimic the original disk.

The acquisition of data from a storage device will allow write-blocking hardware. As the name suggests, write-blocking hardware allows all the content of the subject disk to be read but does not allow any changes to be made to that disk. Similar images may be collected from other types of media such as memory cards or thumb drives. It is now possible to image the random-access memory of a computer.

Very detailed notes are essential. Because the examiner will be working on the original evidence, they must be able to answer potentially challenging questions as to the risks and consequences of their actions. The decision to use this facility is dependent upon the circumstances of the case and the skills of the examiner.

---

1   The dd command captures all files, slack space, and unallocated data. Windows automatically mounts connected storage devices so a write-blocking hardware device must be used. The problem with this is file metadata can be altered when a drive is mounted, changing potentially important evidence.

Forensic Images

A forensic image differs from a mere copy of the data on the original disk in the following ways:

- A forensic image makes the digital evidence amenable to recovery of the deleted or fragmented file.
- The analysis of the image takes place utilizing forensic software.
- The image does not need to be booted.
- It is possible to make any number of restored copies of the original drive without affecting its evidential integrity.

For the commercial sector:

- For purposes of business continuity, it may be totally impractical to take a computer offline for an extended period.
- Most jurisdictions now accept a verified image; therefore, the evidential computer can be returned to the user if the circumstances dictate.
- This allows the machine to continue to function as normal while still retaining the original evidence for the legal process.

A forensic image differs from a mere copy in that every bit and byte on the original disk or card is extracted. Using this methodology, files that have been deleted and fragments of files which previously existed may be recoverable. Through industry standard procedures the image is verified and can identify disparity of even one single bit. From this point onwards, the original machine is locked away while the image is analyzed. The process of analysis for the image takes place using unique forensic software such as EnCase Imager, FTK Imager, Paladin, X-Ways, Axiom, etc. For the examination to take place, the image does not need to be booted. While in its native format, the image is incapable of being so booted. An added bonus, it is possible, when necessary, to make any number of restored copies of the original drive without affecting the integrity of the evidence.

In the commercial sector, the ability to take a verifiable image of a computer at a particular moment in time is essential. This serves the purpose of business continuity, as it may be totally impractical to take a computer offline for an extended period.

In most jurisdictions verified images are accepted; therefore, if the circumstances dictate, the evidential computer can be returned to the user. Although, a better solution may be to retain the original drive and then to restore the forensic image to a new drive which is placed in the subject computer. The machine is then able to continue computer function as normal while retaining the original evidence for the legal process.

THE PRINCIPLES OF COMPUTER-BASED ELECTRONIC EVIDENCE BY ACPO (ASSO-CIATION OF CHIEF POLICE OFFICERS)

- **Rule #1.** No action taken by law enforcement agencies or their agents should change data held on a computer or storage media which may subsequently be relied upon in court.
- **Rule #2.** In circumstances where a person finds it necessary to access original data held on a computer or on storage media, that person must be competent to do so

and be able to give evidence explaining the relevance and the implications of their actions.
- **Rule #3.** An audit trail or other record of all processes applied to computer-based electronic evidence should be created and preserved. An independent third party should be able to examine these processes and achieve the same result.
- **Rule #4.** The person in charge of the investigation, the case investigator, has overall responsibility for ensuring that the law and these principles are adhered to. (Williams, 2012)

Physical Items

Physical items are items on which data objects or information may be stored or through which data objects are transferred.

Data Objects

Data objects are defined as objects or information of potential probative value that are associated with physical items. Data objects may occur in different formats without altering the original information.

As discussed in the previous module, digital evidence is valuable evidence; however, it is often latent in the same sense as fingerprints and DNA evidence. Also, digital evidence is fragile as it can be easily altered, damaged, and destroyed. Furthermore, it is sometimes time-sensitive and can cross borders with ease and speed. For that reason, we should handle digital evidence carefully and in the same manner as forensic evidence is handled—with respect and care in order to preserve its value as evidence. This relates not just to the physical integrity of a device, but also to the digital data it contains.

Original Digital Evidence

Original data evidence refers to physical items and the data objects associated with those items at the time of acquisition or seizure. Digital evidence is categorized into (1) original digital evidence and (2) duplicate digital evidence.

At the time of acquisition or seizure, physical items and the data objects are associated with each other.

Duplicate digital evidence is an accurate reproduction of all data objects contained on an original physical item. To make duplicate digital evidence, the original evidence must be acquired in a manner that protects and preserves the integrity of the evidence.

Four reasons we use duplicate digital evidence include:

- Preventing alterations to the original media and preserving the data/values in their original state;
- Ensuring that the data present on the seized media is available for examination;
- Assisting in defending against allegations of loss or destruction of data caused by examination; and
- Allowing controlled alterations and recovery of the imaged media or data for better examination.

What Is a Forensic Copy (Duplicate)?

The technical term for the end product of a forensics acquisition of a computer's hard drive or other storage device is "forensic copy." This end product is a bit-stream copy (duplicate), which is a bit-for-bit digital copy of a digital original document, file, partition, graphic image, entire disk, or similar object.

It is recommended to make several forensic copies, so if something happens to one copy, another backup is readily available. A drive can be imaged (duplicated) without anyone viewing its contents, thus privacy or confidentiality issues are not at risk.

A bit-stream copy is the basis of all forensic work. Forensics images or clones of the original media should be captured by using hardware or software that can capture a bit-stream copy from the original media to ensure no difference between them. It should be archived to media and maintained consistently according to departmental policy and applicable laws, as should all forensic work.

Whenever using bit-stream copies to make duplicates, we should take the following precautions:

- Hardware or software write-blockers can be used to prevent the evidence from being modified.
- Properly prepared media should be used when making forensic copies to ensure no commingling of data from different cases occurs. Methods of acquiring evidence should be forensically sound and verifiable.

Image 3.1

In Practice: Write Blocking and Protection

It is important to never turn on the PC without equipping write-blocking devices or software. Write-blocking devices prevent any writes to the drives attached and offer very fast acquisition speeds. The mere act of turning on the PC could potentially alter critical data. During the bootup, Windows-based operating systems alter many date/timestamps in the system and the date/timestamps of documents.

Bit-Stream Image File

Generally, there are two methods for making duplicate digital evidence from the original digital evidence. The first method is to make an image file of the original digital evidence. An image file is a digital, sometimes compressed file from which a bit-stream duplicate of an original digital object can be reconstructed. In order to comply with the principles of computer-based electronic evidence, an image should be made of the entire target device. For example, it should be saved to a drive as an image file for later restoration. Sometimes, though, we need to make a logical image file, such as when imaging a server that may be beyond the scope of our hardware or tools. No matter what method or which image file we take, we should verify the integrity of an image file and have documentation that is correct.

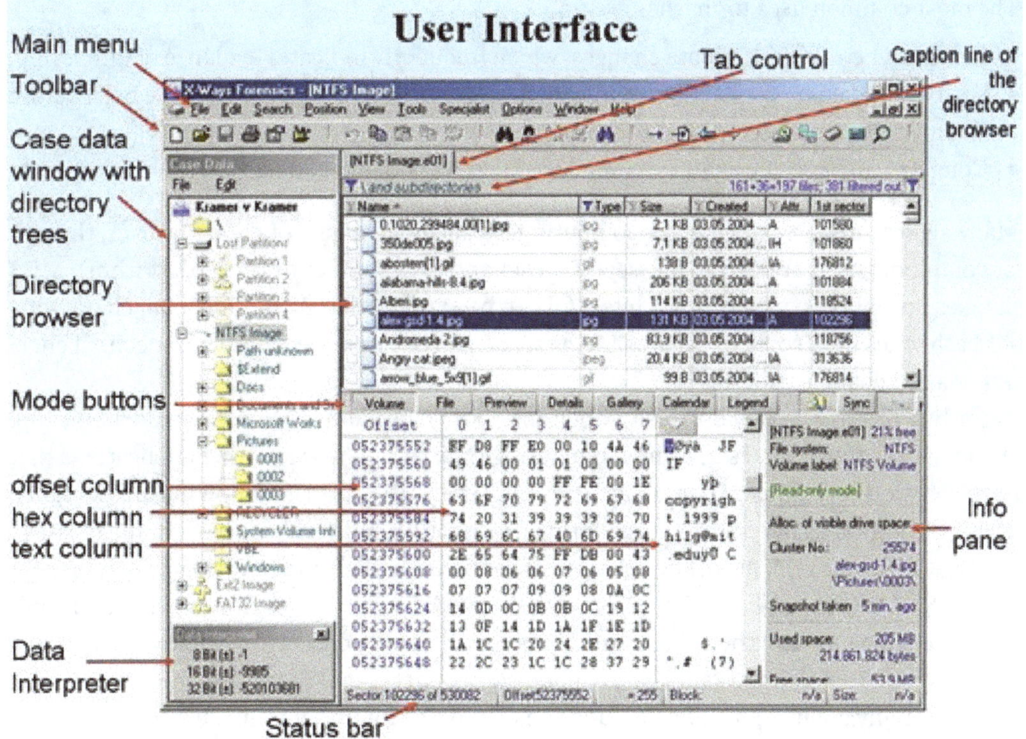

Image 3.2

Bit-Stream Drive Clone or Direct Drive-to-Drive Clone

One option for making duplicate digital evidence from the original digital evidence is to make bit-stream drive clones. Cloning entails making an exact copy of the original drive using hardware or software tools and placing the original drive and the wiped target drive into one computer.

The integrity of the image file should be verified and have documentation that is correct just as an image file does. Essentially, a drive-to-drive copy is when the examiner must make a bootable copy of the drive.

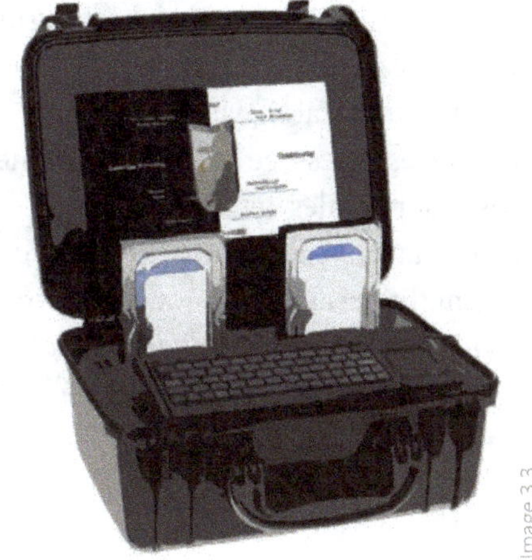

Image 3.3

Hash/Verification

When digital evidence is introduced in legal proceedings, it is difficult to argue against the presence of the proposed evidence. We utilize hash values to prove that the original evidence is or was preserved and that the duplicate worked on is in fact a duplicate. To be as reliable as the original, the integrity of duplicate digital evidence must be verified. Hash verification is a technique that uniquely identifies identical files.

A hash is a digital fingerprint of a file or collection of data, represented as a string of binary data written in hexadecimal notation.

The most common uses for hashes are to:

- Identify when a chunk of data changes, which frequently indicates evidence of tampering;
- Verify that data has not changed, in which case the hash should be the same both before and after the verification; and
- Compare a hash value against a library of known good and bad hashes, seeking a match.

Many algorithms are available for computing the message digest of data; however, the two most commonly used are MD5 (Message Digest) and SHA-1 (Secure Hash Algorithm). MD5 hash takes up 16 bytes, which is 128 bits, and can be expressed as 32 hexadecimal characters. SHA1 hash takes up 20 bytes, which is 160 bits, and can be expressed as 40 hexadecimal characters or as 32 characters (Base32).

According to NIST Policy on Hash Functions (2015), "Federal agencies should stop using SHA-1 for generating digital signatures, generating time stamps and for other applications that require collision resistance. Federal agencies may use SHA-1 for the following applications: verifying old digital signatures and time stamps, generating and verifying hash-based message authentication codes (HMACs), key derivation functions (KDFs), and random bit/number generation. ... SHA-2 (i.e., SHA-224, SHA-256, SHA-384, SHA-512, SHA-512/224 and SHA-512/256): Federal agencies may use these hash functions for all applications that employ secure hash algorithms. NIST encourages application and protocol designers to implement SHA-256 at a minimum for any applications of hash functions requiring interoperability."

Most evidence processors' hash analysis settings allow you to create MD5 and SHA-1 hash values for files so that you can later use them for the reasons specified above. You can check whether to run either or both hashing algorithms when you click the Hash Analysis hyperlinked name displayed in the Edit Settings dialog.

To make digital signatures, the MD5 algorithm is used. It is a one-way hash function, meaning that it takes a message and converts it into a fixed string of digits, also called a message digest. The odds that two files with different contents have the same hash value are roughly 2 raised to the hundred and twenty-eighth power. A "collision" occurs when two different data streams generate the same hash value. No known MD5 or SHA-1 "collision" has occurred outside a lab environment.

This means that if the hash values of two files match, it is quite likely that the file contents match exactly. If it has changed so much as one bit, we will get a completely different value for that media.

Copy

A copy process only gets active files; therefore, ambient data will not be copied. A copy is an accurate reproduction of the information contained on the original physical item, independent of the original physical item. **Slack space and free space are left within a copy, as you only get the data area with a copy.**

Proper Preservation and Presentation of Digital Evidence

The complex methods of recovering digital evidence may seem intimidating, but if the process is followed correctly, the integrity and quality of the evidence can be secured. Without the user's awareness, operating systems and other programs frequently alter the contents of electronic storage, and this change may happen automatically. In an effort to comply with the principles of computer-based digital evidence, a copy image should be made of the entire target device.

Partial copying may be considered when the amount of data to be imaged makes this impracticable. Investigators should be cautious to ensure that all relevant evidence is captured if this approach is adopted. It is essential to display objectivity of evidence in court, as well as the continuity and integrity of the evidence. It is also necessary to demonstrate how the evidence has been recovered, showing each process by which evidence was obtained.

Physical Disks and Logical Volumes

Let's say that we purchase a new 1 Terabyte (TB) external HDD from our favorite electronics retailer. When we get it home, plug it into the computer, and nothing happens. It's not likely that the new HDD is bad. It's more likely that it needs to be prepared for use. We need to partition the HDD into one or more logical volumes. We head on over to disk management and it states that we need to initialize the physical disk, which is just a fancy term for choose and deploy a partition scheme.

A physical disk is a tangible item that we can hold in our hands. We can think of it as a shell of a house. Standing outside the house, we can see how big it is but have no idea how many rooms it is divided into or what the purpose of the rooms is.

A logical volume is simply a room inside the house or a room inside the physical disk. The number of logical volumes that will be present on the physical disk depends on the user and their needs or intended use. A physical disk can be partitioned into 1 or more logical volumes. Each logical volume is treated independently from one another, so each logical volume could contain its own operating system and user data.

How do we go from a physical disk to one or more logical volumes? It's called partitioning or dividing up the physical disk. In computers, there are two common partitioning schemes that can be deployed today. These two partitioning schemes are the old school **Master Boot Record (MBR)** or new school **GUID Partition Table (GPT)**.

The MBR scheme supports up to 2 TB disks and is made up for three components. The first component is the boot code, which is simply some technical information about the disk. The second component is the master partition table (MPT). We can think of the MPT as a map to the physical disk. It tells the computer/OS how many logical volumes are on the physical disk, as well as how big each logical volume is and exactly where those logical volumes are located. The third component is the signature, which is the last two bytes in sector 0. The Signature simply tells the computer/OS that it has reached the end of the MBR/MPT. The Signature is 0x55 0xAA at offsets 510 and 511.

The GUID Partition Table is a newer partitioning style, and the maximum drive size is unreachable. The overall concept of partitioning is consistent: simply apply an organization structure to a physical disk to keep track of logical volumes. There are different components and greater functionality to GPT, but overall, it accomplishes the same goal as MBR. With GPT, there's only one type of partition—a primary partition. This also means that we can have up to 128 volumes, possibly having a bootable operating system on each.How can we tell if we are looking at a physical disk or logical volume? MS Windows typically shows drive letters in file explorer. However, with a little information, we can view physical disk assignments as well. In the command prompt, we can type the below command and it will display all the physical disks attached to the system. As such, the primary internal HDD or SSD will typically be listed as PHYSICALDRIVE 0. In disk manager, it simply states disk 0 or disk 1, etc., depending on how many physical disks are attached. To utilize a volume (drive), the operating system must recognize or support the file system format.

Image 3.4

```
Caption=SAMSUNG MZVLW1T0HMLH-00000
DeviceID=\\.\PHYSICALDRIVE0
Model=SAMSUNG MZVLW1T0HMLH-00000
Partitions=4
Size=1024203640320
```

- Command Prompt

  ```
  wmic diskdrive list brief /
  format:list
  \\.\PHYSICALDRIVE0
  \\.\PHYSICAL DRIVE1
  ```

- Disk Management

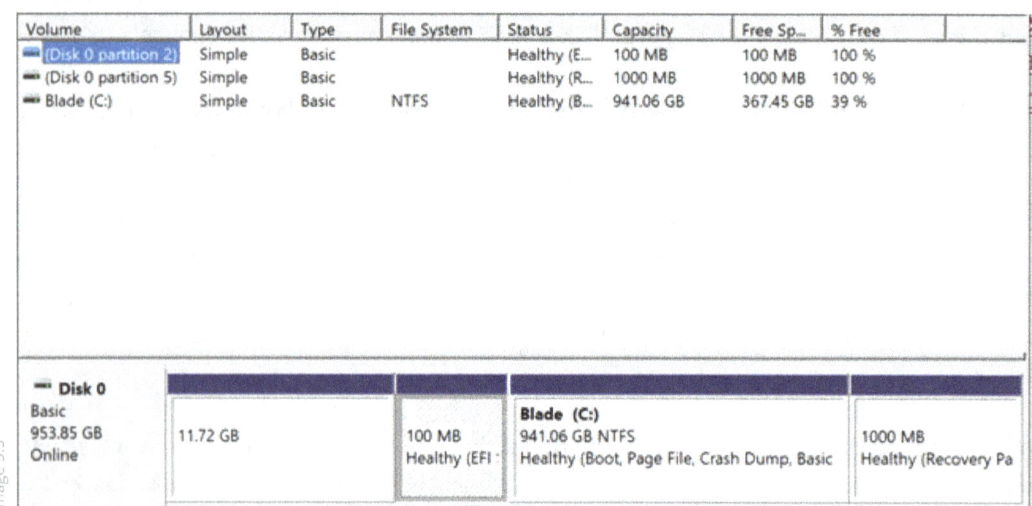

Image 3.5

File Storage: Using Data from Data Files

A data file (also called a file) is a collection of information logically grouped into a single entity and referenced by a unique name, such as a filename. A file can include many data types, such as a document, an image, a video, or an application. The success of processing computer media depends upon the ability to collect, examine, and analyze the files that reside on the media.

This section provides an overview of the most common media types and file systems methods for naming, storing, organizing, and accessing files. It then discusses how files should be collected and how the integrity of the files should be preserved. Various technical issues related to file recovery, such as recovering data from deleted files, will also be outlined. The last portion of the section describes the examination and analysis of files, providing guidance on tools and techniques that can assist analysts. An understanding of the relationship between file systems and investigation is crucial for a digital forensic investigator.

A Stored Computer File Is Just a Collection of Bytes

A stored computer file contains data but no information to identify itself or its attributes. It is no more than an unidentified patch of data on the disk without an indexing system. Operating systems maintain such an indexing system and record the starting position of the file, its length, time attributes, and the name under which it is stored.

The File Itself Does Not Hold Identifying Data

The identifying system holds the key to the identity of the file. If the indexing system becomes badly corrupted or an entry overwritten, then although the data in a file may be recoverable, its name and time data will not.

Information relating to a deleted file may be held in the index system for a considerable time, which is beneficial from an investigative standpoint. If the index entry has not been overwritten by a new one and if the file data has not itself been overwritten, then the file may be recovered together with its identity.

Operating Systems Reuse Disk Space

It is important to grasp the principle of *unallocated clusters.* This occurs if a file is fully deleted from the index (for example, by using the empty recycle bin command on a Microsoft Windows machine) and then the area of the disk is marked as available for reuse.

In many cases, the suspect may have tried to dispose of some of the most crucial evidence. In some circumstances, a document, picture, or other data can be fully recovered, yet the history of that data cannot. Contrary to popular belief, this certainly does not mean that the recovered data has no value.

Certain types of files, notably word processing documents, do contain some internal information relating to them, known as metadata. In turn, recovered emails could contain an abundance of information relating to their previous transmission. Yet, what is not included is their history on the machine from which they were extracted. For example, a child pornography investigation that uncovers relatively few indecent images of children in the live file system would be strengthened by the recovery of hundreds of thousands of images from unallocated clusters.

File Systems

For media to be used to store files, the media must usually be partitioned and formatted into logical volumes. Partitioning is defined as the act of systematically dividing media into portions that function as physically separate units. A logical volume is a partition or a collection of partitions serving as an independent entity that has been formatted with a filesystem. The selected file systems determine the format of the logical volumes.

The way that files are named, stored, organized, and accessed on logical volumes is defined by a filesystem. Various filesystems exist and each provides specialized features and data structures. Yet, some common traits are shared by all filesystems.

First, each file system utilizes the concept of directories and files to organize and store data. The organizational structures that are used to group files together are directories. In addition to files, subdirectories may be contained in the other directories. Next, the location of files on media can be determined from the data structure within the filesystem. Additionally, filesystems store each data file written to media in one or more file allocation units. These are referred to as clusters by some filesystems (e.g., File Allocation Table [FAT], NT File System [NTFS]) whereas other filesystems refer to them as blocks (e.g., UNIX and Linux). A file allocation unit is simply a group of sectors, which are the smallest units that can be accessed on media.

A short list of commonly used filesystems includes the following:

- **FAT12:** FAT12 is used only on floppy disks and FAT volumes smaller than 16 MB. FAT12 uses a 12-bit file allocation table entry to address an entry in the filesystem.
- **FAT16:** MS-DOS, Windows 95/98/NT/2000/XP, Windows Server 2003, and some UNIX OSs support FAT16 natively. FAT16 is also commonly used for multimedia devices such as digital cameras and audio players. FAT16 uses a 16-bit file allocation table entry to address an entry in the filesystem. FAT16 volumes are limited to a maximum size of 2 GB in MS-DOS and Windows 95/98. Windows NT and newer OSs increase the maximum volume size for FAT16 to 4 GB.
- **FAT32:** Windows 95 Original Equipment Manufacturer (OEM) Service Release 2 (OSR2), Windows 98/2000/XP, and Windows Server 2003 support FAT32 natively, as do some multimedia devices. FAT32 uses a 32-bit file allocation table entry to address an entry in the filesystem. The maximum FAT32 volume size is 2 terabytes (TB).
- **NTFS:** Windows NT/2000/XP and Windows Server 2003 support NTFS natively. NTFS is a recoverable filesystem, which means that it can automatically restore the consistency of the filesystem when errors occur. In addition, NTFS supports data compression and encryption, and it allows user- and group-level access permissions to be defined for data files and directories. The maximum NTFS volume size is 2 TB.
- **High-Performance File System (HPFS):** HPFS is supported natively by OS/2 and can be read by Windows NT 3.1, 3.5, and 3.51. HPFS builds on the directory organization of FAT by providing automatic sorting of directories. In addition, HPFS reduces the amount of lost disk space by utilizing smaller units of allocation. The maximum HPFS volume size is 64 GB.
- **Second Extended Filesystem (ext2fs):** ext2fs is supported natively by Linux. It supports standard UNIX file types and filesystem checks to ensure filesystem consistency. The maximum ext2fs volume size is 4 TB.
- **Third Extended Filesystem (ext3fs):** ext3fs is supported natively by Linux. It is based on the ext2fs filesystem and provides journaling capabilities that allow consistency checks of the filesystem to be performed quickly on large amounts of data. The maximum ext3fs volume size is 4 TB.
- **ReiserFS:** ReiserFS is supported by Linux and is the default filesystem for several common versions of Linux. It offers journaling capabilities and is significantly faster than the ext2fs and ext3fs filesystems. The maximum volume size is 16 TB.
- **Hierarchical File System (HFS):** HFS is supported natively by macOS. HFS is mainly used in older versions of macOS but is still supported in newer versions. The maximum HFS volume size under macOS 6 and 7 is 2 GB. The maximum HFS volume size in macOS 7.5 is 4 GB. MacOS 7.5.2 and newer macOSs increase the maximum HFS volume size to 2 TB.
- **HFS Plus:** HFS Plus is supported natively by macOS 8.1 and later and is a journaling filesystem under Mac OS X. It is the successor to HFS and provides numerous enhancements,

Karen Kent, et al., *Guide to Integrating Forensic Techniques into Incident Response: Recommendations of the National Institute of Standards and Technology*, 2006.

such as long filename support and Unicode filename support for international filenames. The maximum HFS Plus volume size is 2 TB.

- **UNIX File System (UFS):** UFS is supported natively by several types of UNIX OSs, including Solaris, FreeBSD, OpenBSD, and Mac OS X. However, most OSs have added proprietary features, so the details of UFSs differ among implementations.
- **Compact Disc File System (CDFS):** As the name indicates, the CDFS filesystem is used for CDs.
- **International Organization for Standardization (ISO) 9660 and Joliet:** The ISO 9660 filesystem is commonly used on CD-ROMs. Another popular CD-ROM filesystem, Joliet, is a variant of ISO 9660. ISO 9660 supports filename lengths of up to 32 characters, whereas Joliet supports up to 64 characters. Joliet also supports Unicode characters within filenames.
- **Universal Disk Format (UDF):** UDF is the filesystem used for DVDs and is also used for some CDs. (Kent et al., 2006)

Other Data on Media

As described above, filesystems are designed to store files on media. However, deleted files or earlier versions of existing files may hold data within the filesystems. Important information can be found within this data.

But how can this data still exist on various media? An outline of how this data can still exist is described below:

- **Deleted Files.** Typically, when a file is deleted, it is not erased from the media; instead, the information in the directory data structure that points to the location of the file is marked as deleted. This means that the file is still stored on the media but is no longer enumerated by the OS. The operating system considers this to be free space and can overwrite any portion of or the entire deleted file at any time.
- **Slack Space.** As noted previously, filesystems use file allocation units to store files. Even if a file requires less space than the file allocation unit size, an entire file allocation unit is still reserved for the file. For example, if the file allocation unit size is 32 kilobytes (KB) and a file is only 7 KB, the entire 32 KB is still allocated to the file, but only 7 KB is used, resulting in 25 KB of unused space. This unused space is referred to as file slack space, and it may hold residual data such as portions of deleted files.
- **Free Space.** Free space is the area on media that is not allocated to any partition; it includes unallocated clusters or blocks. This often includes space on the media where files (and even entire volumes) may have resided at one point but have since been deleted. The free space may still contain pieces of data.

Collecting Files

During the data collection process, the analyst should make multiple copies of the relevant files or filesystems—typically a master copy and a working copy.

The working copy can then be utilized by the investigator without affecting the original files or the master copy. In addition to collecting the files, it is also important to collect the timestamps for the files, such as when the file was last modified or accessed. Later we will discuss timestamps and explain how they can be preserved.

Copying Files from Media

Two different techniques allow files to be copied from media:

- **Logical backup/ Logical Acquisition.** A logical backup copies the directories and files of a logical volume. Not captured in the logical backup is other data that may be present on the media, such as deleted files or residual data stored in slack space.
- **Bit-stream imaging/ Physical Acquisition.** Otherwise known as disk imaging, bit-stream imaging generates a bit-for-bit copy of the original media, including free space and slack space. Bit-stream images require more storage space and take longer to perform than logical backups.

When comparing the acquisition of a physical disk to an acquisition of a logical volume, they are both a bit-by-bit capture. However, the logical volume is a division of the physical disk. Therefore, the data set is limited to the area inside the walls of the room.

In the event evidence may be needed for prosecution or disciplinary actions, the analyst should get a bit-stream image of the original media, label the original media, and store it securely as evidence. Subsequent analysis should then be performed using the copied media to ensure that the original media is not modified and that a copy of the original media can always be re-created if necessary. Each and every step that was taken to create the image copy should be documented. Documentation should allow any analyst to produce an exact duplicate of the original media using the same procedures. In addition, proper documentation can demonstrate that during the collection process evidence was not mishandled. Aside from the steps that were taken to record the image, the analyst should document supplementary information such as the hard drive model and serial number, media storage capacity, and information about the imaging software or hardware that was used (e.g., name, version number, licensing information). The maintenance chain of custody is supported by all these actions.

Digital Forensic Process: Examination, Analysis, and Report

Examining Data Files

After a logical backup or bit-stream imaging has been performed, the backup or image may have to be restored to another media before the data can be examined. This is dependent on the forensic tools that will be used to perform the analysis. Some tools can analyze data directly from an image file, while others require that the backup or image be restored to a medium first. The data should be accessed as read-only to ensure that the data being examined is not modified and that it will provide consistent results on successive runs, regardless of whether an image file or a restored image is used in the examination. As stated, during this process write-blockers can be used to prevent writes from occurring to the restored image. After restoring the backup, the examination of the collected data can begin as the investigator performs an assessment of the relevant files and data by locating all files, including deleted files, remnants of files in slack and free space, and hidden files. Next, the analyst may need to extract the data from some or all of the files, which may be complicated due to obstructing measures such as encryption and password protection. This section describes the processes involved in examining files and data, as well as techniques that can expedite examination.

Locating the Files

The first step in the examination is to locate the files. Many gigabytes of slack space and free space can be captured in a disk image, which could contain thousands of files and file fragments. Manually extracting data from unused space can be a time-consuming and difficult process, as it requires knowledge of the underlying filesystem format. Thanks to several tools, the process of extracting data from unused space and saving it to data files, as well as recovering deleted files and files within a recycling bin, can be automated. The contents of slack space can be displayed by analysts with hex editors or special slack recovery tools.

Extracting the Data

The completion of the examination process involves extracting data from some or all of the files. In order to make sense of the contents of a file, an analyst needs to identify what type of data the file contains. The specific purpose of file extensions is to determine the nature of the file contents; for example, a JPG extension indicates a graphic file and an MP3 extension indicates a music file. Yet, users can assign any file extension to any type of file, such as naming a text file mysong.mp4 or eliminating a file extension. Additionally, some file extensions may be hidden or unsupported on other OSs. Analysts should never assume that file extensions are accurate.

Analysts can more accurately identify the type of data stored in many files by looking at their file headers. A file header contains identifying information about a file and possibly metadata that provides information about the file contents. The figure below shows the file header contains a file signature that identifies the type of data that a particular file contains. The example includes **a file header FFD8**, which indicates that this is a **JPEG** file. The file header could be located in a file separate from the actual file data. A simple histogram is another effective technique for identifying the type of data in a file, showing the distribution of ASCII values as a percentage of total characters in a file. For example, a spike in the 'space,' 'a,' and 'e' lines generally indicates a text file, while consistency across the histogram indicates a compressed file. Files that are encrypted or modified through steganography may be indicated through patterns.

Challenges are often presented to analysts when encryption is present. In an effort to prevent other users from accessing information without a decryption key or passphrase, users might encrypt individual files, folders, volumes, or partitions. A third-party program or the OS may perform the encryption. It can be relatively easy to identify an encrypted file, yet it is usually not so easy to decrypt it. By examining the file header, identifying encryption programs installed on the system, or finding encryption keys (which are often stored on other media) the analyst might be able to identify the encryption method. Upon determining the encryption method, the analyst can better decipher the feasibility of decrypting the file. Often, because the encryption method is strong and the authentication (e.g., passphrase) used to perform decryption is unavailable, decryption may be impossible.

Offset	0	1	2	3	4	5	6	7	8	9	A	B	C	D	E	F	
00000000	FF	D8	FF	E0	00	10	4A	46	49	46	00	01	01	00	00	01	ÿØÿà..JFIF......
00000010	00	01	00	00	FF	DB	00	43	00	08	06	06	07	06	05	08	....ÿÛ.C........
00000020	07	07	07	09	09	08	0A	0C	14	0D	0C	0B	0B	0C	19	12	....
00000030	13	0F	14	1D	1A	1F	1E	1D	1A	1C	1C	20	24	2E	27	20	.... $.'
00000040	22	2C	23	1C	1C	28	37	29	2C	30	31	34	34	34	1F	27	",#..(7).01444.'
00000050	39	3D	38	32	3C	2E	33	34	32	FF	DB	00	43	01	09	09	9=82<.342ÿÛ.C...
00000060	09	0C	0B	0C	18	0D	0D	18	32	21	1C	21	32	32	32	32	.......2!.!2222

image 3.6

Using Forensic Software

Forensic software is used to examine the image that has been created of the original media. What is the right tool for the job? The following list should be considered when selecting an examination tool:

- **Usability:** How easy is it to use?
- **Comprehensive:** How good is it at showing all data?
- **Accuracy:** Has its output been verified?
- **Deterministic:** Repeatable (same steps/same output).
- **Verifiable:** Ability to ensure accuracy.
- **Tested:** Ability to verify that data is accurate and makes no changes on the device.

Keep in mind that while some forensic tools will brag about their features and abilities, no one tool is perfect. The nature of the case determines the procedures that will be employed. The dedicated software allows for many ways in which data can be located and refined.

Various forensic tools that enable examinations and analysis of data, as well as some collection activities, should be accessible to analysts. The variety of forensic products currently available allow analysts to perform a wide range of processes such as analyzing files and applications, collecting files, reading disk images, and extracting data from files. Also, most analysis products offer the ability to generate reports and log all errors that occurred during the analysis. Although these products are invaluable in performing analysis, it is crucial to comprehend which processes should be run to answer particular questions about the data. In regard to the collected data, the analyst may need to provide a quick response or just answer a simple question.

In these cases, it may not be necessary or even feasible for a complete forensic evaluation. The forensic software should contain applications that can accomplish data examination and analysis in many ways and can be run quickly and efficiently from a forensic workstation. An analyst should be able to perform the following processes with a variety of tools.

Using File Viewers

Using viewers instead of the original source applications to display the contents of certain types of files is an important technique for scanning or previewing data and is more efficient because the analyst does not need native applications for viewing each type of file. Various tools are available for viewing common types of files, and there are also specialized tools solely for viewing graphics. If available file viewers do not support a particular file format, the original source application should be used; if this is not available, then it may be necessary to research the file format and manually extract the data from the file.

TABLE 3.1 Most Commonly Used Forensic Software Suites

Autopsy	Autopsy is a free open-source tool provided by Basis Technology. It is actually a GUI for The Sleuth Kit (TSK).	 Image 3.7a
AXIOM	AXIOM began as an enhanced version of the original Internet Evidence Finder (IEF), but now has become a complete forensic suite. During transition from IEF to AXIOM, Magnet purposely kept the interface the same as IEF for ease of use. AXIOM can also be used individually for smartphone/computer/cloud examinations.	 Image 3.7b
EnCase	EnCase is traditionally used in forensics to recover evidence from seized hard drives. EnCase allows the investigator to conduct in-depth analysis of user files to collect evidence such as documents, pictures, internet history, and Windows Registry information.	 Image 3.7c
Forensic Toolkit (or FTK)	Forensic Toolkit, or FTK, is computer forensics software made by AccessData. It scans a hard drive looking for various information. It can, for example, locate deleted emails and scan a disk for text strings to use them as a password dictionary to crack encryption.	 Image 3.7d
X-Ways	X-Ways Forensics is an advanced work environment for computer forensic examiners. X-Ways Forensics is based on the WinHex hex and disk editor and part of an efficient workflow model where computer forensic examiners share data and collaborate with investigators that use X-Ways Investigator.	 Image 3.7e

Uncompressing Files

Compressed files may contain files with useful information, as well as other compressed files. Therefore, it is important that the analyst locate and extract compressed files. Files should be uncompressed early in the forensic process to ensure that the contents of compressed files are included in searches and other actions. However, analysts should keep in mind that compressed files might contain malicious content, such as compression bombs, which are files that have been repeatedly compressed, typically dozens or hundreds of times. Compression bombs can cause examination tools to fail or consume considerable resources; they might also contain malware and other malicious payloads. Although there is no definite way to detect compression bombs before uncompressing a file, there are ways to minimize their impact. For instance, the examination system should use up-to-date antivirus software and should be stand-alone to limit the effects to just that system. In addition, an image of the examination system should be created so that, if needed, the system can be restored.

Graphically Displaying Directory Structures

This practice makes it easier and faster for analysts to gather general information about the contents of media, such as the type of software installed and the likely technical aptitude of the user(s) who created the data. Most products can display Windows, Linux, and UNIX directory structures, whereas other products are specific to iOS directory structures.

Identifying Known Files

The benefit of finding files of interest is obvious, but it is also often beneficial to eliminate unimportant files, such as known good OS and application files, from consideration. Analysts should use validated hash sets, such as those created by the NIST National Software Reference Library (NSRL) project or personally created hash sets that have been validated, as a basis for identifying known benign and malicious files. Hash sets typically use the SHA-1 and MD5 algorithms to establish message digest values for each known file.

Performing String Searches and Pattern Matches

String searches aid in examining large amounts of data to find keywords or strings. The analyst can reduce the volume of information to review by developing concise sets of search terms for common situations.

This methodology provides the benefit that keywords may be found in hidden locations or in the remnants of deleted files. It is necessary for the investigator to examine that document using human eyes and brain to establish whether the hit is relevant to the case when a keyword is located in a document.

A reasonable example would be a case involving the manufacture of illegal drugs. Searching a computer for the words "drugs" or "narcotics" is likely to produce a vast amount of false hits. However, searching the names of the chemicals used in the manufacture of drugs or the current slang names for drugs used by people (users or suppliers) is likely to lead to far better results. When a computer searches for a word, it can identify results as only a collection of letters; the computer cannot establish any meaning nor context. Often, the collection of letters that form a short word may be found within longer words thus resulting in many false positives when searched. My native language is English, so it is difficult to provide examples that may be meaningful to people who speak other languages. To overcome this problem, it is possible to utilize special search techniques, but skill and experience are required to do so effectively.

Particularly in cases involving illegal pictures, forensic software is very useful in identifying all the picture files on the disk. Results will include all those that have been hidden or deleted. In order to establish whether an image is illegal, though, the examiner must look at the pictures one by one.

Accessing File Metadata

File metadata provides details about any given file. For example, collecting the metadata on a graphic file might provide the graphic creation date, copyright information, description, and

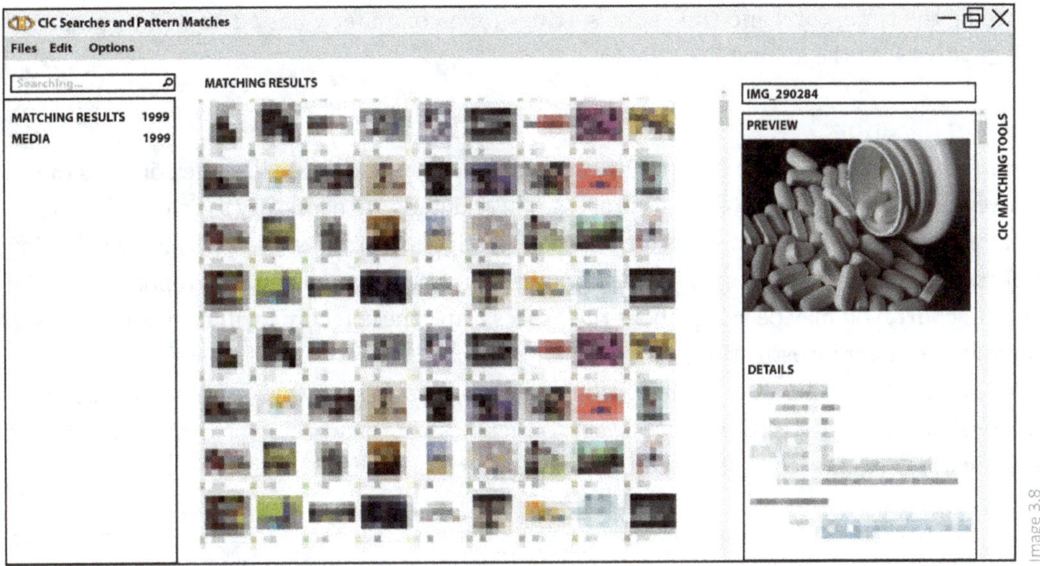

Image 3.8

the creator's identity. Metadata for graphics generated by a digital camera might include the make and model of the digital camera used to take the image, as well as F-stop, flash, and aperture settings. For word processing files, metadata could specify the author, the organization that licensed the software, when and by whom edits were last performed, and user-defined comments. Special utilities can extract metadata from files.

TABLE 3.2

SkyDrive executable information (using X-Ways 16.5).

Name	SkyDrive.exe	SkyDrive.exe
Path	\Users\cloud\AppData\Local\Microsoft\SkyDrive	\Users\cloud\AppData\Local\Microsoft\ SkyDrive
Size	290 KB (296,672)	290 KB (296,672)
Created	27/05/2012 17:07:59	09/06/2012 12:46:34
Int. creation	16/04/2012 09:18:43	25/05/2012 16:40:09
Hash	D943C6ADF16045E7FE00233FF9C15376	4F7C80E5A420E47B584055EB1AC61562
Metadata	signed true CA Microsoft Timestamping PCA Signing date 16/04/2012 18:50:38	signed true CA Microsoft Timestamping PCA Signing date 26/05/2012 02:12:21

Analysis

The next step, after the examination has been completed is to perform analysis of the extracted data. The process of analysis is the process of assigning meaning to the results of the examination process and to trying to find the related information for the arrest of the suspect. This process generally includes both forensic examiners and field investigators discussing the meaning of the documentation produced after the examination process.

As reviewed, in analysis of different types of data there are many forensic software tools that can be helpful. Knowing when an incident occurred, a file was created or modified, or an email was sent can be critical to forensic analysis. For example, such information can be used to reconstruct a timeline of activities. Although this may seem like a simple task, it is often complicated by unintentional or intentional discrepancies in time settings among systems.

Knowing the time, date, and time zone settings for a computer whose data will be analyzed will greatly assist the analyst.

Writing a Report

As the analyst is writing the report, they should include details regarding the contents of the recovered data as well as the analysis process. These details are necessary to ensure the appropriateness of the actions and steps during the investigation process. During the investigation process the contents which are recorded and have not been included in the report should be kept. In court, you may be asked about the process, and these records will be useful for testimony and for later investigation.

The investigators involved during the investigation process may eventually be called on to testify about the investigation process, appropriateness of the working process, and whether investigators are entitled to the specific working process.

Summary

The original files should not be examined; the analysts should examine copies of files. Multiple copies of the desired files or file systems should be made during the collection phase; typically, a master copy allows for a working copy. The analyst can then work with the working copy of the files without risking the original files or the master copy. A bit-stream image should be created if evidence may be needed for prosecution or disciplinary actions, or if preserving file times is important.

Analysts should preserve and verify file integrity. During a backup, using a write-blocker and imaging prevents a computer from writing to its storage media. The integrity of copied data can be verified by computing and comparing the message digests of files. Whenever possible, backups and images should be accessed as read-only; write-blockers can also be used to prevent writing to the backup or image file or restored backup or image file.

To identify file content types, analysts should rely on file headers not file extensions. Because users can assign any file extension to a file, analysts should not assume that file extensions are accurate. By looking at their file headers, analysts can identify the type of data stored in many files. Although people can alter file headers to conceal actual file types, this is much less common than altering file extensions.

Analysts should have a forensic tool kit for data examination and analysis. The tool kit should provide various tools that allow quick reviews of data as well as in-depth analysis. The tool kit should provide applications that can be executed quickly and efficiently from removable media or a forensic workstation.

Lab 3: Hash Verification Practice

Download Hashcalc from the link below:
https://www.slavasoft.com/hashcalc/

On your desktop, right click and select "New" then "Text Document". Leave the default name. Open notepad and type "**This is a test.**" (Capitalization and punctuation matter.) Save the file and from the desktop browse to the saved file from hashcalc. If typed correctly, your hash value should be as shown on the right side.

Re-open the document and change the period to an exclamation point: **"This is a test!"** Save and close the document again.

Change the filename and recalculate. Please notice that the hash value does not change after renaming the file.

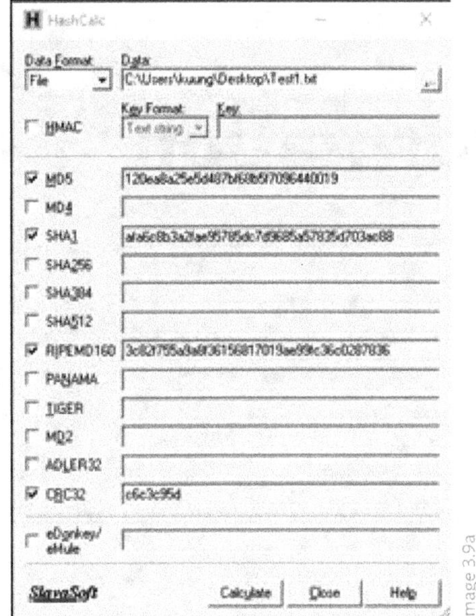

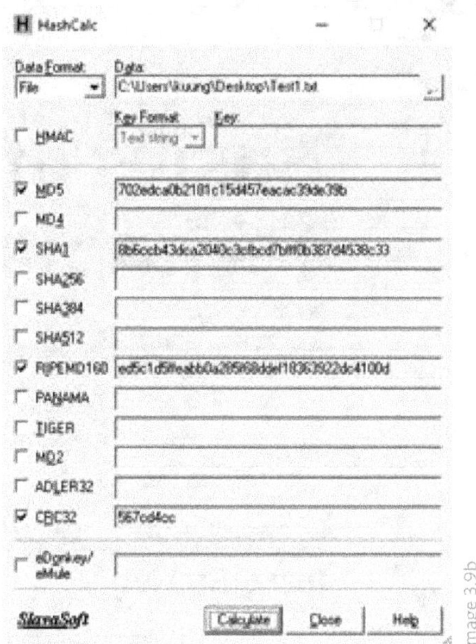

References

Kent, K., Chevalier, S., Grance, T., & Dang, H. (2006). Guide to integrating forensic techniques into incident response. *NIST Special Publication, 10*(14), 800–886. https://csrc.nist.gov/publications/detail/sp/800-86/final

Kipling, R. J. (1902). The elephant's child. *Just so stories*. Macmillan.

National Institute of Standards and Technology (2015, August 5). *NIST policy on hash functions*. https://csrc.nist.gov/Projects/Hash-Functions/NIST-Policy-on-Hash-Functions

Williams, J. (2012). ACPO good practice guide for digital evidence. *Association of Chief Police Officers*.

Credits

Basics of Information Technology

Sinchul Back

Introduction

In line with the overview of digital evidence and the process of digital forensics, it is noteworthy to explore important foundations in information technology in order to become a highly effective cyber-forensics investigator. The evidence that digital forensic examiners collect must be coupled with a comprehensive understanding of the operating systems and file systems that exist on the cyber-crime suspect's digital devices. Due to the significance of operating systems, digital forensic investigators have to digest this knowledge and accurately apply it to the collection and examination of digital evidence. Due to the advancement of technology, cybercriminals use more and more diverse computing devices. Desktop workstations, laptop computers, tablets, smartphones, smartwatches, Internet of Things (e.g., baby monitors, printers, smart TVs, and refrigerators) are all examples of computing devices. Thus, digital forensics investigators need to understand these computing devices in order to properly collect and seize electronic evidence.

We now introduce some terminology that will be useful throughout the book to enhance our understanding of digital forensic investigation. We will begin with the concept of a computer system, which can consist of hardware, software, data, and communication facilities and networks (Stallings & Brown, 2015). Hardware contains computer systems and other data processing, file storage, and data communications components. Software includes the computer operating system, system utilities, and applications. Data consist of files and databases, information security data (e.g., password-protected files and encrypted information). Communication facilities and networks consist of local network communication links, bridges, and routers (Stallings & Brown, 2015). In this sense, this chapter reviews some concepts and details (i.e., file storage system, file systems, networks, and peripherals) that were covered in Chapters 2 and Chapter 3 in the hope that it may help readers have a landscape view of information security components.

Networks and Network Components

First and foremost, this section gives you a thorough overview of computer networks and networking components. We will define a network and the basic concepts/terms. What *is* a network? A computer network is defined as a combination of computer hardware, cabling, network devices, and computer software used together to enable computers to communicate with each other (Odom & Knott, 2006). Both (1) "a small network" and (2) "a very large network (i.e., the internet)" are examples of networks. A small network enables two computers, A and B, to communicate via a networking cable. A very large network, that is, the internet, is a global network that enables almost every computer on Earth to communicate with other electronic devices around the world. To make the internet connect, a firm referred to as an internet service provider (ISP) builds the necessary network.

As stated previously, there are four networking components that create a computer network: computer hardware, cables, networking devices, and computer software.

Computer Hardware

As a component of computer hardware, personal computers (PCs)—which include desktop computers and laptop computers—are commonly used by a single person at a time. A PC generally consists of the following components.

Processor (called a central processing unit, or CPU). A computer processor, or CPU, acts as the computer's brain. The CPU's work includes creating the image that is displayed on the computer's screen, taking input from the keyboard or mouse, sending data over a network, and processing data for software running on the computer.

Microprocessor. A silicon chip that contains a CPU. A typical PC has several microprocessors, including the CPU.

Temporary memory (called random-access memory, or RAM). RAM is a volatile memory that stores the data, instructions, the results of the program currently being executed by the processor, and temporary data. RAM is the computer equivalent of the papers and notes you might keep on your desk when studying. The CPU can quickly and easily access the data stored in RAM, and that data typically pertains to something the PC is actively processing. Note that the contents of RAM are lost when the computer is powered off.

Read-only memory (ROM). ROM is a type of computer memory in which data has been prerecorded. After data has been written onto a ROM chip, it cannot be removed; it can only be read. PCs can re-record information into another type of ROM, called electronically erasable programmable read-only memory (EEPROM). The basic input/output system (BIOS) in most PCs is stored in EEPROM.

Permanent memory (such as disks). Computers need long-term memory to store data that might be needed later, even after a computer is powered off. Permanent memory typically consists of a type of device called a disk drive or hard disk.

Input/output devices (I/O). To interact with humans, the computer must be able to know what the human wants it to do and provide the information to the human. Humans tell a computer what to do by manipulating an input device. For example, this occurs when the human types on the PC keyboard or moves/clicks with a mouse. For output, the computer uses a video display, audio speakers, and printers (Odom & Knott, 2006, pp. 8–9).

Network interface cards (NICs). For a PC to use a network, NICs provide interface to a network. It is also known as a network card. Each PC on a network should have a NIC.

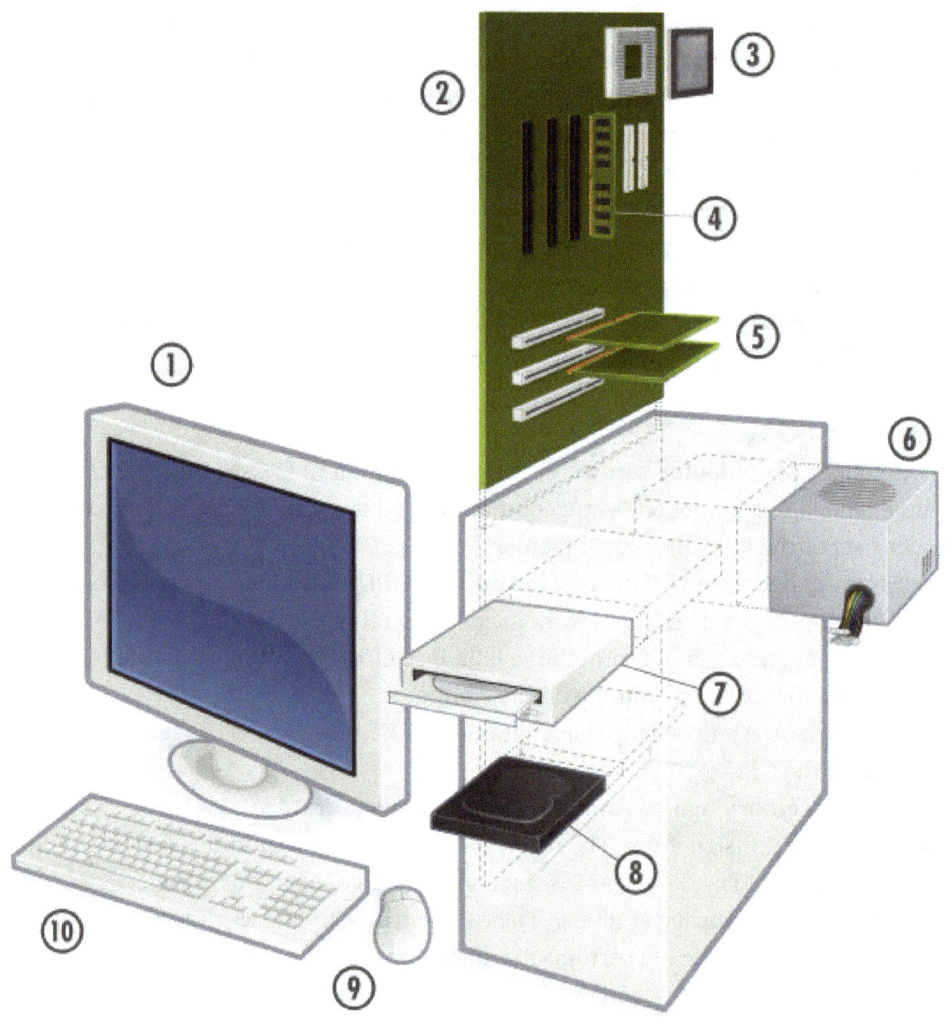

FIGURE 4.1 Personal computer

Also, digital forensic investigators need to be familiar with PC motherboards, which contain many of the PC's most important components. The following list details some of the essential components of the motherboard:

Printed circuit board (PCB). A thin plate on which chips (integrated circuits) and other electronic components are placed. Examples include the motherboard and various expansion adapters.

Transistor. A device that amplifies a signal or opens and closes a circuit. Microprocessors can have millions of transistors.

Integrated circuit (IC). A device made of semiconductor material. It contains many transistors and performs a specific task. The primary IC on the motherboard is the CPU. ICs are often called chips.

Memory chips. Another name for RAM, memory chips are integrated circuits whose primary purpose is to temporarily store information that is processed by the CPU.

Resistor. An electrical component that is made of material that opposes the flow of electric current.

Capacitor. An electronic component that stores energy in the form of an electrostatic field. It consists of two conducting metal plates separated by insulating material.

Connector. A port or interface that a cable plugs into. Examples include serial, parallel, USB, and disk drive interfaces.

Light-emitting diode (LED). A semiconductor device that emits light when a current passes through it. LEDs are commonly used as indicator lights.

Parallel port. An interface that can transfer more than 1 bit at a time. It connects external devices, such as printers.

Serial port. An interface used for serial communication in which only 1 bit is transmitted at a time. The serial port can connect to an external modem, plotter, or serial printer. It can also connect to networking devices, such as routers and switches, as a console connection.

FIGURE 4.2 Computer motherboard

Mouse port. A port designed for connecting a mouse to a PC.

Keyboard port. A port designed for connecting a keyboard to a PC.

Power connector. A connector that allows a power cord to be connected to the computer to give electrical power to the motherboard and other computer components.

Universal Serial Bus (USB) port. This interface lets peripheral devices, such as mice, modems, keyboards, scanners, and printers, be plugged and unplugged without resetting the system. PC manufactures may one day quit building PCs with the older parallel and serial ports completely, instead using USB ports.

Firewire. A serial bus interface standard that offers high-speed communications and real-time data services.

Networking Components

This section introduces the general terms of networking components to create networks. Dordal (2020) defines local area networks (LANs) as the physical networks that provide the internet connection between computers. Also, he explains that Internet Protocol, or IP, "provides an abstraction for connecting multiple LANs into the internet; while TCP deals with transport and connections and actually sending user data" (Dordal, 2020, p. 13). In this regard, networking layers are composed of these three components: LANs, IP, and TCP.

It is critical that digital forensic investigators understand devices that have logging capabilities, since they likely hold historical evidence for a crime scene. In this sense, the investigator should provide information routed through several networking servers and all corresponding server timestamps. The following are some of the network devices that have logging capabilities.

Web servers. A web server stores and serves up HTML documents and related media resources in response to client requests. As such, when a suspect types a URL into the web browser and then presses Enter, a data packet is sent to that web server that includes the suspect's IP address. That web page is then sent to the suspect's computer and stored in a cache. This cache can provide the digital forensic examiner important information about a suspect's internet activity.

Proxy servers. A computer that relays a request for a client to a server computer. This is important to remember because a suspect might use another computer or multiple computers as an intermediary to request information from a server. For example, investigators might seize a web server hosting illicit photographs of minors, but the client requests to download pages or images might actually show IP addresses of unsuspecting individuals who had their computers compromised by hackers.

Dynamic Host Configuration Protocol (DHCP) servers. DHCP is a standard for enabling a server to assign IP addresses and configuration to hosts on a network. The DHCP server assigns a unique IP address. Then after a client leaves the network, that IP address is released. To that end, a DHCP server offers an IP address, a subnet mask, and a default gateway.

Subnet mask. Facilitates the communication between segregated networks.

Default gateway. A system that connects two network segments running different protocols.

Simple Mail Transport Protocol (SMTP) server. Used to send email for a client; the email is then routed to another SMTP server or other email server. As a cybercrime investigator, you

0. Simple Web Stack

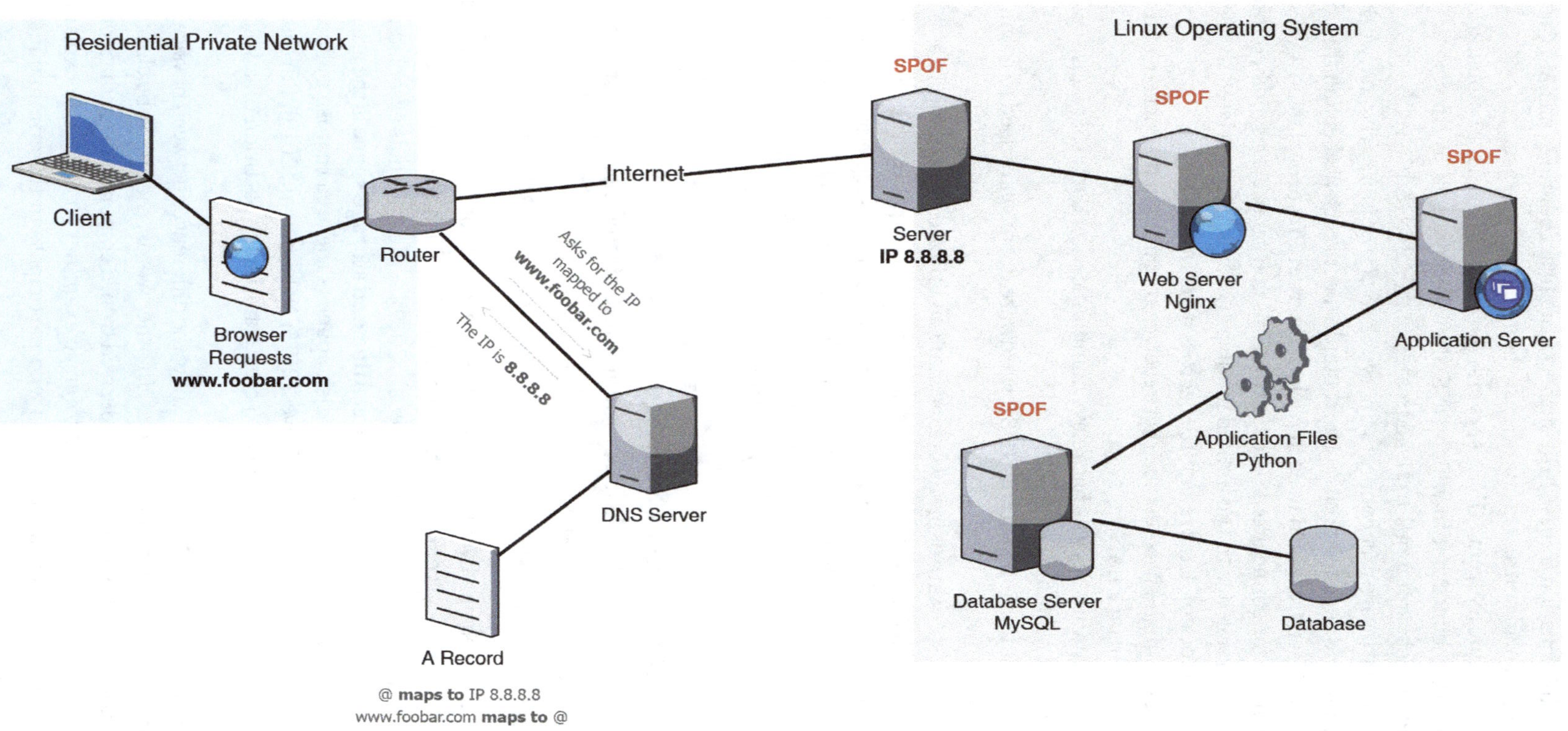

FIGURE 4.3 Computer network

can obtain information from (1) the suspect's email account, (2) an SMTP server of the sender and recipient email addresses, and (3) the body of the message.

Domain Name System (DNS) servers. A naming system for computers and other devices connected to the internet in which DNS allows computers to convert domain names (e.g., https://centercicboston.org/) to IP addresses (e.g., 3.34.116.101).

Routers. Hardware that connects a network to one or more other networks and directs data packets from one node to another. The router inspects each packet and then directs that packet based on the IP address held in each packet. Internet Protocol version 4 (IPv4) has been commonly utilized for connectionless data transmission on packet-switched internetworks. IPv6 is the latest version due to the limited number of IP addresses related to IPv4.

Packet. A block of data transmitted across a network.

Switches. A high-speed device that receives incoming data packets and redirects them to their destination on a LAN.

Hubs. A hardware device that relays communication data; that is, it sends data packets to all devices on a network.

Intrusion-detection system (IDS). Hardware or software used to monitor network traffic for malicious activity.

Firewalls. A software or hardware mechanism utilized to examine data packets on a network and determine, based on its set of rules, whether each packet can be admitted through. For instance, an information security officer prohibits some suspicious inbound/outbound traffic from IP addresses that originate from China. Then Chinese hackers adapt and get through using other U.S. IP addresses for command and control so that they can access a network without being blocked by the firewall (Hayes, 2015).

Physical and Logical Storage

Prior to collecting digital evidence, cybercrime investigators need to ask themselves the following:

- Where will this search be conducted?
- Which parts of the electronic devices will need to be investigated to find electronic evidence?

This section begins by outlining the basic justifications of file systems, storage space, and partition(s). A **file system** refers to the manner in which files are stored and provides the OS a road map to data on a disk (Nelson et al., 2019). A file system is associated with determining allocated and unallocated storage space. **Allocated storage space** is the area on a volume where a file is stored, whereas **unallocated disk space** is the area that isn't allocated to a file which contains data deleted previously (Hayes, 2015; Nelson et al., 2019). Furthermore, unallocated storage space is pervasively utilized to create a primary partition on a volume. A **partition** is a logical drive on a disk.

Digital forensic investigators should know the differences between physical and logical storage when they investigate cybercrime cases. In other words, they must be clear whether the digital evidence comes to file storage or files derived from a computer or media storage device (Hayes, 2015). Also, it is crucial for a digital forensic examiner to be able to explain this information to courts and nontechnical audiences who participate in the criminal justice system. In the context of computer systems, a cybercrime investigator must understand the fundamental list of file storage concepts as follows (Hayes, 2015, pp. 73–76):

- **Byte.** Composed of eight bits, this is the smallest addressable unit when it comes to memory.
- **Sector.** On a magnetic hard disk, this represents 512 bytes; on an optical disk, it represents 2048 bytes.
- **Bad sector.** An area of the disk that can no longer be used to store data.
- **Cluster.** A logical storage unit on a hard disk that contains contiguous sectors.
- **Tracks.** Thin, concentric bands on a disk that comprise sectors where data is stored.
- **File slack.** Refers to the remaining unused bytes in the last sector of a file.
- **Physical file size.** The actual disk space used by a file.
- **Logical file size.** The amount of data stored in a file.
- **Platter.** A circular disk that stores data magnetically.
- **Spindle.** Found at the center of the disk, this is powered by a motor and used to spin the platters.
- **Actuator arm.** Contains a read/write head that modifies the magnetization of the disk when writing to it.
- **Cylinder.** The same track number on a collection of stacked platters, spanning all platters in a hard drive.
- **Disk geometry.** Refers to the structure of a hard disk in terms of platters, tracks, and sectors.

Consult Figures 4.4 and 4.5 for an illustrative example of some of these storage subcomponents and structures as they relate to memory.

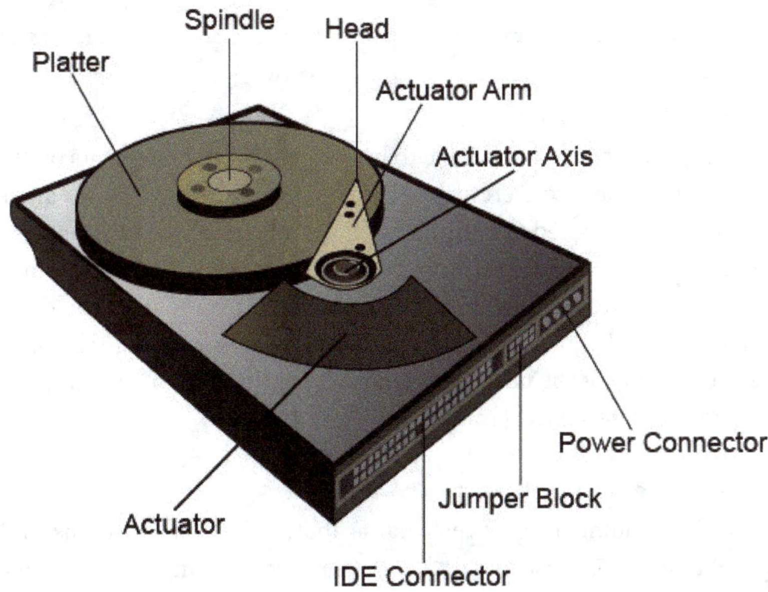

FIGURE 4.4 Mechanical components of a typical hard disk drive

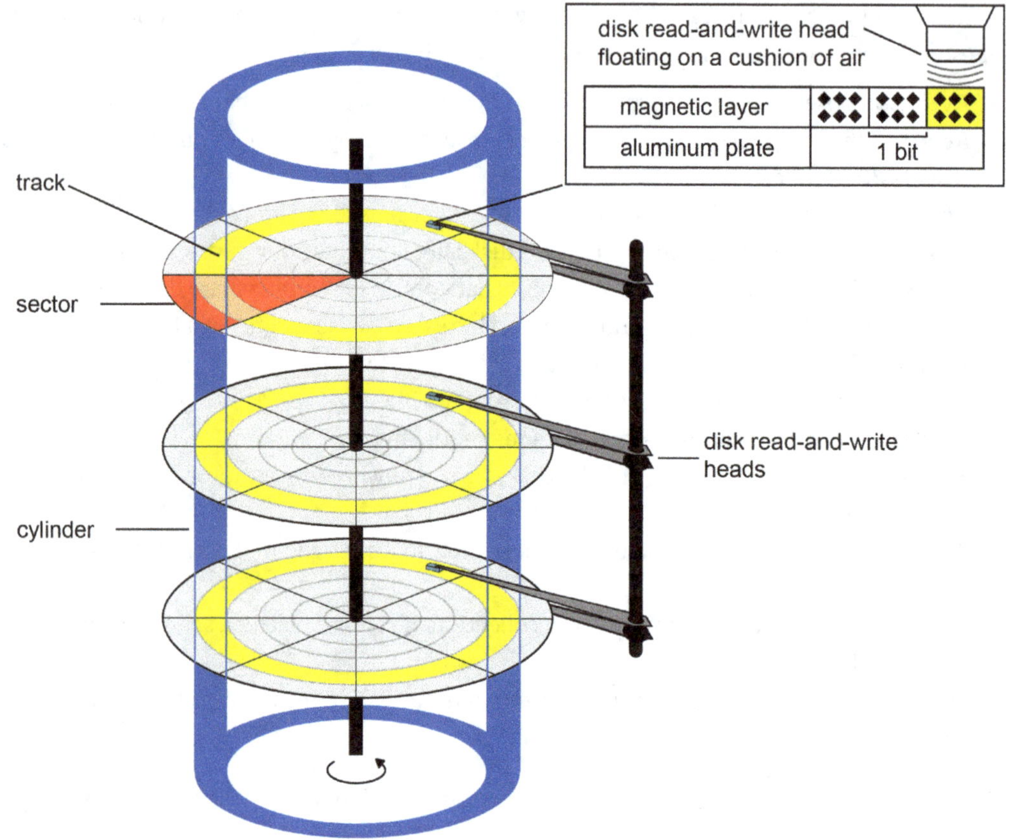

FIGURE 4.5 Basic structures and operations of a hard disk drive

Data Organization

Binary-to-Decimal Conversion

Digital forensic investigators need to be able to investigate different formats of data. For interpreting information, they must convert files or data on a computer into binary and hexadecimal formats. Humans are accustomed to working with decimal numbers. In contrast, computers utilize binary numbers which means that they consist of 0s and 1s. A decimal number is a series of symbols, which, in Figure 4.6, is in the rightmost column and has a value of 1, and the next column to the left has a value of 10 (Hayes, 2015). As we can see in Figure 4.6, we are able to calculate the decimal value of the number by multiplying the symbol in each column with the column's value and augmenting the products.

Hexadecimal-to-Decimal Conversion

Hexadecimal is another numbering system that is analyzed by digital forensics tools such as FTK or X-Ways (Hayes, 2015; Nelson et al., 2019). Hexadecimal employs 16 symbols (base 16), which include the numbers 0 to 9 and letters A to F. Most digital forensics tools (e.g., hex editors) enable a forensic investigator to read the entire contents of a file. Table 4.1 demonstrates the conversion from binary to decimal to hexadecimal.

Decimal-to-Binary Conversion

Base^{Exponent}	2^7	2^6	2^5	2^4	2^3	2^2	2^1	2^0
Place Value	128	64	32	16	8	4	2	1
Example: Convert decimal 35 to binary	0	0	1	0	0	0	1	1

$$35 = \qquad\qquad 2^5 \quad + \qquad\qquad 2^1 + 2^0$$
$$35 = \qquad\qquad (32 * 1) \quad + \qquad\qquad (2 * 1) + (1 * 1)$$
$$35 = 0 + 0 + 1 + 0 + 0 + 0 + 1 + 1$$
$$35 = 00100011$$

FIGURE 4.6 Binary-to-decimal conversion

Hexadecimal-to-ASCII Conversion

ASCII is the American Standard Code for Information Interchange. Understanding how to convert hex to ASCII is very important because it enables a forensic examiner to change unreadable files to readable contents (Nelson et al., 2019). Table 4.2 illustrates the conversion from hexadecimal to ASCII. In Table 4.2., two hexadecimal values can be equivalent to one ASCII character (e.g., Hex "41" = ASCII character "A").

For instance, an uppercase "M" has the hexadecimal value 4D, and a lowercase "r" has the hexadecimal value 72. We use Table 4.3 to convert the sentence, *'Mr. Evil Noodle!'*

Operating Systems

An **operating system** (**OS**) is a set of programs that coordinate and manage a computer's hardware and system resources (Hayes, 2015). Currently, there are three main computer operating systems:

- Windows
- Linux or UNIX
- Apple Mac (macOS)

TABLE 4.1 Binary/Decimal/Hexadecimal Conversion

BINARY	DECIMAL	HEXADECIMAL
0000	00	0
0001	01	1
0010	02	2
0011	03	3
0100	04	4
0101	05	5
0110	06	6
0111	07	7
1000	08	8
1001	09	9
1010	10	A
1011	11	B
1100	12	C
1101	13	D
1110	14	E
1111	15	F

TABLE 4.2 Hexadecimal-to-ASCII Conversion Table

ASCII Table

Dec	Hex	Oct	Char	Dec	Hex	Oct	Char	Dec	Hex	Oct	Char	Dec	Hex	Oct	Char
0	0	0		32	20	40	[space]	64	40	100	@	96	60	140	`
1	1	1		33	21	41	!	65	41	101	A	97	61	141	a
2	2	2		34	22	42	"	66	42	102	B	98	62	142	b
3	3	3		35	23	43	#	67	43	103	C	99	63	143	c
4	4	4		36	24	44	$	68	44	104	D	100	64	144	d
5	5	5		37	25	45	%	69	45	105	E	101	65	145	e
6	6	6		38	26	46	&	70	46	106	F	102	66	146	f
7	7	7		39	27	47	'	71	47	107	G	103	67	147	g
8	8	10		40	28	50	(	72	48	110	H	104	68	150	h
9	9	11		41	29	51	)	73	49	111	I	105	69	151	i
10	A	12		42	2A	52	*	74	4A	112	J	106	6A	152	j
11	B	13		43	2B	53	+	75	4B	113	K	107	6B	153	k
12	C	14		44	2C	54	,	76	4C	114	L	108	6C	154	l
13	D	15		45	2D	55	-	77	4D	115	M	109	6D	155	m
14	E	16		46	2E	56	.	78	4E	116	N	110	6E	156	n
15	F	17		47	2F	57	/	79	4F	117	O	111	6F	157	o
16	10	20		48	30	60	0	80	50	120	P	112	70	160	p
17	11	21		49	31	61	1	81	51	121	Q	113	71	161	q
18	12	22		50	32	62	2	82	52	122	R	114	72	162	r
19	13	23		51	33	63	3	83	53	123	S	115	73	163	s
20	14	24		52	34	64	4	84	54	124	T	116	74	164	t
21	15	25		53	35	65	5	85	55	125	U	117	75	165	u
22	16	26		54	36	66	6	86	56	126	V	118	76	166	v
23	17	27		55	37	67	7	87	57	127	W	119	77	167	w
24	18	30		56	38	70	8	88	58	130	X	120	78	170	x
25	19	31		57	39	71	9	89	59	131	Y	121	79	171	y
26	1A	32		58	3A	72	:	90	5A	132	Z	122	7A	172	z
27	1B	33		59	3B	73	;	91	5B	133	[	123	7B	173	{
28	1C	34		60	3C	74	<	92	5C	134	\	124	7C	174		
29	1D	35		61	3D	75	=	93	5D	135	]	125	7D	175	}
30	1E	36		62	3E	76	>	94	5E	136	^	126	7E	176	~
31	1F	37		63	3F	77	?	95	5F	137	_	127	7F	177	

TABLE 4.3 Conversion of Mr. Evil Noodle! to Hex

ASCII	M	R	.		E	V	I	L		N	o	O	D	L	E	!
Hex	4D	72	2E	20	45	76	69	6C	20	4E	6F	6F	64	6C	65	21

A former FBI agent and current cyber instructor for the FBI National Academy located at Quantico, Virginia, Pedro D. Cordero (2017, pp. 6–7) asserted that "each of these three operating systems has different and specific vulnerabilities that are exploited by cyber-threat actors. It is important to ensure that the organization secure all operating systems to reduce these vulnerabilities; if the organization does not, it risks being exploited, possible data loss, and compromising the organization's IT infrastructure." From the perspective of computer forensics, it is essential that digital forensic examiners are knowledgeable in computer operating systems. This is because each of these three operating systems have differing and specific vulnerabilities, which can become the primary targets of cyber adversaries or major sources of digital evidence related to a cybercrime incident. In other words, digital forensic examiners should recognize how OSs store and manage data on a potential suspect's computer.

As stated previously, a file system provides an OS with a road map to data that is on a disk. When forensic examiners need to access a suspect's computer to acquire data relevant to their investigation, they should examine the computer's OS and file system so that they are able to access and modify system settings when necessary (Nelson et al., 2019).

In the worldwide market, the Windows system is the most pervasively utilized computer operating system. The newest version of Windows operating system is Windows 10, and older Windows operating systems are Windows 8.1., Windows 8, Windows 7, Windows Vista, and Windows XP (Cordero, 2017). Apple OS for MacBook, MacBook Air, and MacBook Pro is the second leading operating system. The Linux or UNIX operating system was introduced in the 1970s, which included Silicon Graphics, Inc. (SGI) IRIX, Santa Cruz Operation (SCO) UnixWare, Sun Solaris, IBM AIX, and HP-UX (Nelson et al., 2019).

In the context of computer operating systems, a digital forensic investigator must understand the fundamental concepts of operating systems as follows (Hayes, 2015, pp. 73–76; Nelson et al., 2019):

- **Kernel.** At the core of the operating system, a kernel is responsible for communication between applications and hardware devices, including memory and disk management.
- **Basic Input/Output System (BIOS).** Starts an operating system by recognizing and initializing system devices, including the hard drive, CD-ROM drive, keyboard, mouse, video card, and other devices.

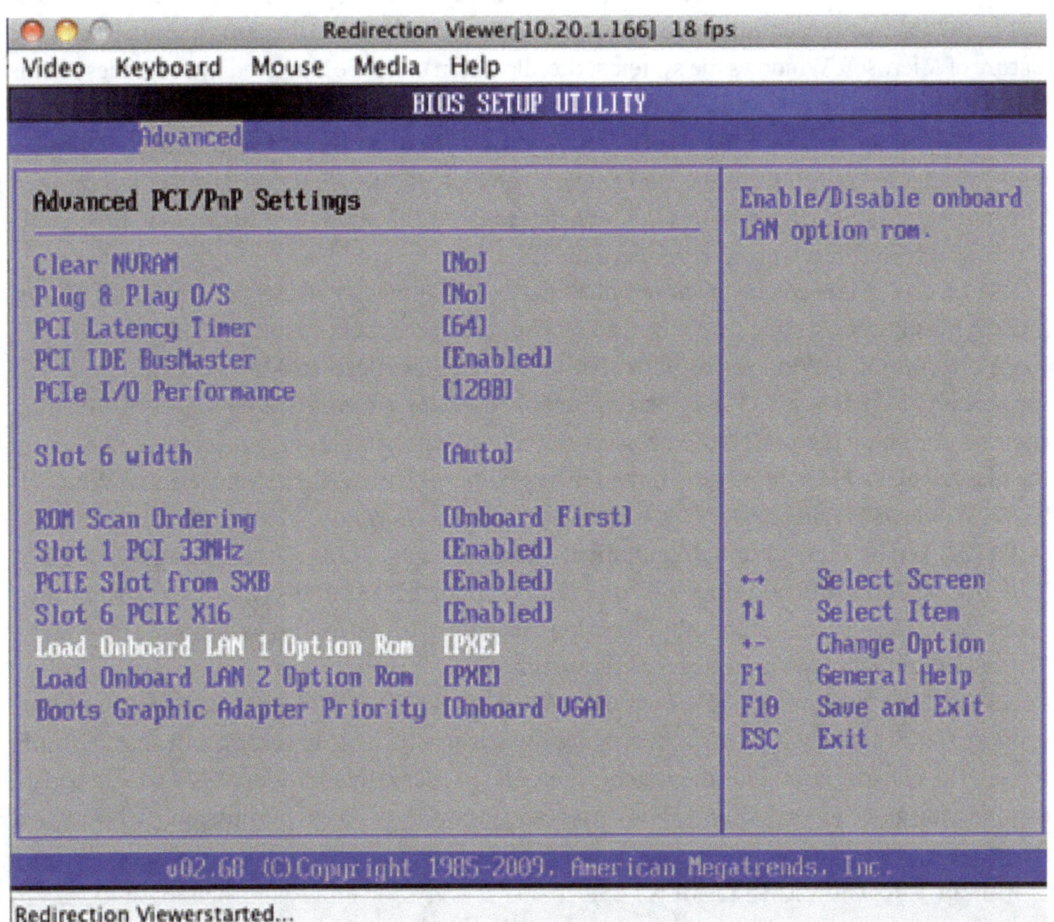

FIGURE 4.7 Viewing the BIOS

- **Bootstrapping.** The process of running a small piece of code to activate other parts of the operating system during the boot process. The bootstrap process is contained in the ROM chip.
- **Read-Only Memory (ROM).** Nonvolatile storage that is generally not modified and is used during the boot process.
- **Random Access Memory (RAM).** A computer's main memory is its temporary workspace for storing data, code, settings, and so forth.
- **Master Boot Record (MBR).** Involved in the boot process and stores information about the partitions on a disk, including how many exist and their location.
- **Master Partition Table.** Contains descriptions about the partitions on a hard disk.
- **Master Boot Code.** Code used by the BIOS to start the boot process.
- **Disk Signature.** Identifies the disk to the operating system.
- **End of Sector Marker.** A two-byte structure found at the end of the MBR.

File Systems

Microsoft Windows supports five main file systems such as NTFS, FAT64, FAT32, FAT16, and FAT12. **NTFS (New Technology File System)** replaced FAT (File Allocation Table). Due to the feature of Microsoft Windows file systems, the allocation units are clustered by some filesystems such as NTFS and FAT. A file allocation unit is a group of sectors—the smallest units that can be accessed on media. Successful forensic examiners need to understand the various file systems in order to collect, examine, and analyze the files that reside on the suspect's computer. The following lists the file systems that are commonly utilized (Hayes, 2015, p. 46; Kent et al., 2006, pp. 35–36; Nelson et al., 2019, pp. 313–324):

FAT12. FAT12 is used only on floppy disks and FAT volumes smaller than 16 MB. FAT12 uses a 12-bit file allocation table entry to address an entry in the filesystem.

FAT16. MS-ODS, Windows 95/98/NT/2000/XP, Windows Server 2003, and some UNIX OSs support FAT16 natively. FAT16 is also commonly used for multimedia devices such as digital cameras and audio players. FAT16 uses a 16-bit file allocation table entry to address an entry in the filesystem. FAT16 volumes are limited to a maximum size of 2 GB in MS-DOS and Windows 95/98. Windows NT and newer OSs increase the maximum volume size for FAT16 to 4 GB.

FAT32. Windows 95 Original Equipment Manufacturer (OEM) Service Release 2 (OSR2), Windows 98/2000/XP, and Windows Server 2003 support FAT32 natively, as do some multimedia devices. FAT32 uses a 32-bit file allocation table entry to address an entry in the filesystem. The maximum FAT32 volume size is 2 terabytes (TB).

NTFS (New Technology File System). Windows NT/2000/XP and Windows Server 2003 support NTFS natively. NTFS is a *recoverable filesystem*, which means that it can automatically restore the consistency of the filesystem when errors occur. In addition, NTFS supports data compression and encryption, and allows user- and group-level access permissions to be defined for data files and directories. The maximum NTFS volume size is 2 TB.

Master File Table (MFT). In NTFS, the MFT keeps file and folder metadata in NTFS, including the filename, creation date, location, size, and permission for every file and folder. Compression or encryption files are stored in the MFT. The MFT can also maintain deleted file records and track the reallocated space.

High-Performance File System (HPFS). HPFS is supported natively by IS/2 and can be read by Windows NT 3.1, 3.5, and 3.51. HPFS builds on the directory organization of FAT by providing automatic sorting of directories. In addition, HPFS reduces the amount of lost disk space by utilizing smaller units of allocation. The maximum HPFS volume size is 64 GB.

Second Extended Filesystem (ext2fs). ext2fs is supported natively by Linux. It supports standard UNIX file types and filesystem checks to ensure filesystem consistency. The maximum ext2fs volume size is 4 TB.

Third Extended Filesystem (ext3fs). ext3fs is supported natively by Linux. It is based on the ext2fs filesystem and provides journaling capabilities that allow consistency checks of the filesystem to be performed quickly on large amounts of data. The maximum ext3fs volume size is 4 TB.

Fourth Extended Filesystem (ext4fs). ext4fs is supported natively by Linux. It added support for partitions larger than 16 TB, improved management of large files, and offered a more flexible approach to adding file system features. Because these changes affected the way the Linux kernel interacts with the file system, adoption of ext4fs was slower in some Linux distributions, but it is now considered the standard file system for most distributions.

ReiserFS. ReiserFS is supported by Linux and is the default filesystem for several common versions of Linux. It offers journaling capabilities and is significantly faster than the ext2fs and ext3fs filesystems. The maximum volume size is 16 TB.

Hierarchical File System (HFS). HFS is supported natively by macOS. HFS is mainly used in older versions of macOS but is still supported in newer versions. The maximum HFS volume size under macOS 6 and 7 is 2 GB. The maximum HFS volume size in macOS 7.5 is 4 GB. MacOS 7.5.2 and newer macOSs increased the maximum HFS volume size to 2 TB.

HFS Plus. HFS Plus is supported natively by macOS 8.1 and later and is a journaling filesystem under Mac OS X. It is the successor to HFS and provides numerous enhancements, such as long filename support and Unicode filename support for international filenames. The maximum HFS Plus volume size is 2 TB.

Logical/Physical EOF (end of file). The HFS and HFS+ file systems have two descriptors for the end of a file (EOF)—the logical EOF and the physical EOF. The *logical EOF* is the actual ending of a file's data, so if file B has 510 bytes of data, byte 510 is the logical EOF. The *physical EOF* is the number of bytes allotted on the volume for a file; for file B, its byte size is 1023.

Plist Files. Plist files are preference files for installed applications on a macOS, usually stored in /Library/Preferences.

UNIX File System (UFS). UFS is supported natively by several types of UNIX OSs, including Solaris, FreeBSD, OpenBSD, and Mac OS X. However, most OSs have added proprietary features, so the details of UFSs will differ among implementations.

Compact Disc File System (CDFS). As the name indicates, CDFS is the filesystem used for CDs.

International Organization for Standardization (ISO) 9660 and Joliet. The ISO 9660 filesystem is commonly used on CD-ROMs. Another popular CD-ROM filesystem, Joliet, is a variant of ISO 9660. ISO 9660 supports filename lengths of up to 32 characters, whereas Joliet supports up to 64 characters. Joliet also supports Unicode characters within filenames.

Universal Disk Format (UDF). UDF is the filesystem used for DVDs and is also used for some CDs.

As stated previously, filesystems are allowed to store files on media. However, filesystems can retain data from deleted files to existing files. These files can also provide significant data for digital evidence. Files can be categorized as follows (Kent et al., 2006, pp. 36–37):

Deleted Files. When a file is deleted, it is typically not erased from the media; instead, the information in the directory's data structure that points to the location of the file is marked as deleted. This means that the file is still stored on the media but is no longer enumerated by the OS. The operating system considers this to be free space and can overwrite any portion of or the entire deleted file at any time.

Slack Space. As noted previously, filesystems use file allocation units to store files. Even if a file requires less space than the file allocation unit size, an entire file allocation unit is still reserved for the file. For example, if the file allocation unit size is 32 kilobytes (KB) and a file is only 7 KB, the entire 32 KB is still allocated to the file, but only 7 KB is used, resulting in 25 KB of unused space. This unused space is referred to as *file slack space*, and it may hold residual data such as portions of deleted files.

Free Space. Free space is the area of media that is not allocated to any partition; it includes unallocated clusters or blocks. This often includes space on the media where files (and even entire volumes) may have resided at one point but have since been deleted. The free space may still contain pieces of data.

Windows Registry

From a digital investigator's perspective, the *Registry* can contain valuable evidence. As a digital forensic examiner, you need to investigate the Registry of all Windows systems. The Registry is a hierarchical database that stores hardware and software configuration information, network connections (e.g., internet searches and websites accessed), user preferences (including usernames and passwords), and setup information. To discourage corrupting the system and possibly making it unbootable, the examiner should pay careful attention to a live system which could be modifying any Registry setting (Nelson et al., 2019).

The Registry consists of two elements: keys and values. Keys function similarly to a folder, and most keys hold subkeys (or folders). In addition, the Windows Registry has five basic hives (Nelson et al., 2019, pp. 239–240):

- **HKEY_CLASSES_ROOT.** A symbolic link to HKEY_LOCAL_MACHINE\SOFTWARE\CLASSES; it provides file type and file extension information, URL protocol prefixes, and so forth.
- **HKEY_CURRENT_USER.** A symbolic link to HKEY_USERS; it stores settings for the currently logged-on user.
- **HKEY_LOCAL_MACHINE.** Contains information about installed hardware and software.
- **HKEY_USERS:** It stores information for the currently logged-on user; only one key in this—HKEY_CURRENT_USER.
- **HKEY_CURRENT_CONFIG.** A symbolic link to HKEY_LOCAL_MACHINE\SYSTEM\CurrentControlSet\Hardware ProfileVxxxx (with xxxx representing the current hardware profile); it contains hardware configuration settings.

Cryptography

Encrypted files are encoded to protect data on electronic devices against unauthorized access. To protect an encrypted file, a password or passphrase is implemented for authorized users. This is a great benefit to end-point users to protect against information theft or an unauthorized data breach. Since most electronic devices use disk encryption, collecting encrypted data from a suspect's computer and/or mobile device is becoming more significant to the practice of digital forensics. In other words, without the password or passphrase, acquiring the contents of encrypted files can be extremely challenging for digital forensic examiners. Some encryption technologies are very sophisticated and can cost digital forensic examiners a lot of time and effort to crack—days, weeks, or even years via password-cracking tools (e.g., brute-force attack). In line with that, this chapter briefly introduces you to cryptography technology.

Cryptography can be defined as the art and science of protecting data (Taylor et al., 2019). Encryption can be defined as the technique of using an algorithm—a series/sequence of instructions—to transform information and make it unreadable to unauthorized access. The information can then be retrieved by a user possessing a secret code, called a **key** (Cordero, 2017; Taylor et al., 2019). Therefore, this encoded data may only be decrypted with a certain granted key for the authorized user. Key distribution can be important to exchange data without allowing others to see the key. Key distribution is available and can be achieved in numerous ways. For two parties, such as Mr. Evil Noodle and Mr. Evil Noodle's co-offender (a.k.a. Joker), the following could happen:

- A key could be selected by Mr. Evil Noodle and physically delivered to Joker;
- A third party could select the key and physically deliver it to Mr. Evil Noodle and Joker;
- If Mr. Evil Noodle and Joker have previously and recently used a key, one party could transmit the new key to the other, encrypted using the old key;
- Mr. Evil Noodle and Joker may each have an encrypted connection to a third party, "C", who could deliver a key on the encrypted links to Mr. Evil Noodle and Joker; or
- If Mr. Evil Noodle signs a message for a hacking operation with a private key, the Joker can only verify that Mr. Evil Noodle sent the message and that the message has not been altered—for verification, the public key of Mr. Evil Noodle should be executed (see Figure 4.8).

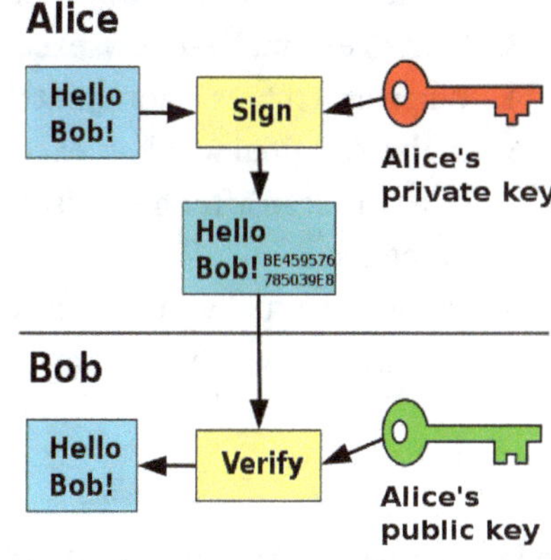

FIGURE 4.8 Public-key cryptography

Summary

Local area networks (LANs) are the physical networks that provide the internet connection between computers. Internet Protocol, or IP, offers an abstraction for connecting multiple LANs to the internet, while TCP deals with transport and connections and actually sending user data.

It is essential that digital forensic investigators understand devices that have logging capabilities and, therefore, hold historical evidence for a crime scene. In this sense, the investigator should provide information routed through several networking servers and all corresponding server timestamps.

Digital forensic investigators should know the differences between physical and logical storage when they investigate cybercrime cases in order to effectively acquire valid digital evidence.

An **operating system (OS)** is a set of programs that coordinate and manage a computer's hardware and system resources (Hayes, 2015). Currently, there are three main computer operating systems: Windows, Linux or UNIX, and Apple Mac (macOS).

From a digital investigator's perspective, the *Registry* can contain valuable evidence. As a digital forensic examiner, you need to investigate the Registry of all Windows systems. The Registry is a hierarchical database that stores hardware and software configuration information, network connections (e.g., internet searches and websites accessed), user preferences (including usernames and passwords), and setup information.

Lab 4.1: Reconnaissance on a Network

The objective of this project is to practice how to conduct reconnaissance on a network in order to better understand network systems as well as testing potential vulnerabilities in a computer network. In addition, for a digital forensic examiner, information gathered during footprinting exercises is very useful for discovering targets for active reconnaissance. To conduct reconnaissance on a network, please execute the following commands:

1. Start Kali Linux → Open the Terminal window.
2. In the Terminal window, execute "nmap" command to discover hosts and services.
3. In the Terminal window, execute "man nmap" command to discover hosts and services.
4. When asked press h or q in the Terminal window, press q to quit the manual.
5. In the Terminal window, execute "nmap -sn 10.0.0.1/24" command to order nmap not to do a port scan after host discovery, and only print out the available hosts that responded to the scan.
6. In the Terminal window, execute "nmap -O -v 10.0.0.1" command to run an operating system discovery scan against the server and router.

Lab 4.2: Identifying a Suspicious Account on the System User Groups

In this lab, you will learn how to identify a suspicious account on the System User Groups. Through this practice, you can have a better understanding of the User Groups feature in the Window operating system. Please execute the following commands:

1. Open Computer Management from Start menu.
2. Go to System Tools and in the System Tools tree navigate to Local User and Groups > Users.

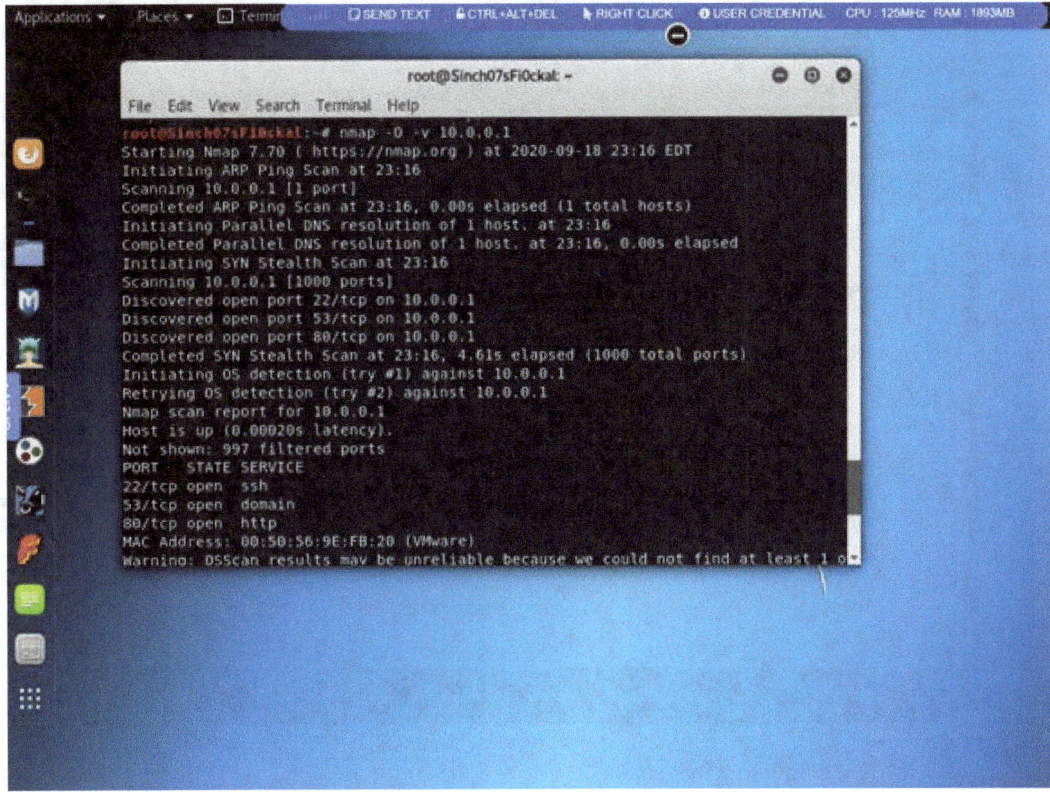

FIGURE 4.9 Nmap for operating system discovery scan

3. Please verify that several accounts are listed in the middle pane.
4. Please open the Properties window for the ●●●●● account (see Fig. 4.10).
5. On the Member Of tab, please verify that this account is a member of the Administrators group.

Lab 4.3: Monitoring a Denial-of-Service (DoS) Attack

In this lab, you will learn how to monitor a denial-of-service (DoS) attack. A DoS attack is a cyberattack in which the cyber adversary seeks to make a computer or network resource unavailable to its intended users by temporarily or indefinitely disrupting services of a host connected to the internet. Please run the following commands:

1. Open a VM for Windows.
2. On the desktop, open Notepad++ and enter the following text:

   - :loop
   - ping 169.254.94.56 -1 65500 -w 1 -n 1
   - goto:loop

3. Go to the File tab on Notepad++, and then click Save As.

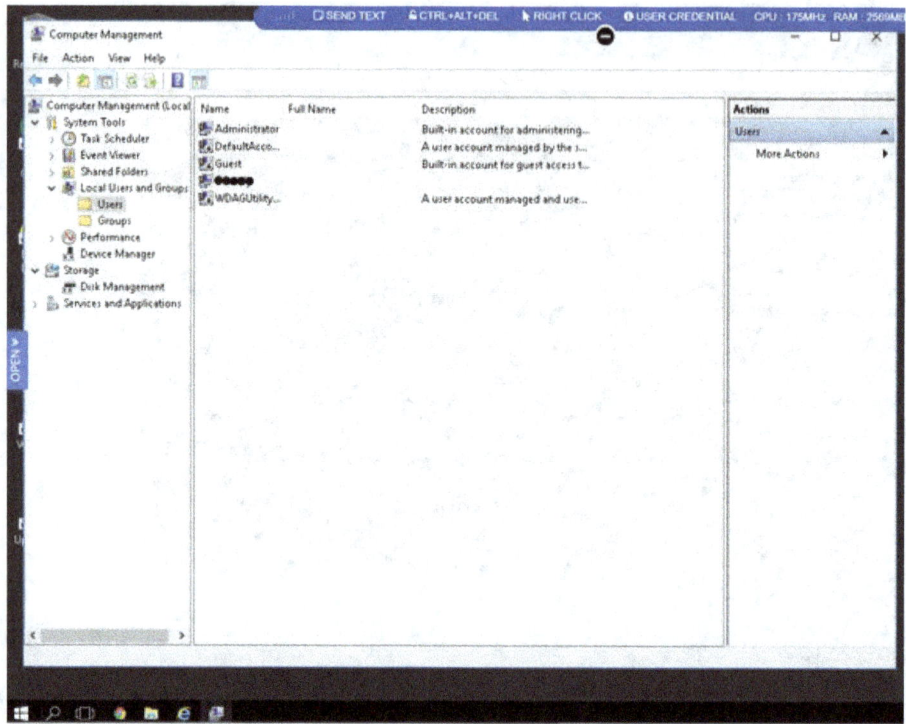

FIGURE 4.10 Computer management

4. Select Desktop, and in the File name textbox, type **Dos.bat**, and then click Save.
5. On the desktop, double-click dos.bat.
6. In the C:\WINDOWS\system32\cmd.exe terminal window, observe the results.
7. On the taskbar, right-click and select Task Manager.
8. In the Task Manager window, go to the Performance tab, and observe the results.

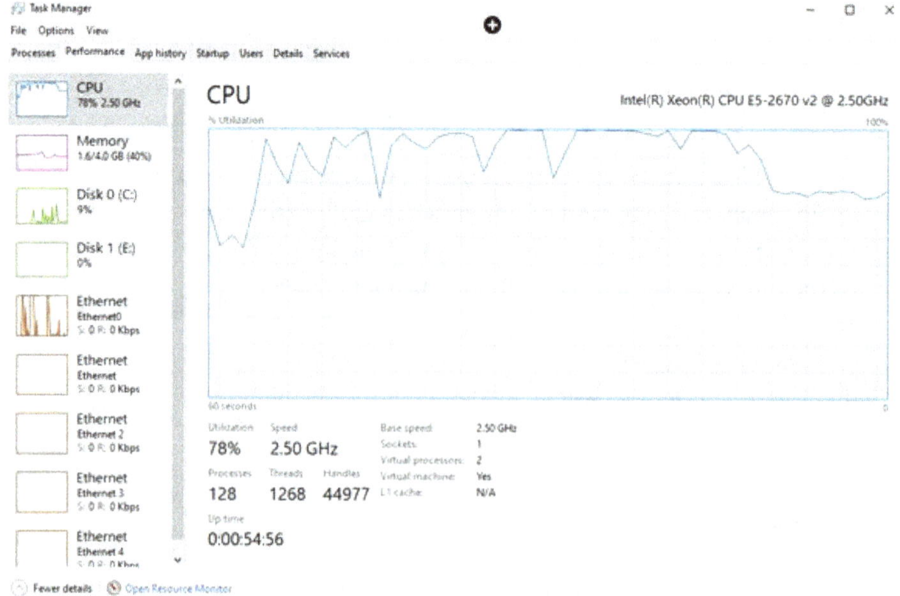

FIGURE 4.11 Task management window

Lab 4.4: Encrypting and Decrypting a File Using AES Crypt

In this project, you will learn how to encrypt and decrypt files using AES Crypt. AES Crypt is a file encryption software available on several operating systems that uses the industry standard Advanced Encryption Standard (AES) to easily and securely encrypt files. For this project, please run the following steps:

1. Open a VM for Windows.
2. On the desktop, right-click and navigate to New > Folder.
3. Rename the folder as **Encryption & Decryption**.
4. Double-click Encryption & Decryption and right-click and navigate to New > Text Document.
5. Double-click New Text Document.
6. In New Text Document, type **Secret communication channel for distribution plan of drugs from Mexico** and press Ctrl+S.

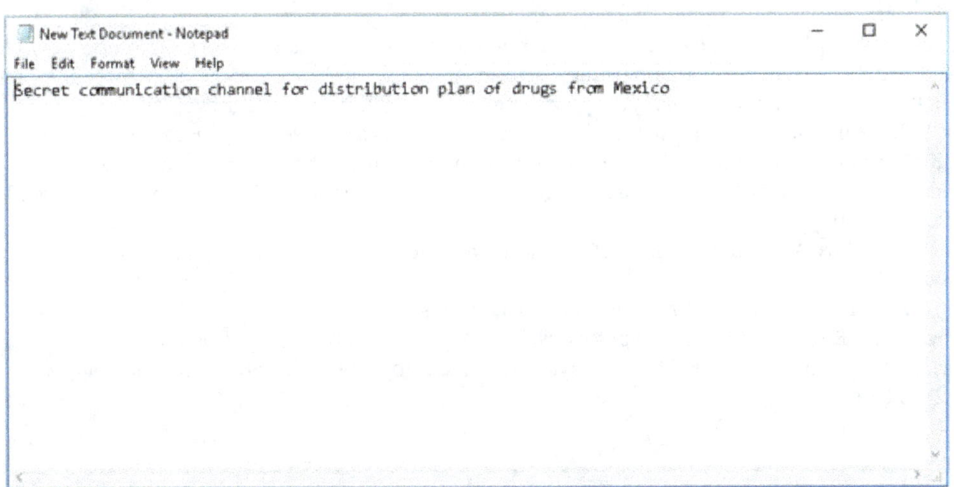

FIGURE 4.12 Example of secret communication

7. Close the New Text Document window.
8. Right-click New Text Document and click AES Encrypt.
9. In the AES Crypt Password window, in the Enter password and Confirm password textboxes, type **mexicoline12345**, and click OK.
10. In the Encryption & Decryption window, right-click New Text Document.txt and select Cut.
11. Navigate to This PC > Documents.
12. In the Documents window, right-click and select Paste.
13. Double-click New Text Document.txt.
14. In the AES Crypt Password window, in the Enter password textbox, type **mexico-line12345**, and click OK.

15. In the Documents window, double-click New Text Document and observe the decrypted data.

AES Crypt Password

Enter password:

Confirm password:

OK

Cancel

FIGURE 4.13 AES Crypt password window

References

Cordero, P. D. (2017). *Hacking the cyber threat: A cybersecurity primer for law-enforcement leaders and executives*. CreateSpace Independent Publishing Platform.

Dordal, P. L. (2020). *An introduction to computer networks*. Loyola University Chicago.

Hayes, D. R. (2015). *A practical guide to computer forensics investigations*. Pearson Education.

Kent, K., Chevalier, S., Grance, T., & Dang, H. (2006). Guide to integrating forensic techniques into incident response. *NIST Special Publication*, 10(14), 800–886. https://csrc.nist.gov/publications/detail/sp/800-86/final

Nelson, B., Phillips, A., & Steuart, C. (2019). *Guide to computer forensics and investigations (Loose-leaf version)*. Cengage Learning.

Odom, W., & Knott, T. (2006). *Networking basics*. Cisco Press.

Stallings, W., & Brown, L. (2015). *Computer security: Principles and practice*. Pearson Education.

Taylor, R. W., Fritsch, E. J., Liederbach, J., Saylor, M. R., & Tafoya, W. L. (2019). *Cyber crime and cyber terrorism*. Pearson.

Credits

Mobile Forensics

Sinchul Back

Introduction

Mobile devices such as smartphones and tablet PCs have been pervasively used in both individual and professional lives. For example, individuals utilize mobile devices for texting and email, taking photos, surfing the internet, online banking, online shopping, and playing games as well as making phone calls. Given that, mobile devices have become essential devices which hold a tremendous amount of information pertaining to a person's life and movements. Therefore, many digital forensic examiners are faced with investigating components of diverse mobile devices.

In the aftermath of the ISIS-inspired terrorist attack in San Bernardino, California, in 2015, the Federal Bureau of Investigation (FBI) attempted to extract information from an Apple iPhone used by one of the alleged terrorists. The smartphone was protected by Apple security measures. The FBI, with a court order, wanted to crack the iPhone's PIN code to access its contents for evidence; however, they were initially unable to break through Apple's security barriers, as these consisted of a two-step verification/public-key encryption system. Unfortunately, Apple refused to aid the FBI in breaking the security measures, claiming that it was unconstitutional based on the First Amendment's protection against compelled speech. Eventually, the FBI was able to extract information from the suspect's iPhone through a third-party provider (Bay, 2017; Zetter, 2016). The Apple/FBI case of 2016 highlights how extracting and analyzing evidence from a mobile device became an important part of crime investigation procedure. In addition to all the challenges of digital forensics, the field of mobile forensics presents new challenges such as unbreakable encryption technologies, a wide diversity of platforms, limited investigation tools, and additional legal issues.

This chapter begins with an overview of mobile devices and then explains mobile forensic procedures. You will learn how to identify and reconstruct file fragments, analyze file headers, and repair damaged file headers.

This chapter also explores the forensic tools used for acquiring data from mobile devices. In addition, this chapter explains and applies the procedures of the validation, preservation, acquisition, examination, analysis, and reporting of digital information from a mobile device. Its purpose is to inform digital forensic examiners of the various technologies engaged and the potential ways to approach them from a digital forensic perspective. The intended audience for this chapter is varied—from cyber incident response team members managing a cybersecurity incident to information security officials investigating an employee-associated situation to forensic examiners involved in cybercrime investigations.

Overview of Mobile Devices

Mobile phones are highly mobile communication devices that execute an array of functions ranging from that of a simple digital organizer to that of a low-end personal computer (Ayers et al., 2014). Mobile phone technology has advanced rapidly in the past three decades.

The Cellular Network

By the end of 2020, mobile phones had gone through five generations of cellular networks (Nelson et al., 2019): analog, digital personal communications service (PCS), third-generation (3G), fourth-generation (4G), and fifth-generation (5G). 3G introduced some advanced services such as the multimedia broadcast and multicast service (MBMS). 4G applied a user-centric system, providing higher data rates but also a clear and tangible advantage in users' lifestyles. The International Telecommunication Union (ITU) suggested the criteria for mobile devices to be considered 4G are those using Orthogonal Frequency Division Multiplexing (OFDM), Mobile WiMAX, Ultra Mobile Broadband (UMB), and Long-Term Evolution (LTE) (Nelson et al., 2019). 5G cellular networks with greater bandwidth, giving higher download speeds (i.e., 10 gigabits per second), were finalized in 2020.

A mobile phone signal's data moves from one cell to another cell. A radio transceiver device communicates with mobile phones, and the controller manages the transceiver equipment and performs channel assignment in alignment with the base transceiver station (BTS), the base station controller (BSC), and the mobile switching center (MSC). The National Institute of Standards and Technology explains that the BTS is the equipment at the cell site that enable communications between the mobile phone user and the carrier's network; the BSC manages the radio signals for the BTS; and the MSC facilitates switching data packets from one network path to another on a cellular network (Ayers et al., 2014).

Forensically, it is essential to know how a cellular network is built so that the examiner is able to identify the type of evidence that can be derived from the carrier's network, even without access to the suspect's device (Hayes, 2015). Cell site records provided by a carrier for a user suspected of criminal activity show information as to where a user has been and when they were there.

Mobile Devices

Mobile devices include simple phones, smartphones, tablets, and smartwatches. NIST's Guidelines on Mobile Device Forensics indicates that most mobile devices consist of hardware (e.g., microprocessor, ROM, RAM, a digital signal processor, a radio module, a microphone and speaker, hardware interfaces) as well as removable memory cards, Bluetooth, and Wi-Fi. In the context of the mobile device, a digital forensic examiner must understand the fundamental concepts of mobile station (Hayes, 2015, pp. 73–76; Nelson et al., 2019; Reiber, 2018).

FIGURE 5.1 Apple iPhone (front)

FIGURE 5.2 Apple iPhone (back)

Mobile Device Security Systems

Understanding mobile device security can help digital forensic examiners determine the proper course of action during and after device collection. If a device is powered on without security settings, you are able to examine the device at a computer forensics lab with more ease. Also, you need to ask the owner to determine the type of security enabled and any ways to bypass it. If these options are not available to the cybercrime scene investigator, the digital forensic examiner(s) will face the challenge of multiple security settings, including the device password or PIN, a subscriber identity module (SIM), an encryption password, and a password for the device's backup. As a result, I want to provide the reader with the general guidelines to recognize the type of device and respond with the appropriate method(s), which will be important for the follow-up collection process.

Apple iOS System

Apple mobile devices (the iPhone series) are more personalized and often interconnected to an Apple environment so that the same evidence can be duplicated from multiple devices. The iOS system is the mobile operating system developed by Apple, and it has been applied on the iPhone, iPad, iPod, and Apple TV (Apple Inc., 2015). This operating system works with a case-sensitive file system that also employs journaling. The iOS 4 system added full disk encryption, which meant that any unallocated space on the device remains fully encrypted even if the password is released. The iOS 7 system includes a Control Center feature, which is used to facilitate an investigative method for forensic examiners to activate airplane mode and disable cellular and Wi-Fi connections (Hayes, 2015; Reiber, 2018). iOS 8, released in 2014, has significant implications for investigators because Apple modified the method of device encryption. For example, the passcode of an iOS device used to encrypt that device disables Apple itself from recovering the user data stored on the device if the passcode is not released. Before iOS 8, cybercrime investigators were able to send locked iOS devices with appropriate court documents to Apple security officers, and the investigators were able to acquire a disc image of the user partition for forensic analysis. With the post-iOS 8 system modification, it is no longer possible to perform these types of recoveries. Following the release of Apple's iOS 8 security system with the iOS 11, there are continued difficulties with employing these methods of recovery. Forensics examiners should know many different elements relevant to the iOS security systems.

iOS Protection Methods

iOS protects its data in several different ways such as Passcode, Touch ID/Face ID, minimal protection, strong protection, password on iTunes, and Apple ID/password on iCloud (Tipton et al., 2014). The default length of the passcode changed from four digits to six digits for all devices with a Touch sensor when iOS 9 was unveiled. Touch ID/Face ID is a type of biometric platform used as a passcode to access the device. All files in iOS 8 and/or above versions of the iOS system are encrypted with minimal protection. Usually, picture and video files fall into this class of protection. Strong protection is implemented to encrypt files with higher protection methods, and investigators are unable to decrypt those files without the user passcode. A password on iTunes protects iTunes backup data, which is encrypted with a *key* generated from the password. Apple ID/password on iCloud protects an iCloud backup to prevent unauthorized access.

Handling iOS Devices

When starting a computer forensic investigation on an iOS device, as an examiner, you need to check whether the device is on or off, locked or unlocked, passcode set or not. You can isolate the device along with executing airplane mode if the device is on and unlocked. If the passcode is set, we need to maintain the device charged or extract the data from the device as quickly as possible. If the device is on but locked, you can bypass the passcode to access to the data. If the device is off, take out the SIM card. You can boot it into the recovery/DFU mode later for analyzing the data on this device.

Bypassing the Passcode

Bypassing the passcode is not always available; however, for iOS 4 and later versions, forensic examiners have a few options to solve the problem of not having the passcode of the suspect's mobile device. If the mobile device is locked, forensic examiners need to copy it to the forensic workstation to access the locked device and execute a backup through iTunes without the password. An important aspect of this process is the Lockdown certificate, which is a pairing/sync certificate. It is a plist file named with a unique device identifier (UDID). On Mac, it is stored at /var/db/lockdown by default. On a Windows system,

FIGURE 5.3 iPhone passcode

it is preserved at C:\Program\Data\Apple\Lockdown. You need to utilize the "sudo" command in the terminal. In newer versions, a trust alert will prompt you on the screen at first in order to request an authorized passcode when the forensic workstation connects to the suspect's device. (Figure 5.3 has been accessed by the forensic workstation so it already has this trust relationship.)

If you are unable to unlock this device after a few attempts of typing in the password, you may need to execute a brute-force attack or dictionary attack (Epifani & Stirparo, 2016; Gangula, 2019). A brute-force attack thoroughly checks all possible passwords until the genuine one is detected. In fact, cracking a four-digit passcode takes around 20 minutes with a brute-force attack, a four-character alphanumeric passcode takes around 20 days, and a six-character alphanumeric passcode can possibly take more than 2 years to crack. Dictionary attacks systematically examine all words in a dictionary to find the passwords. Dictionary attacks may attempt a preset dictionary with common passcodes. If you want to increase the possibility of cracking the passcode, you can customize the dictionary based on information you gathered about the target. Customized dictionaries can speed up the cracking process. Figure 5.4 demonstrates a brute-force attack.

iOS Acquisition

As a process of iOS logical acquisition, a forensic examiner can extract the data from the copy of a backup on iOS devices. If the examiner has privileged access to the device, the examiner can acquire a physical bit-by-bit copy of the device storage (Epifani & Stirparo, 2016; Gangula, 2019).

```
root@Sinch07sFi0pent: ~
File   Edit   View   Search   Terminal   Help
root@Sinch07sFi0pent:~# unshadow /etc/passwd /etc/shadow > ~/Documents/pwd.txt
root@Sinch07sFi0pent:~# john --wordlist=~/Documents/crack.lst ~/Documents/pwd.tx
t
Created directory: /root/.john
Warning: detected hash type "sha512crypt", but the string is also recognized as
"crypt"
Use the "--format=crypt" option to force loading these as that type instead
Using default input encoding: UTF-8
Loaded 1 password hash (sha512crypt, crypt(3) $6$ [SHA512 128/128 AVX 2x])
Press 'q' or Ctrl-C to abort, almost any other key for status
ucertify        (root)
1g 0:00:00:00 DONE (2020-10-29 22:41) 50.00g/s 250.0p/s 250.0c/s 250.0C/s abcd..
uC@123456
Use the "--show" option to display all of the cracked passwords reliably
Session completed
root@Sinch07sFi0pent:~# john --show ~/Documents/pwd.txt
root:ucertify:0:0:root:/root:/bin/bash

1 password hash cracked, 0 left
root@Sinch07sFi0pent:~#
```

FIGURE 5.4 Example of brute-force attack

Jailbreaking the device is, therefore, an essential step that gives examiners a chance to access the device data.

A logical acquisition can be performed via iTunes backup. The iTunes software creates a backup copy of device contents, which is a very useful source for the digital forensic investigator; however, the data is encrypted and protected with a password. It means that examiners must decrypt it for forensic analysis.

Encrypted Backup Cracking

If the iTunes backup data is encrypted, the forensic examiner needs to bypass the encryption through methods such as a brute-force attack or dictionary attack (Epifani & Stirparo, 2016; Gangula, 2019). First, if we can successfully crack the encrypted data in the iTunes backup, we are able to acquire keychain data that is unavailable in the unencrypted backup. However, the keychain file is encrypted and cannot be decrypted offline or reactivated on a different device. As a result, the forensic examiner needs to crack the backup password through tools such as Elcomsoft Phone Breaker, Passware Kit Forensic, and iPhone Backup Unlocker. The backup file is stored with the default backup folders on the computer: Mac: ~/Library/Application Support/MobileSync/Backup. There is a subfolder for each iOS device, which includes a folder containing every file copied from the iOS device, titled sha1(Domain[subdomain-]fullpath/filename.ext).

For instance, the folder has a sha1 file named AppDomain-com.skype.skype-Library/Preferences/com.skype.skype.plist. In contrast, there are several standard files created by the device provider which provide general information regarding the collected device and its data. In other words, these devices and their data may always be examined no matter whether the backup data is encrypted. It may be a good starting point for a mobile forensic investigation. The following list provides some of the most useful files to help forensic investigators initiate a cybercrime investigation (Epifani & Stirparo, 2016; Gangula, 2019):

- **Info.plist:** Plain text data about the device such as timestamps, device name, model, phone number, iTunes versions, etc.
- **Status.plist:** Completion status of backup and whether it was successful or not.
- **Manifest.plist:** List of applications installed on the backed-up device and their information such as name, version, backup date, backup type.
- **Manifest.mbdb:** Descriptions of all the other files in the backup directory. The file information includes domain, path, link target, UID and GID, MAC (last Modified, Accessed and Changed) time, file size, file permission, file hash.

Jailbreaking

Jailbreaking is the process of removing software restrictions on iOS devices, enabling forensic examiners to gain root access to the iOS file system and allowing them to execute additional applications not available in the Apple App Store (Ali et al., 2019; Varenkamp, 2019). Jailbreaking is the only way to make a physical acquisition for iOS 4 and higher; therefore, it is a very important method for forensic examiners seeking evidence from a suspect's iPhone. However, the jailbreaking process might alter the device, which can cause damage to the digital data on iOS devices.

FIGURE 5.5 Jailbreaking iPhones

After jailbreaking, Cydia (an alternative to Apple's App Store) can be installed on a jailbroken iPhone allowing forensic examiners access to apps that aren't available through the Apple Store (Johansen, 2020). In this sense, most jailbreaking tools install Cydia during the jailbreaking process. In summary, a forensic examiner can use software to jailbreak an iPhone and install Cydia as well as other recommended tools (apt, make, wget, sqlite, etc.). The default root password is *alpine* and you need to change it through the passwd command. In this stage, the forensic examiner needs to access all the data and decode it with encryption methods. At the same time, the examiner can decrypt email and stored passwords through a jailbreaking technique. You can use an imaging tool to examine all the image data. However, many forensic examiners will be faced with encrypted image data and may need to decrypt the image data and the keychain to produce a readable image.

File System Acquisition

File system acquisition is recognized as logical acquisition. Also, the NIST indicates that file system acquisition is still a type of logical acquisition along with the NIST's mobile device tool classification system (Ayers et al., 2014). The file system acquisition process enables the forensic examiner to obtain the whole file system (all files and folders); this method can acquire more diverse data than a backup. However, to execute this file system acquisition process the forensic examiner must have privileged access to the device. Similar to physical acquisition, you need to use the jailbreaking technique to bypass the security restriction. Then it is possible to copy the whole file system.

iOS Data Analysis

This section will explore the iPhone file system and data. This section will give us a chance to familiarize the reader with the iPhone file and directory structure as well as the iPhone backup file and folder structure. This section will focus on the analysis of iOS data.

On iDevices, property list (plist) files and SQLite databases are two main formats used to store data. These files are utilized through Plist Editor and SQLite database browser. These tools can analyze time stamps, geolocation data, phone call data, messages, photos, etc.

There are numerous tools available to extract data from an iTunes backup. iPhone Backup Extractor, Cellebrite UFED, iPhone Backup Analyzer (iPBA), and iPhone Analyzer are the main extraction tools (Zdziarski, 2012). This section will introduce the use of iPhone Backup Extractor and iPhone Analyzer.

The iPhone Directory Structure

As the first step of the investigation for a suspect's iPhone, a forensic examiner needs to explore the iPhone directory structure. You need to open the iPhone Backup Extractor application. iPhone Backup Extractor will provide the plist file and load the backup file. Figures 5.6 and 5.7 demonstrate iPhone Backup Extractor and the iPhone Directory Structure. The iPhone Backup Extractor locates and extracts the AddressBook.sqlitedb, Calendar.sqlitedb, and call_history. db files from the iPhone's iTunes backup file.

iOS Forensic Process

This section will explore the iOS forensics process using SQLite. SQLite is an open source that implements a self-contained, zero configuration, and trinational SQL database engine. A single SQLite file holds multiple tables, triggers, and views. SQLite Browser can analyze the extracted data to identify and examine important case data such as contacts, call logs, SMS images, audio and video files, web history, passwords, and application data. Also, deleted data can be recovered

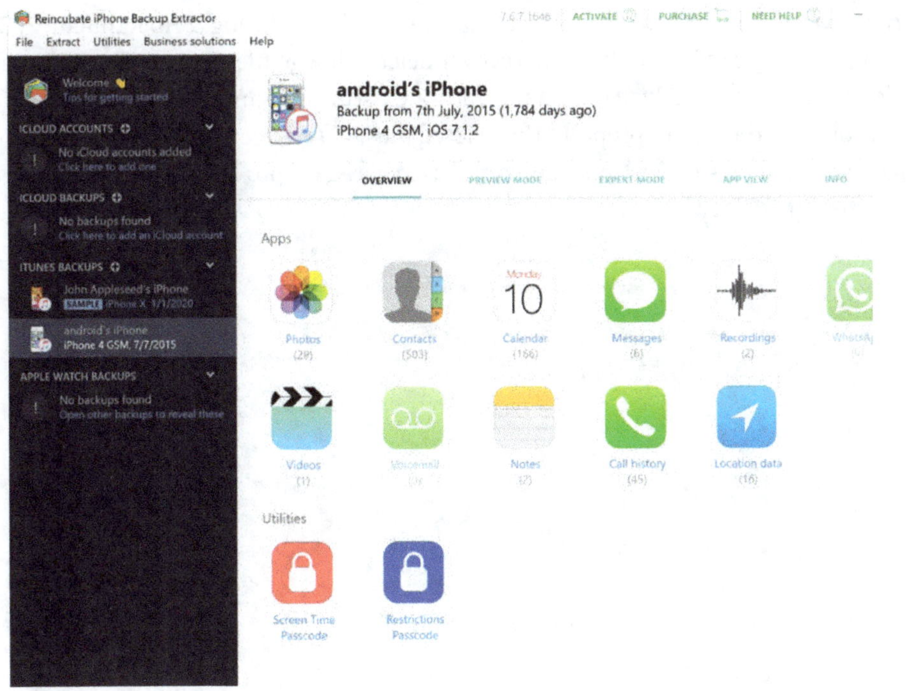

FIGURE 5.6 iPhone Backup Extractor

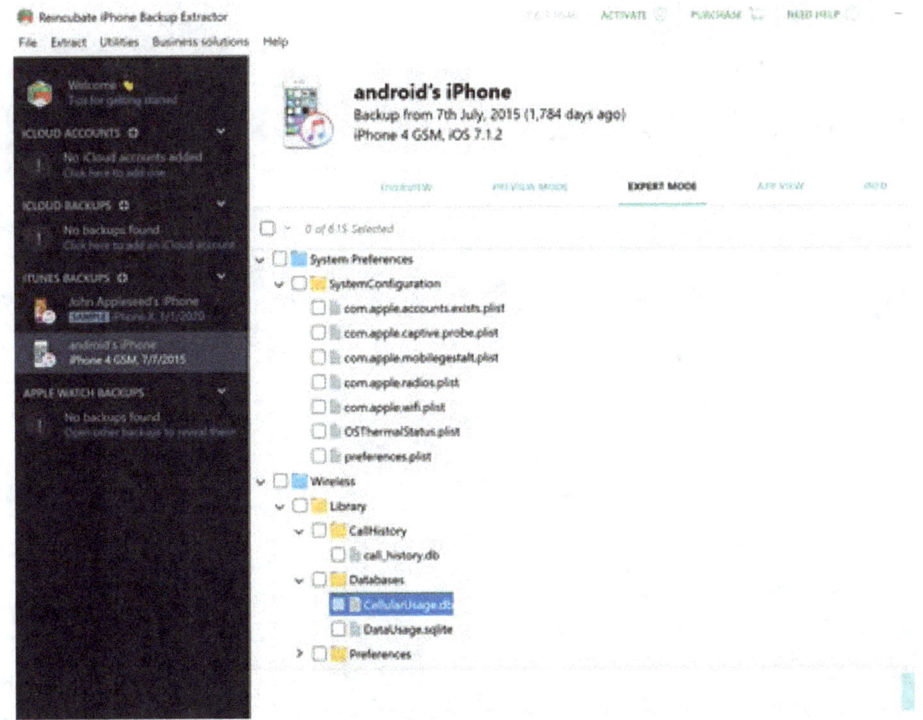

FIGURE 5.7 iPhone directory structure

from the physical image of the iOS device (Cahyani et al., 2017; Ong & Ab Rahman, 2020). Hex editor and Autopsy can be employed to recover deleted image files from a physical image with the following steps: (1) analyze the catalog file for existing files and the journal files for deleted files and compare the two to pinpoint the deleted files; (2) extract encrypted and deleted files; (3) analyze metadata such as timestamps and associated crypto keys; (4) decrypt the deleted file with the per-file key.

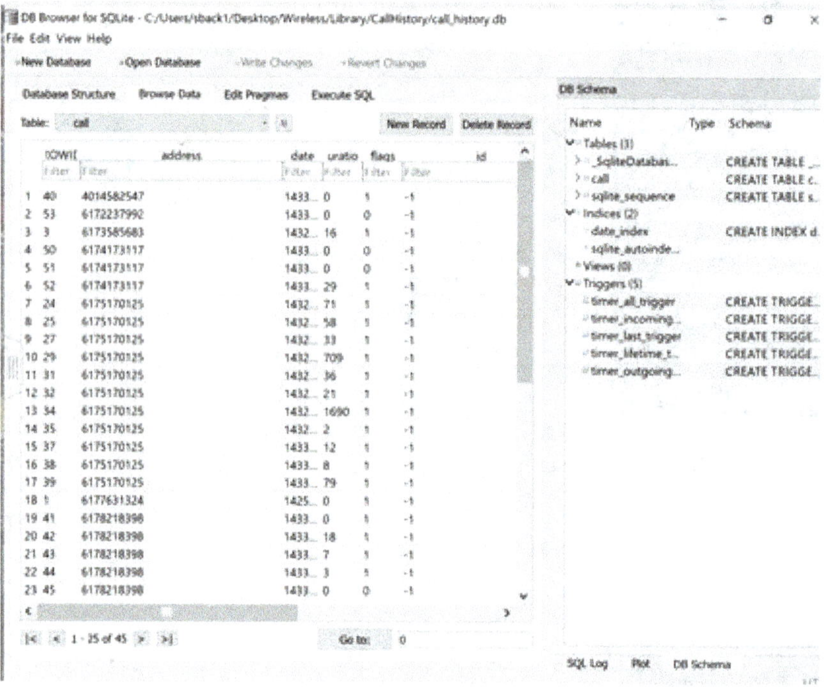

FIGURE 5.8 DB browser for SQLite

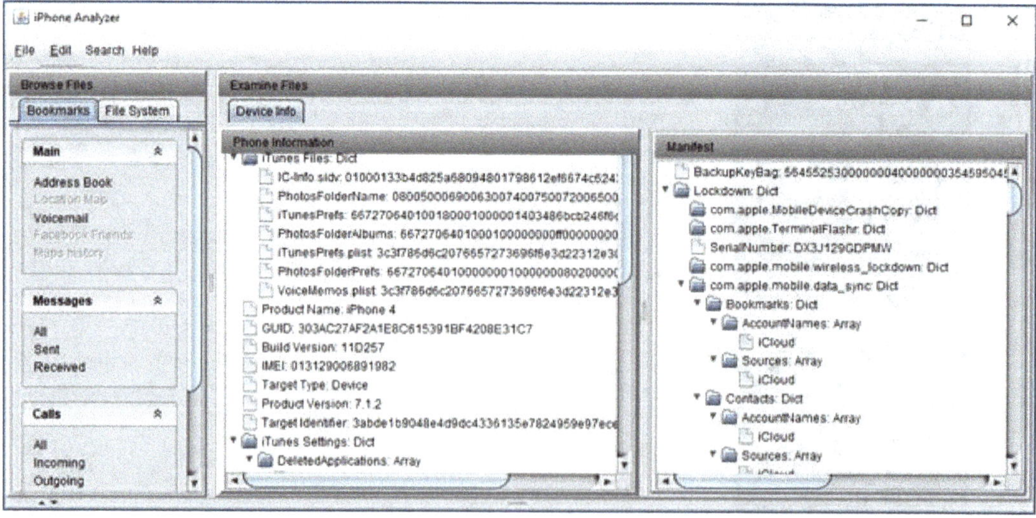

FIGURE 5.9 iPhone Analyzer

Another useful tool is iPhone Analyzer, which can explore the internal file structure of iDevices. Specifically, iPhone Analyzer has features including iPhone backup browsing, viewing plist and sqlite files, searching including regular expressions, ssh access for jailbroken phones, recovering backups, viewing all iPhone photos, examining address books, and finding/recovering passwords.

Android System

The Android operating system is used for Android devices, which are manufactured by original equipment manufacturers (OEMs) such as Samsung, HTC, ASUS, Huawei, ZTE, and various other mobile device manufacturers. Google leads the way in developing the Android operating system, which is now maintained through the Android Open Source Project (AOSP). As an open-source software stack, Android has been applied on a wide range of mobile devices. Android consists of the Linux kernel code with Android's open nature so that mobile phone producers can access it without restrictions as well as freely modify and use the software (Source, 2020). Even though Android systems can be modified, they still contain some permanent source software components, including bootloaders, firmware, DRM, and apps (Source, 2020).

Android Platform

Hardware

We have already explored the hardware in mobile systems and workstations in the previous sections. In this section, we will focus on several hardware components and features that forensic examiners commonly need to be aware of for their cybercrime investigations. These include components such as RAM, the volatile storage unit; NAND flash, which is the built-in memory of a device; SD card slots to support SD memory cards; the Universal Integrated Circuit Card (UICC), which is the hardware portion of the smart card (storing data and applications); and the SIM card or the subscriber identity module, which is the software application included on the card (Reiber, 2018). SIM cards were invented to enable portability and to store information to enable authentication on the cellular network.

Android Architecture and Kernel

Layered architecture is applied as a basic Android software stack (see Figure 5.13). It consists of System apps, Java API

FIGURE 5.10 Samsung Galaxy (back)

FIGURE 5.11 Samsung Galaxy (front)

FIGURE 5.12 SIM card

framework, Java C/C++ libraries, Android Runtime, hardware abstraction layer (HAL), and Linux kernel. A hardware abstraction layer clarifies a standard interface for various hardware models so that Android can be agnostic about lower-level driver implementations. The Linux-based kernel controls basic system resources (e.g., CPU, memory, storage, battery, and networking). Due to the low-level process-based isolation, a separate virtual address space is used for each process. In this regard, these system resources are unable to access each other's memory. Therefore, inter-process communication (IPC) channels are major communication platforms between processes. System apps are pre-installed formats to demand higher privileges for access. The key components of an Android system are presented below (Google Inc., 2020; Source, 2020):

- **Manifest file.** The file AndroidManifest.xml is provided by the app developers for each app. It contains device version information, component definition, permission definition, external library information, and shared UID information.
- **Intent filter.** An expression in an app's manifest file that specifies the type of intents (information requests) it is willing to receive and the criteria of the sender.
- **Activity.** This is an application feature that provides a UI such as dialing the phone, taking a photo, sending an email, or viewing a map.
- **Service.** A service is an Android component running in the background.
- **Android Software Development Kit (SDK).** Android SDK is a tool kit that allows developers to build, test, and debug Android applications through the provision of software libraries, APIs, reference material, and an emulator.
- **Native Development Kit (NDK).** The NDK enables developers to write code in C/C++ and compile directly for the CPU. It is very useful for game engines, signal processing, and physics simulation(s).
- **Android Debug Bridge (ADB).** ADB is a command-line utility that allows communication between a computer and an Android device.

- **YAFFS2 (Yet Another Flash File System).** It is an open-source file system that works with NAND flash devices, which includes the following features. a log-structured file system, built-in wear leveling and error correction capable of handling bad blocks, a single-threaded file system, and a small footprint in RAM.
- **EXT4 (Extended File System).** A de facto filesystem for Linux. EXT4 uses regular block devices along with Universal Flash Storage (UFS) instead of the NAND flash memory. EXT4 includes features such as supporting full permissions and semantics, dual-core systems, and supporting encryption natively.

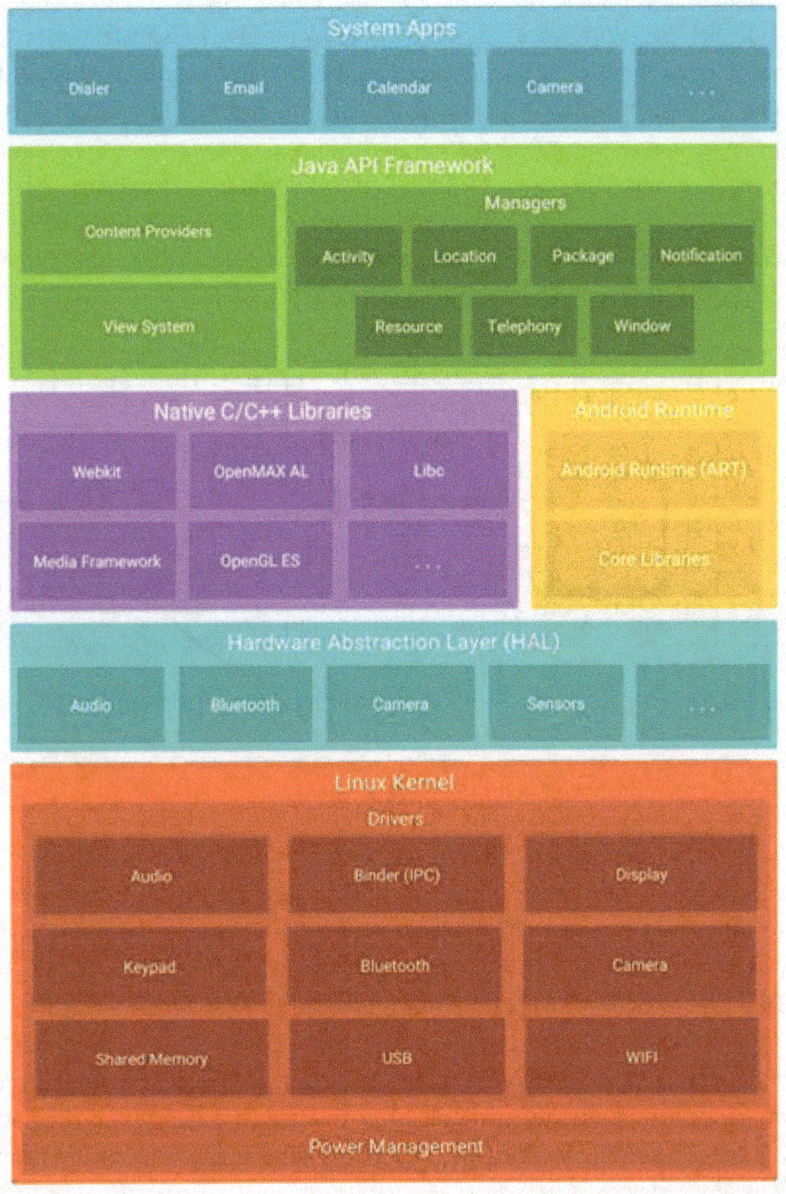

FIGURE 5.13 Android architecture and kernel

Android Security and Forensics

Android security systems differ from the iOS security system. In this section, you will learn about the various techniques and tools used to execute the Android forensic process. Some components of these concepts and techniques overlap with those used in iOS forensics. I will briefly discuss other specifics and techniques that forensic examiners can implement during Android mobile forensic investigations.

Data Protection and Encryption

Android systems normally employ an encryption technique to provide user data security through the encryption of all user data in the kernel (Google, 2020). There are two methods for device encryption: full-disk encryption and file-based encryption. Full-disk encryption uses a single key protected by the user-set device passcode. Once a device is executed by a full-disk encryption technique, all user-created data is automatically encrypted. The full disk encryption is based on dm-crypt, an encryption subsystem in the kernel that works at the block device layer for EXT4. It uses a 128-bit AES (Advanced Encryption Standard) algorithm with cipher-block chaining (CBC), which is used to encrypt all data (see Figure 5.14). Android devices generate a random 129-bit master key upon first boot, and are encrypted with another key that is based on the user password in order to protect this master key.

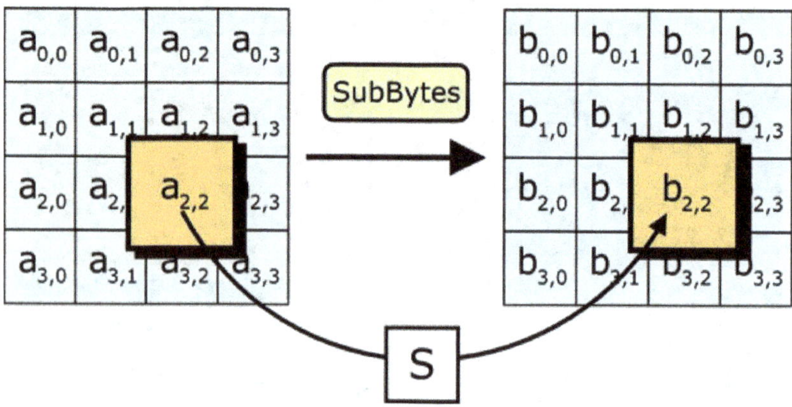

FIGURE 5.14 Advanced Encryption Standard (AES)

Android 7 and later versions support file-based encryption (FBE), which allows files to be encrypted with different keys that can be unlocked independently. In other words, it enables a new feature called Direct Boot, which allows encrypted devices to boot straight to the lock screen. FBE improves usability while securing private user information. In this sense, EXT4 supports FBE at the directory level and allows Android to leverage this feature to formulate different storage locations (directories) to applications for each user, along with credential encrypted (CE) storage and device encrypted (DE) storage. Credential encrypted storage is a default storage location and only available after the user has decrypted the device. Device encrypted storage is a storage location available both during Direct Boot mode and after the user has unlocked the device.

In addition, the Android system applies device access control in accordance with a user-set password, which prevents unauthorized access to an Android device. Also, users can utilize a lock screen pattern to protect the Android device. The restriction(s) on PINs, patterns, and passwords should be set through device management tools (e.g., Smart Lock).

Data Acquisition from Android Devices

The forensic investigator can use Android Debug Bridge (ADB) commands to extract the data from the mobile device. ADB is a command-line utility that enables a forensic examiner to communicate with Android mobile devices. Also, ADB commands can be employed to operate applications and alter devices. Android forensic tools—Cellebrite, XRY—can be utilized to execute ADB to communicate with the device and extract data from it.

Rooting and Rooting Process

Rooting is the process of gaining root privilege to Android devices. It is required for more advanced and potentially dangerous operations. There are two types of rooting techniques that need to be explored. Temporary rooting acquires the root privilege and is then lost after a reboot, while permanent rooting acquires root privilege that remains after a reboot. Therefore, forensic examiners wish to gain this rooted privilege so that they can determine whether the suspect modified or removed some files from his or her Android device. That said, when Library is written by the superuser (su) into the system partition, the rooting is permanent and remains available after a boot; without systematic root, it is also possible to fast boot from a temporary boot image, resulting in temporary rooting. Before reboot, a list of root exploits available include Wunderbar/asroot, Volez, Exploid, Zimperlich and Zysploit, and GingerBreak.

To execute a root exploit on Android, the Common Vulnerabilities and Exposure (CVE) system can be utilized. The CVE is the numbered reference system employed to catalog disclosed vulnerabilities and exploits (Distortion, 2018). The Linux kernel operating system is a primary target for root exploit. This is because Linux kernel, as the main OS on an Android device, provides the privileged position that can help malicious actors achieve full control over a targeted Android device. For instance, TowelRoot (CVE 2014-3153) exploits the 'futex syscall bugs' to acquire root privilege. Vendors provide vendor-customized device drivers for camera or sound. Therefore, when malicious actors execute code inside the kernel space it subsequently leads to full control over the device. To that end, the odds of having security vulnerabilities in these devices (e.g., Samsung smartphones) can be high. Libraries layer is another main target for the root exploit. Exploits at the libraries layer can facilitate the Android libraries attack. For example, the ZergRush exploit (CVE-2011-3874) in Android leads to root privilege escalation that came along with overflow vulnerability. Libraries layers can be embedded by multiple programs with a malicious code and root exploit.

Sun, Cuadros, and Beznosov (2015) assert that malicious actors are limited to exploit a system or kernel vulnerability without rooting. It means that any app may easily obtain the root privilege with Run-time.exec ("su") on a rooted device (Sun et al., 2015). For example, once root privilege is granted, malware or spyware could obtain unauthorized permission to access and to intercept any sensitive data on the rooted device. An Android security report revealed that there were numerous devices from major Chinese corporations that planted rooting exploits which allow root access by default (Cecelia, 2015; Google Inc., 2015).

Forensic Acquisition and Android Data Analysis

After acquiring data via ADB command, content providers show user data such as contacts, SMS, and call logs without rooting the device. Also, physical acquisition allows access to the storage media, which can allow forensic examiners a look at raw images in the device. Android devices contain various types of files such as SQLite DB and XML files. SQLite database stores persistent data for the user and app (Skulkin et al., 2018). XML is eXtensible Markup Language, which stores preference data (e.g., Chrome settings data) along with settings, passwords, and protocols. SQLite DB files are located in the *databases/subdirectory* under each application data folder */data/data/<appname>/*. SQLite browser can be utilized to extract and analyze these data; moreover, a hex editor can be utilized to find out the type of binary file. XML files are located in the shared_pref/ subdirectory under each application's data folder */data/data/<appname>/*. The cat command can be utilized to extract and analyze these data, which can provide forensic examiners important data in order to move to the next steps for digital forensic procedures.

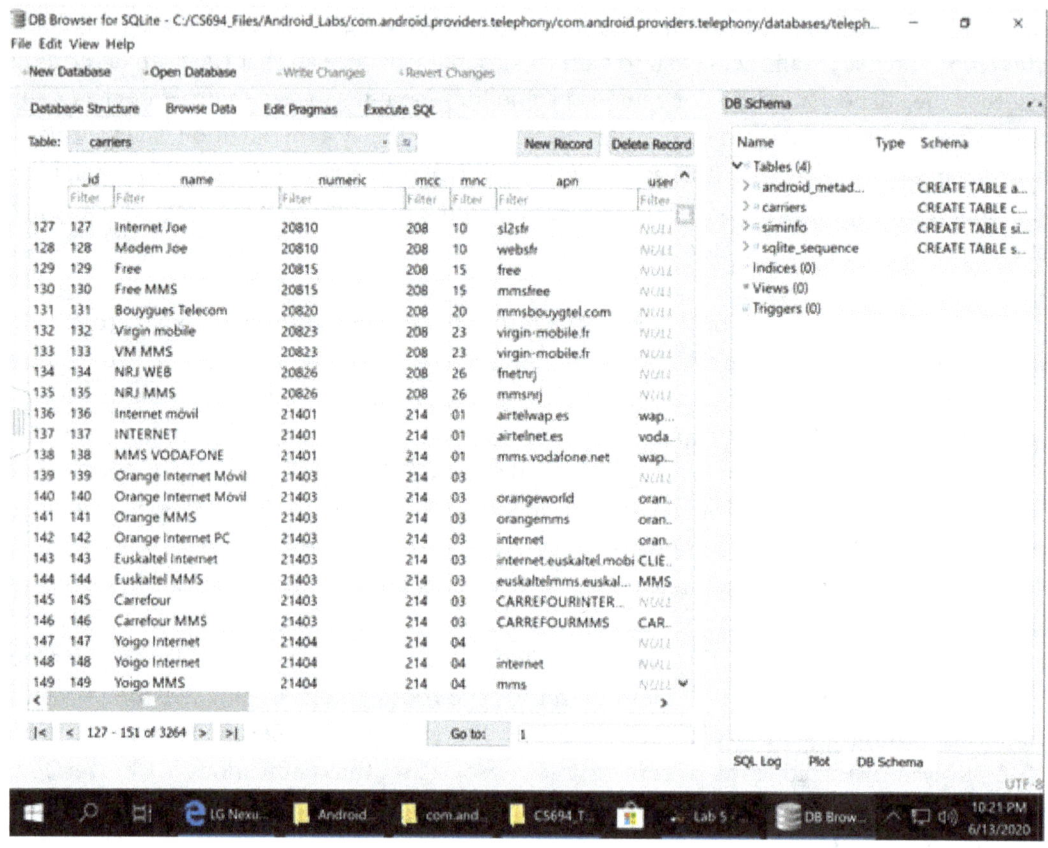

FIGURE 5.15 Acquisition data from an Android device

The timeline analysis in an Android system is similar to the iOS system, which involves extracting timestamps from the file metadata and displaying them in human-readable format. Android uses UNIX epoch time in the EXT4 file system, which represents the number of seconds. There are many tools, including the Linux command date, which can be utilized to convert the UNIX timestamp to human readable format.

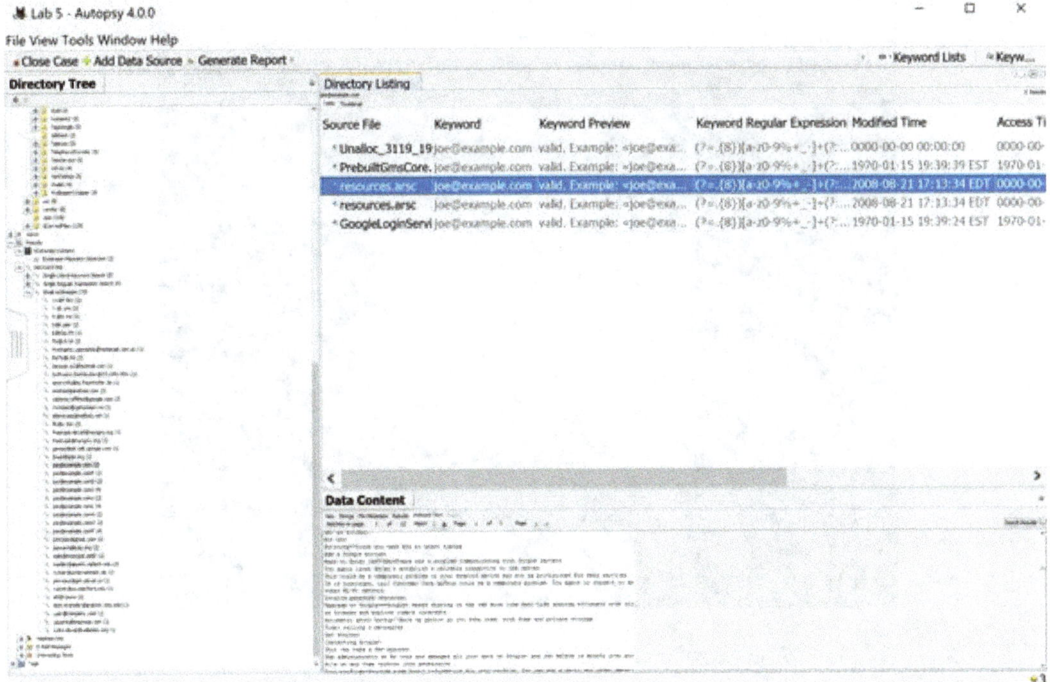

FIGURE 5.16 Timeline analysis with Autopsy

Using the Autopsy application and SQLite browser, you can analyze an image file and locate the corresponding information from its database. Also, you are able to load the database files from these .zip files (contact, call logs, and calendar apps) into the SQLite browser (see Figure 5.15).

Summary

Mobile devices are rich sources of information such as calls, text messages, photos and music/video files, and address books. A suspect's data can be acquired from volatile memory, SIM cards, and compact flash cards. As a process of iOS logical acquisition, a forensic examiner can extract data from a backup on iOS devices or acquire a physical bit-by-bit copy file from the device storage media. In this regard, jailbreaking the device is an essential step in accessing device data. The forensic investigator can use ADB commands to extract the data from the mobile device. Android Debug Bridge (ADB) is a command-line utility that enables a forensic examiner to communicate with Android mobile devices. Also, ADB commands can be employed to operate applications and alter devices. Android forensic tools—Cellebrite, XRY—can be utilized to execute ADB utilities allowing communication with the device and data extraction.

Lab 5.1: Use of the Mount Utility

In this project, as a forensic investigator, you obtained Mr. Evil Noodle's iPhone XS device to investigate a drug trafficking case. You will utilize Santoku Linux to create an Android virtual device and then gather data from it. This hands-on experience will give you a sense of how to investigate a cybercrime scene on mobile devices.

FIGURE 5.17 Creating an AVD in Santoku Linux

Download Santoku Linux from the website link http://santoku-linux.com/download/.

1. Open VirtualBox, operate a virtual machine, and execute Install Santoku 15.03. When prompted, restart the virtual machine.

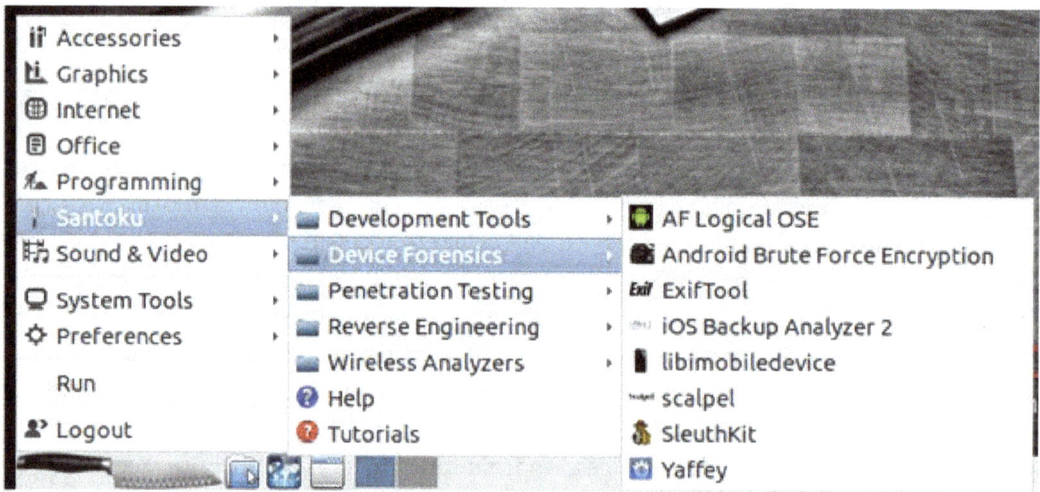

FIGURE 5.18 Santoku Device Forensics menu

2. You can find mobile forensics tools by clicking the Santoku knife, selecting Santoku, and Device Forensics.
3. To open a command-line terminal, click the Santoku knife, select Accessories, and LXTerminal.

4. Show your output when you run the cat/proc/cpuinfo command.

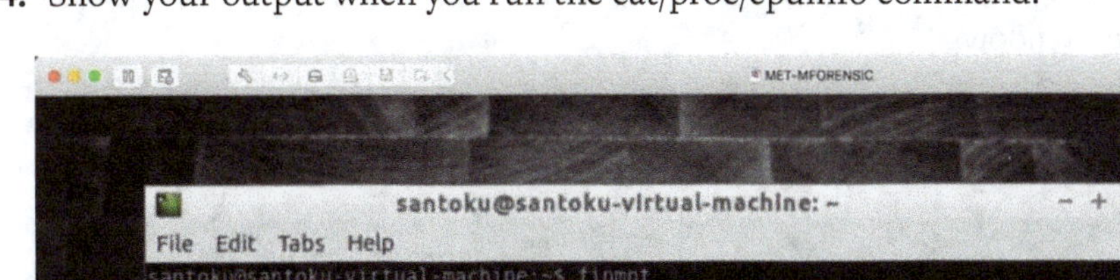

FIGURE 5.19 Use of the mount utility

5. Show your output when you run the cat /proc/meminfo command.
6. Show your output when you run the df command.
7. Use the mount utility to list the mounted filesystems and compare them to the results you see from cat /proc/filesystems. It displays the target mount point, the source device, file system type, and relevant mount options for each filesystem, whereas proc filesystems show only file style.

Lab 5.2: Explore the iPhone Backup Structure

Create an Assignment\Chap5-2\Projects folder on your computer before starting these projects; it is referred to as your "Assignment folder" in steps.

In this project, as a forensic investigator, you obtained Mr. Evil Noodle's iPhone XS device to investigate a drug trafficking case. You will utilize the iPhone Backup Extractor application to explore the backup structure. This hands-on experience will give you a sense of what it is to investigate a cybercrime scene on mobile devices.

1. Extract the contents of the iPhone Xs_Chuck_backup.zip file. In Windows, you can right-click on the file and choose Extract All.
2. Open the iPhone Backup Extractor application.

3. Beneath the iTunes Backup section, click the *No backups found* option to display the Preferences window.

FIGURE 5.20 No backups found

4. To add an iTunes backup location, click the plus symbol (+).
5. Next, browse to your extracted iPhone backup location. Make sure you select the folder 7jhghahag5hahhahr65767glhalhahag9ljhah234hahh83. Once located, iPhone Backup Extractor will read the plist file and load the backup.
6. Click the appropriate item from the iTunes Backups section.
7. Go to the *Expert Mode* tab and examine each folder.

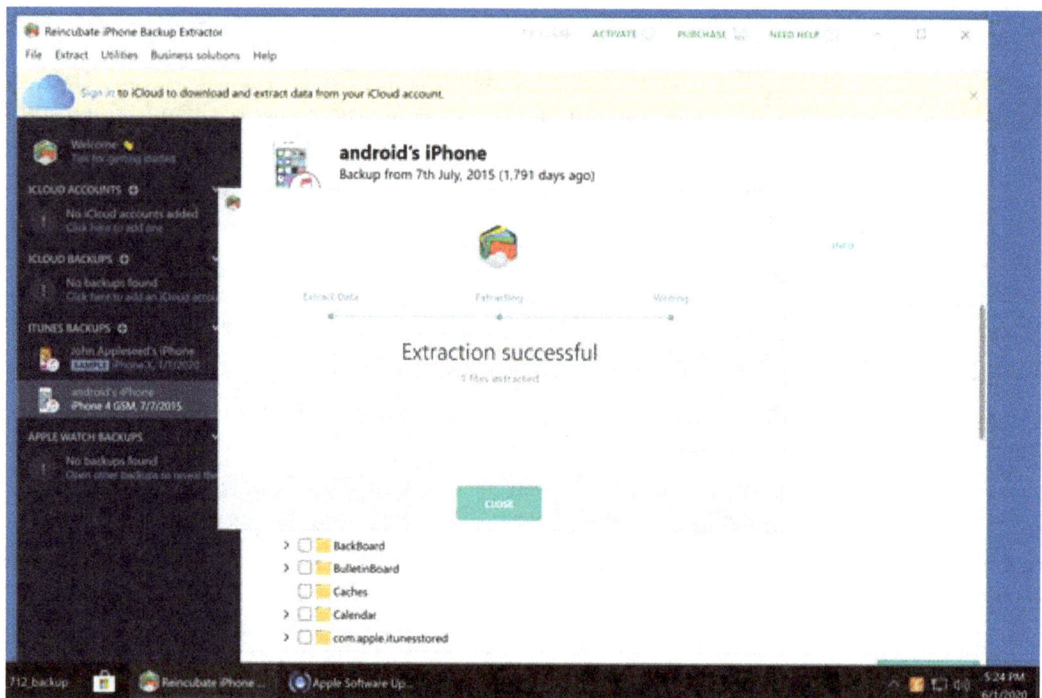

FIGURE 5.21 iPhone backup extraction

Lab 5.3: Android Data Analysis Using Autopsy

Create an Assignment\Chap5-3\Projects folder on your computer before starting these projects; it is referred to as your "Assignment folder" in steps.

In this project, as a forensic investigator, you obtained Mr. Evil Noodle's LG Nexus 5 smartphone to investigate an internet child pornography case. You will utilize the Autopsy application to analyze an image file and locate the corresponding information from its data source.

1. Open the Autopsy application.
2. Click Create New Case from the Welcome window.
3. Enter **Lab 5-3** as the Case Name and select your Documents folder to store the case files. Click the Next button to continue.
4. Use **1** for the Case Number and enter your name as the Examiner. Click the Finish button to end the case setup process.
5. When prompted to Add Data Source, locate the nexus5_system_061615_bs64.img file and click the Next button to continue.
6. Ensure all ingest modules are selected and click the Next button to continue.

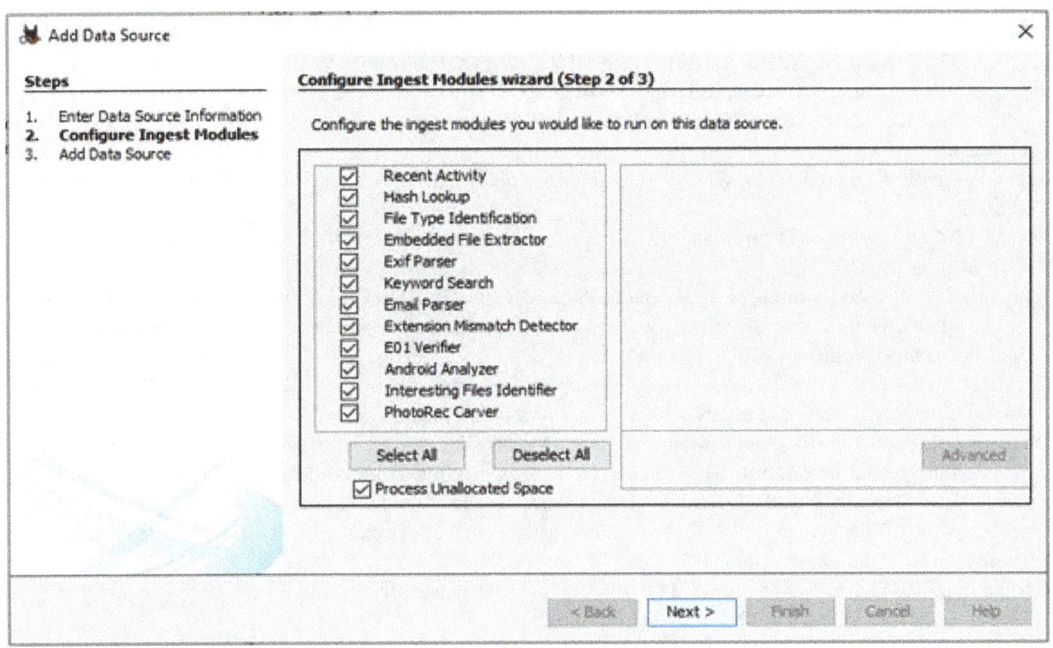

FIGURE 5.22 Configure Ingest Modules wizard

7. When prompted, click the Finish button to allow Autopsy to analyze the image. Autopsy will display status indicators.
8. Once Autopsy finishes all processing, explore the results at the left of the display.

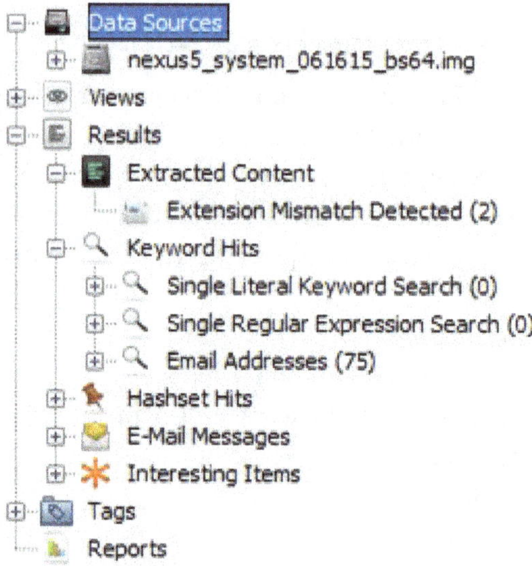

FIGURE 5.23 Autopsy results

References

Ali, A. A., Cahyani, N., & Jadied, E. (2019). Digital forensic analysis on iDevice: Jailbreak iOS 12.1.1 as a case study. *Indonesia Journal on Computing*, 4(2), 205–218.

Apple Inc. (2015). iOS security (iOS 9.0 or later). https://www.apple.com/business/docs/iOS_Security_Guide.pdf

Ayers, R., Brothers, S., & Jansen, W. (2014). Guidelines on mobile device forensics. *NIST Special Publication*, 1(1), 85.

Bay, M. (2017). The ethics of unbreakable encryption: Rawlsian privacy and the San Bernardino iPhone. *First Monday*, 22(2).

Cahyani, N. D. W., Ab Rahman, N. H., Glisson, W. B., & Choo, K. K. R. (2017). The role of mobile forensics in terrorism investigations involving the use of cloud storage service and communication apps. *Mobile Networks and Applications*, 22(2), 240–254.

Cecelia. (2015, April 10). *80% China's mobile users rooted smartphones in 2014*. China Internet Watch. https://www.chinainternetwatch.com/12926/80-china-smartphone-users-rooted/

Distortion. (2018). *Top 10 exploit databases for finding vulnerabilities*. WonderHowTo.com https://null-byte.wonderhowto.com/how-to/top-10-exploit-databases-for-finding-vulnerabilities-0189314/

Epifani, M., & Stirparo, P. (2016). *Learning iOS forensics*. Packt Publishing Ltd.

Gangula, M. R. (2019). Overcoming forensic implications with enhancing security in iOS. *Culminating Projects in Information Assurance, 77*.

Google Inc. (2015). *Android security 2014 year in review*. https://source.android.com/security/reports/Google_Android_Security_2014_Report_Final.pdf

Google Inc. (2020). *Android enterprise security [White paper]*. https://static.googleusercontent.com/media/www.android.com/en//static/2016/pdfs/enterprise/Android_Enterprise_Security_White_Paper_2019.pdf

Hayes, D. R. (2015). *A practical guide to computer forensics investigations*. Pearson Education.

Johansen, A. G. (2020, March 22). Is jailbreaking legal and safe? NortonLifeLock. https://us.norton.com/internetsecurity-mobile-is-jailbreaking-legal-and-safe.html

Nelson, B., Phillips, A., & Steuart, C. (2019). *Guide to computer forensics and investigations (Loose-leaf version)*. Cengage Learning.

Ong, W. S., & Ab Rahman, N. H. (2020). A forensic analysis visualization tool for mobile instant messaging apps. *International Journal on Information and Communication Technology*, (60)2, 78–87.

Reiber, L. (2018). *Mobile forensic investigations: A guide to evidence collection, analysis, and presentation*. McGraw Hill Professional.

Skulkin, O., Tindall, D., & Tamma, R. (2018). *Learning Android forensics: Analyze Android devices with the latest forensic tools and techniques*. Packt Publishing Ltd.

Source. (2020). *Set up for Android development*. https://source.android.com/setup

Sun, S. T., Cuadros, A., & Beznosov, K. (2015, October). Android rooting: Methods, detection, and evasion. In *Proceedings of the 5th Annual ACM CCS Workshop on Security and Privacy in Smartphones and Mobile Devices* (pp. 3–14).

Tipton, S. J., White II, D. J., Sershon, C., & Choi, Y. B. (2014). iOS security and privacy: Authentication methods, permissions, and potential pitfalls with touch id. *International Journal of Computer and Information Technology, 3*(03).

Varenkamp, P. (2019). *iPhone acquisition using jailbreaking techniques* [Master's thesis, NTNU].

Zdziarski, J. (2012). iOS forensic investigative methods [Technical draft]. https://www.zdziarski.com/blog/wp-content/uploads/2013/05/iOS-Forensic-Investigative-Methods.pdf

Zetter, K. (2016, February 18). Apple's FBI battle is complicated. Here's what's really going on. *Wired*. https://www.wired.com/2016/02/apples-fbi-battle-is-complicated-heres-whats-really-going-on/

Credits

Email Evidence and Analysis

Kyung-Shick Choi

Introduction

In this chapter, we will first discuss email evidence and then move on to a discussion involving email analysis using an actual computer crime case. One of the highest volumes of digital evidence available today is email. Free email accounts such as Microsoft Hotmail, Google Gmail, and Yahoo! Mail can easily be set up within a couple of minutes, making it possible for almost everyone to have an email account.

Before we discuss email evidence analysis in further detail, let's review the Enron scandal.

Applied Case: Enron

Enron was the largest energy company in the 1990s and was rated the most innovative large company in America according to *Fortune's* Most Admired Companies survey. The Enron scandal is one of the largest white-collar crime cases in U.S. history. In mid-2000, Enron executives sold their shares after the stock prices superficially raised, and then the stock price plunged to less than $1 by the end of November 2001. On December 2, 2001, Enron filed one of the biggest bankruptcies in U.S. history.

The Enron scandal is a great example of white-collar crime and computer crime, highlighting how computer technology can be used to deceive not only the public but also every government agency in the nation.

During the Enron investigations, one piece of crucial evidence was an internal email at JPMorgan Chase that described one of these disguised loans as a "prepay." The email chain of evidence also included, "Enron loves these deals as they are able to hide funded debt from their equity analysts because they book it as deferred or bury it in their trading liabilities."

The substantial financial losses and the reputation of damages impacted JPMorgan Chase and Citigroup. As a consequence of the scandal, the Sarbanes–Oxley Act was created to increase penalties for destroying, altering, or fabricating records in federal investigations or for attempting to defraud shareholders (Brown, 2003).

TABLE 6.1

✓	Ⓐ	R	★★★★★	File Name/Subject	■	A	📎	Extension
✓	T	R		External QA.doc				doc
✓	T	R		Internal QA.doc				doc
✓	T	R		Internal QA.doc				doc
✓	T	R		FW: my unfair treatment at Enron-please HELP			☐	
✓	T	R		FW: my unfair treatment at Enron-please HELP			☐	
✓	T			**Enron Files Chapter 11 Reorganization**			☐	
✓	T	R		Re: Clinton Energy Vacation Policy & Request			📎	
✓	T	R		Clinton Energy Vacation Policy & Request			☐	

Through the Nuix company, the Enron data set was made available and intensively studied. The Enron data set consists of 1.3 million email messages and attachments from former Enron staff.

The Nuix investigation found the following:

- 60 items containing credit card numbers, including department contact lists that each contained hundreds of individual credit cards
- 572 items containing Social Security or other national identity numbers
- 292 items containing individuals' dates of birth
- 532 items containing information of a highly personal nature such as medical or legal matters
- Through email accounts, much more information was exposed

TABLE 6.2

TYPE OF INFORMATION	NUMBER OF ITEMS CONTAINING THIS TYPE OF INFORMATION
Credit card number	60
Date of birth	292
Highly personal information	532
National identity number	572
Personal contact details	6,237
Résumés containing substantial personal contact details	3,023

Types of Email: Web-Based, User-Based, and Server-Based Email

Although there are numerous email addresses and accounts in use today, they are categorized into three major types. The first category is web-based email, which is the most commonly used email service today.

Web-Based Email

Seemingly, the most popular type of email service is free web-based email including Microsoft's Hotmail, Yahoo!, or Google's Gmail. Special programs are not required in order to send and retrieve email with these services. Any computer with internet access can facilitate access to mail stored on servers provided by the service provider, making these services extremely portable.

Image 6.1

People often believe that there will be no record stored on the local computer when using a web-based email account via the internet browser. However, even with this type of account, bits and pieces of information from web-based email may in fact be recoverable as HTML documents, in the form of temporary internet files or drive space that has not been overwritten by new files. Web-based email is cached to the local hard drive as a web page. Meaning, these web pages can be located in the internet cache or in unallocated space. In the event that the email itself cannot be recovered, the individual messages read or emails sent and received can be recovered in the user's mailbox. Furthermore, a more feasible option may be to get email directly from the service provider via subpoena. A web-based account allows an email to be traced back to the originating ISP, resulting in the possibility of determining the IP address of the machine that connected when the account was created. This process is impacted by whether the service provider maintained those records for any specific period of time. It is extremely difficult to retrieve the email if the email service purges deleted emails due to inactivity of the accounts by service providers. Importantly, the investigator must check with the custodian of records to find out the retention policy of the individual service (Daniel & Daniel, 2012).

Corporate companies tend to use the second category: mail server–based email.

Mail Server–Based Email

Corporate email accounts are typically hosted on a mail server which is either owned or leased by the company. These mail servers can be Microsoft Exchange, Lotus Notes, Novell Group-Wise, or another type of server. Corporate mail servers are typically backed up on a regular basis and backups could be stored on site on disk. Remote location may have additional backups via off-site storage applications. It is still possible to retrieve the email from the backups by duplicating the missing server on a new computer or in a virtual environment in cases of older backups where the physical server is no longer available.

Free email accounts are hosted by companies. These companies are in the business of marketing via the internet, with Microsoft and Google dominating the free email marketplace. Depending on the free email account provider, emails may be stored for only a short time, especially if an account goes inactive for a certain period; however, the record of the account creation can be stored for a very long time.

Internet service providers (ISPs) also provide email accounts as part of their service when you sign up for an account.

An email system that resides on the local user's personal computer is the third category. These email systems run with user-based email storage. However, email on mobile devices such as cell phones and tablet computers has become dominant.

User-Based Email (Computer)

One of the most common user-based email systems is Microsoft's Outlook or Outlook Express, which runs on the user's computer. The setup can be easy if the account holder has an active internet connection. Emails sent and received through this type of account will be stored locally on the user's computer. In the event that this type of email program is used to generate and send illegal communications, it is likely that evidence of those communications could be recovered from the personal computer.

TABLE 6.3 The Common Formats of Computer User-Based Email Storage

EMAIL CLIENT	EXTENSION	TYPE OF FILE
Microsoft Outlook Express	.dbx	OE mail database
	.dgr	OE fax page
	.email	OE mail message
	.eml	OE electronic mail
Microsoft Outlook	.pab	Personal address book
	.pst	Personal Storage Table
	.ost	Offline Storage Table
Apple macOS	.mbox	X mail
GroupWise	.MLM	Saved email
Lotus Notes	.NSF	Notes database

User-Based Email (Phone)

Email functions with their own format are also accessible on smartphones. As an example, if the owner of a phone connects the iPhone to a computer running iTunes, the phone will want to perform an automatic backup of the phone's contents. All of the user's emails currently on the phone will be backed up to the computer and these backups can be found in the iTunes backup folder for the phone (Daniel & Daniel, 2012).

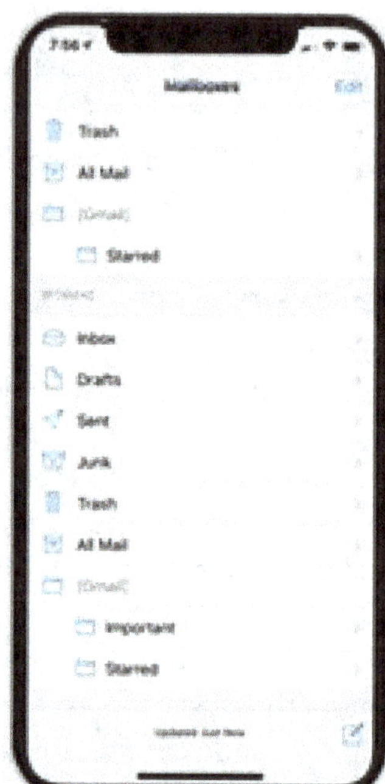

Image 6.3

SCENARIO: DO FACTORY SETTINGS ON A SMARTPHONE REMOVE ALL TRACES OF EMAIL EVIDENCE?

In a case where a large company wanted to know if an employee had sent an email containing a nude picture to the board of directors, the employee's iPhone and company laptop computer were submitted for examination. When the phone was examined, it was discovered that it had been factory reset, wiping all data from the phone. This ruled out checking the phone itself for evidence. Next the computer was examined, and it did not have any pictures or emails of interest. However, there was also evidence that the computer had been cleaned up using a file-wiping program. In some instances, this would have been the end of any possibility of discovering what really happened. However, what the clever employee missed was the backups of the iPhone that were still on the computer in the iTunes directory. The picture was recovered from the iTunes backup folder to prove that the picture originated from the employee's iPhone and email (Daniel & Daniel, 2012, p. 243).

Email Evidence and Investigation

Today, gathering email evidence is more important than ever due to the increasing number of people who send and receive email on numerous devices.

There are normally four things that we need to uncover with email evidence:

- The content of the email
- Who sent it
- When it was sent
- Where it was sent from

Determining the origination of an email can be quite complicated. The evidence of the origination of email is embedded within the email itself in the form of an email header. However, portions of an email header can be easily manipulated by inserting false records. Additionally, spammers often fake the "sent from" email address as well as the internet address of the sending server in efforts to fool antispam software.

Authenticating Email Messages and Other Digital Evidence

In order to be authenticated, either direct evidence or circumstantial evidence is used for a physical document. Examples of circumstantial evidence would be the paper document's appearance, content, or substance. The same circumstantial evidence the courts use to authenticate physical documents crosses over to email messages.

In order to authenticate email messages, Rule 901 requires that the person (proponent) who introduces the message provides "evidence sufficient to support a finding that the email message is what its proponent claims."

The reliability of the digital evidence itself and the reliability of the methods and procedures used must be established. Rule 901 generally can be satisfied by proof that

- The computer equipment is accepted in the field as standard and competent and was in good working order;
- Qualified computer operators were employed;
- Proper procedures were followed in connection with the input and output of information;
- A reliable software program and hardware were used;
- The equipment was programmed and operated correctly; and
- The exhibit is properly identified as the output in question.

Proof must be provided for all six of these issues or for all issues that apply to the handling of the evidence. It is not a surprise that opposing counsel will challenge the authentication of the digital evidence. In fact, evidence should be challenged to ensure that it accurately and fully represents the truth (Volonino et al., 2007, pp. 414–415).

ILOVEYOU Virus Case

On May 3, 2000, many people received an email from a spouse, a friend, a co-worker, or even from a school or government official with the subject line "ILOVEYOU." Without knowing that this was one of the most vicious viruses ever, the people opened this message and helped spread the disastrous ILOVEYOU virus.

The ILOVEYOU virus was a fast-infecting virus that changed Windows Registry settings and then emailed copies of itself to everyone in the Microsoft Outlook Express address book.

Opening the email revealed text reading, "Kindly check the attached LOVELETTER coming from me." An icon below had the name "LOVE-LETTER-FOR-YOU.TXT.VBS". The ".txt.vbs" extension gave the attachment the appearance of a text document, but the ".vbs" portion caused the attachment to be executed as a Visual Basic script, which accounted for the worm-type attacks. Thus, just by clicking on the icon, the virus was activated. Activating the virus forwarded the email to each address contained in the affected computer's Outlook address book. Then, the VBS searched for passwords contained within the host computer and emailed the stolen passwords to an address, mailme@super.net.ph, in the Philippines. Once the ISP, Sky

Internet, discovered that the passwords were being sent to an email address at their domain, the virus was quickly disabled.

The suspect was revealed by the Philippines National Bureau of Investigation (NBI) when they traced the ISP, Sky Internet, which unknowingly housed the password-stealing program WIN-BUGFIX.exe. The cybercriminal was Onel de Guzman. He was a 23-year-old student at AMA, a computer college in the Philippines. The main motivation for committing the computer crime was to obtain free internet access. Guzman studied the virus program as his senior thesis at AMA College, and school officials rejected his proposals many times because they always focused on illegal activities such as stealing and retrieving accounts from victims' computers. Guzman was also a member of a virus-writing club, GRAMMERsoft, which frequently vandalized the school administrative computer system with viruses.

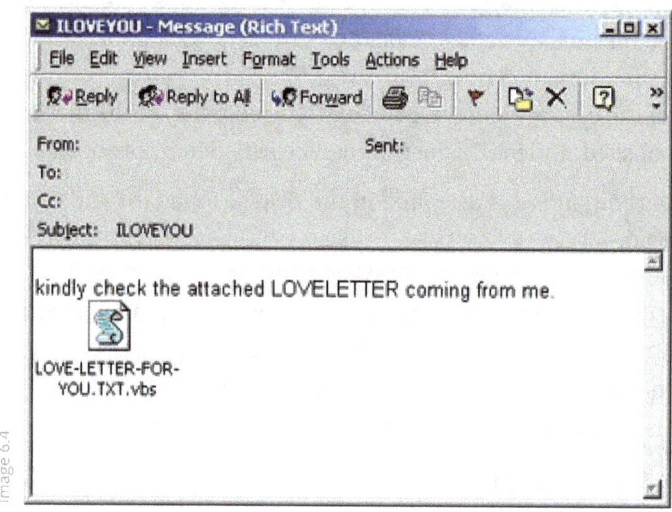

Image 6.4

To investigate computer crimes like the ILOVEYOU virus, the methods for gathering and dealing with electronic evidence must be learned. A step-by-step approach to the investigation, seizure, and evaluation of computer evidence is required. A sample case using one of the ILOVEYOU virus victim's emails as digital evidence is demonstrated below.

ILOVEYOU Virus Investigation

In order to investigate computer crimes like the ILOVEYOU virus, first, the methods for gathering and dealing with electronic evidence must be learned and practiced. This requires a step-by-step approach to the investigation, seizure, and evaluation of computer evidence. A sample case below demonstrates how to investigate the crime by using an email from one of the ILOVEYOU virus victims as digital evidence.

The first step in investigating the ILOVEYOU virus is to determine who has been notified of the infection and if a crime has in fact been committed. The victim (William Farewell, wfarwell@lynx.dac.neu.edu) should have notified mail-server security personnel of the infection, and the server's security system should have detected the infection and immediately reported it. The suspect appeared to be using the internet and email to infect the victim's computer system; based on the computer code, the victim's local internet service provider is Mail1.dac. neu.edu, the Northeastern University server.

Review of the Code

Below is a brief review of the computer code (although, this is not the entire code), which indicates clues as to how the virus infected the user's computer, who allegedly sent the virus, and when and where the infection occurred. Importantly, the header should not be overlooked as it contains the sender and receiver, as well as the dates and times of all computers that handled the message. If the sender, receiver, location, dates, and times in the header do not match, it would indicate this email is a forged message.

In order to deceive a law enforcement investigator, cybercriminals often use email forging skills by making an alibi. In this code, the sender's and receiver's email addresses, location, and all dates and times match; therefore, the investigator would proceed to the next step of the investigation.

```
;Return-Path: <CAddario@MDFA.State.MA.US>
;Received: from bosmail1.mdfa.state.ma.us (bosmail1.mdfa.state.ma.us
[146.243.41.125])
        ;by mail1.dac.neu.edu with ESMTP id KAA18880
        ;for {wfarwell@lynx.dac.neu.edu}; Thu, 4 May 2000 10:29:31 -0400 (EDT)
;Received: by bosmail1.mdfa.state.ma.us with Internet Mail Service
(5.5.2448.0)
        ;id {J576MW5H}; Thu, 4 May 2000 10:29:23 -0400
;Message-ID: {85E48FF36731D3119F6600508B12140349E79B@bosmail1.mdfa.state.
ma.us}
;From: "Addario, Carmen" {CAddario@MDFA.State.MA.US}
;To: "Farwell, William" {wfarwell@lynx.dac.neu.edu}
;Subject: ILOVEYOU
;Date: Thu, 4 May 2000 10:29:19 -0400
;MIME-Version: 1.0
;X-Mailer: Internet Mail Service (5.5.2448.0)
;Content-Type: multiSection/mixed;
;Status:
;This message is in MIME format. Since your mail reader does not
understand
;this format, some or all of this message may not be legible.
;Content-Type: text/plain
kindly check the attached LOVELETTER coming from me.
;Content-Type: application/octet-stream;
        ;name="LOVE-LETTER-FOR-YOU.TXT.vbs"
;Content-Transfer-Encoding: quoted-printable
;Content-Disposition: attachment;
        ;filename="LOVE-LETTER-FOR-YOU.TXT.vbs"
;rem barok -loveletter(vbe) {i hate go to school}
;rem                by: spyder / ispyder@mail.com / @GRAMMERSoft Group / =
;Manila,Philippines
;On Error Resume Next
;dim fso,dirsystem,dirwin,dirtemp,eq,ctr,file,vbscopy,dow
```

```
;eq=3D""
;ctr=3D0
;Set fso =3D CreateObject("Scripting.FileSystemObject")
;set file =3D fso.OpenTextFile(WScript.ScriptFullname,1)
;vbscopy=3Dfile.ReadAll
```

Seeking a Search Warrant: Indicating Criminal Violations

In order to attain a search warrant in efforts to obtain evidence of the distribution of a virus, such as the ILOVEYOU virus, the investigator must indicate that the suspect in the case has engaged in criminal activity.

State computer crime laws punish anyone who, without authorization, do the following:

- Accesses a computer;
- Uses a computer (i.e., steals computer time);
- Denies or disrupts computer services (e.g., inserts a virus or worm which shuts down the computer system);
- Alters, damages, or deletes data stored within the computer (which may include inserting a virus as well as committing vandalism);
- Takes, copies, or makes use of data stored within the computer (this conduct may also violate information theft laws); or
- Engages in illegal wiretapping and electronic eavesdropping.

This definition of computer crime would include virus distribution, in turn making it a crime to send the virus if and when done intentionally. In the event a computer is used in the manner of a weapon in a criminal act, additional charges or a heightened degree of punishment could follow. When a computer is infected with a virus, it fits into this additional category as well.

The investigator needs to check the internet connections and what type of computer system was accessed. By checking with computer services, the victim's "logon" identification can be found, since this is what the ILOVEYOU virus steals, and held as evidence against the suspect. (In the event the suspect possesses the victim's logon name and password, then the suspect is likely a knowing originator of the virus.)

Collecting Information About the Virus Intrusion

As investigators collect information about the virus intrusion, specific information is needed: when, how the intrusion began, etc. To preserve evidence, the victim should keep all records pertaining to the virus in a secure area of the computer and stop using the computer immediately. No one involved in the case should discuss it. The investigator should make an exact copy of all virus-related data on the computer (e.g., the ILOVEYOU virus codes in the victim's computer system). Additionally, a hard-copy record of the damages and infection needs to be kept in order to be displayed as evidence at trial.

If the suspect is using the internet, the investigator should seek assistance from the victim's internet service provider. The service provider may be able to provide information helpful in tracking the suspect's computer system. The investigator can then arrange to capture and examine the suspect's data packets for source as well as code/destination information (though this may require a warrant or cooperation from the owner).

The investigator should examine whether the source of the intrusion as reported by the trap and trace or internet service provider is the actual location of the suspect. Someone knowledgeable in computers could have routed the calls through many different phone companies before reaching their target. If the location returned by the trap and trace is an institution (i.e., a company or a university), contact that institution and seek assistance. If it is a residence, obtain records, such as utility bills, identifying the occupants of that residence. Investigators should consider checking whether the local school or police department is familiar with a juvenile living in the residence.

Even though an investigator sometimes needs to serve the warrant immediately due to exigent circumstances, such as the possibility of a suspect fleeing or causing further damage, it is generally a good idea for the investigator to wait to serve the search warrant. By waiting, extra evidence in a case may be obtained (either from police resources or from the suspect's actions) which could assist in conviction at trial. Following step-by-step instructions, investigators are able to draft the warrant. In dealing with search warrants for computer-related items, it is very important that the warrant be detailed and extensive.

The investigator may find drafting the warrant easier by intentionally including the following information:

- Internet service used by the suspect
- Any passwords to the victim's computer system that may be known or used by the suspect (the victim needs to change those passwords before the warrant is filed)
- The name of the account used by the suspect
- Information unique to the victim's computer system which you expect the suspect to have downloaded, in particular, sensitive information or business critical data, the victim's name, email lists, or name of the victim's computer (i.e., any information that could point to the suspect having unauthorized contact with the victim's computer system)
- Messages or commands sent by the suspect to the victim's computer system (e.g., like other VBS_LOVELETTER cases, this email has the **subject "I LOVEYOU"; body: "kindly check attached LOVELETTER coming from me"**)

Searching the Intruder's Residence

The final step involves obtaining a search warrant in order to search the intruder's residence. As previously discussed, the goal is to seize the intruder's computer or related items, which may contain evidence of the intrusion. The suspect may have the victim's file directories, which will provide clues as to the type and location of information stored on the computer. Computers can keep track of the time of day when users log on and log off and, in many cases, can monitor the intruder's activities. The header contained the information that the suspect (Addario, Carmen/ Caddario@MDFA.State.MA.US) sent the VBScript virus-containing email to the victim (Farewell, William/wfarwell@lynx.dac.neu.edu) on Thursday, the 4th of May 2000, 10:29:19-0400.

The investigator should also consider how to prove which occupant of that location is indeed the primary suspect (e.g., which sibling or employee) while securing the warrant and collecting evidence. For example, even though the virus was generated and distributed from Carmen Addario's location or email address (Caddario@MDFA.State.MA.US), it does not specify who specifically used the computer; perhaps a roommate or co-worker could have sent the virus during Carmen Addario's absence. Thus, the investigator should not jump to conclusions.

When obtaining a description of the residence to include in the search warrant, it is important to drive by the residence and check information provided by the Internet Service Provider (ISP) to make sure that it is not connected to an adjacent residence occupied by the suspect. Additionally, the investigator needs to identify any third parties connected with the suspect by using a carefully drafted warrant. The investigator then needs to arrange for a magistrate to sign the warrant. The investigator should include specific statements in the probable cause section to justify searching and seizing certain property. To justify such requests, present evidence that the target had no right and no reason to possess any of the victim's data.

Serving the Search Warrant

Prior to serving the warrant, investigators should undoubtedly consider the following questions:

- Do we have enough officers to allow the investigating officer to interview the suspect (after providing appropriate Miranda warnings)?
- Are we better off serving the warrant when the suspect is not at home? If we are planning to "turn" the suspect into an informant and are going to serve the warrant when the suspect is not at home, can we determine the suspect's whereabouts in advance?

Typically, it is difficult to describe information within a computer investigation since the investigator's description of the items sought must be narrow enough to not make the warrant too broad. When the investigator writes the warrant, it is important to narrate broad requests when seeking stolen information, and use narrow requests for seeking tangible items, which can be described easily. Additionally, the investigator should not disclose trade secrets in the affidavit as there is no guarantee that the magistrate will protect the trade secrets.

During the search, investigators should not ignore the following items, which may appear in plain view:

- Printouts containing the internet service, credit card numbers, or any string of numbers, which may be access codes. Also, look for names of bulletin boards (BBSs), which may reveal data caches.
- Pads of paper.
- Financial documents.
- Passwords.
- Evidence identifying the user of the computer (e.g., Carmen Addario). Look for names inside manuals or on any labels or folder names affixed to physical disks or portable devices.
- Evidence of confederates.
- Magazines relating to cracking (e.g., Sample of I LOVEYOU virus codes in the computer magazines).
- Computer manuals for the computer used by the victim.

Rules

As a recap of the previous discussion, there are a variety of rules that tie together any seizure process. One of the most important rules is investigators must be able to explain what steps were taken to arrive at a particular conclusion. If presented with an exact replica of the scene, the investigator should be able to refer to the notes of the case and do everything exactly the same from arriving on scene, to collecting the evidence, to walking out the door. In order to achieve this level of detail, there are two essential considerations.

Document everything. Have one person process the scene while the other writes down each and every single step. If you are working alone in the seizure process, consider using a voice recorder to narrate each step for later transcription. This step becomes crucially important if the target computer is manipulated (i.e., moving the mouse to deactivate the screensaver or initiating a shutdown sequence). Seek assistance from someone more qualified. If you do not feel comfortable with what is being done.

The second rule is that **whatever is seized as having potential evidentiary value must be authenticated by the court before it can be admitted into the case**. The ability of the court to authenticate the evidence is a significant issue related to digital evidence. Authentication is governed by the Federal Rules of Evidence, Rule 901 (28 U.S.C.), which states, "The requirement of authentication or identification as a condition precedent to admissibility is satisfied by evidence sufficient to support a finding that the matter in question is what its proponent claims."

Subject to Miranda, interview the suspect. Ask the suspect whether the computer is rigged. If the suspect is to cooperate in assisting police investigations of friends, secure the suspect's cooperation immediately. A long delay (more than a day) before the "turned" suspect returns online may warn confederates that the suspect is no longer their ally.

The latest *Searching and Seizing Computers and Obtaining Electronic Evidence in Criminal Investigations* manual published by the Department of Justice supports the proposition that the seizure of digital evidence should be an incremental process, based both on the situation and the training level of the responder. The manual provides the following steps in its incremental approach:

- Upon arriving on scene, agents will attempt to identify a system administrator, or similar person, who is willing to assist law enforcement in identifying, copying, and/or printing out copies of the relevant files or data objects as defined by the warrant.
- In the event that there are no company employees available to assist the agent, the agent will ask a computer expert to attempt to locate the computer files described in the warrant and will attempt to make electronic copies of those files. It is assumed that if the agent is an expert, the agent will be able to proceed with the retrieval of the evidence.
- If the agent or expert is unable to retrieve the files, or if the onsite search proves infeasible for technical reasons, then the next best option is to create an image of those sections of the computer that are likely to store the information described in the warrant.
- If imaging proves impractical or impossible for technical reasons, the agent should then seize those components of storage media that the agent reasonably believes includes the information described in the warrant.

Investigation After Serving the Search Warrant

When executing the search warrant, it is important to not arrest anyone unless a suspect is a flight risk because most suspects will post bail immediately (Rosenblatt, 1995).

Should you follow the search warrant with a subpoena? If your state law allows it, consider using a subpoena to back up your search. By serving the target with a subpoena or by subpoenaing the evidence you can force the suspect/evidence to appear in court.

Because breaking a subpoena is against the law, the suspect is caught in a catch-22. If the search is ruled inadmissible but they bring the evidence, the evidence may still be valid under the argument that it would have been discovered through the subpoena (doctrine of independent discovery), but disobeying the subpoena also convicts the suspect of a crime.

Since judges often rule evidence in computer cases inadmissible due to vague search warrants or improper searches, this backup method increases the chances that the case can continue (Rosenblatt, 1995, p. 220).

Summary

Email evidence can be among the most crucial evidence to find because often email contents provide a suspect's characteristics, intentions, and motivation of crime commission. This module covered email as evidence and where this evidence can be located. Three different types of email and the methods that can be used for digital forensic investigation were also discussed. Additionally, the process of digital forensic investigation was demonstrated using an ILOVEYOU virus victim's email as an example.

Lab 6.1: Evidence Files

Lab evidence files should be placed onto a second physical hard drive or a second partition. This hard drive or partition should be wiped, verified, and documented as sanitized from any previous data being stored. This eliminates the cross-contamination argument. Lab evidence files may be placed onto a central file server, which gives multiple examiners access to the evidence files.

As a part of the investigation, you will be involved with case management. One part of case management is maintaining the evidence file in a location that is both convenient and secure. At the onset of your investigation, you are likely to acquire some type of digital evidence into evidence files using various hardware and software tools. Once the media is acquired, you will usually find yourself with the data within the evidence file format, dd image, or others.

The EnCase evidence files and other files relating to your case should be placed onto a separate physical hard drive that has been previously wiped. Any potential data that resided on the hard drive will be overwritten by the wiping process. Once wiped, the wiping process should be verified and documented. By placing the EnCase evidence files on a separate sanitized hard drive, you will eliminate any arguments of cross-contamination of data that had resided there before. In addition, the evidence files may be placed onto a central file server. This will give multiple examiners access to the evidence files from one location.

Folders for case organization:

- Case folder
  - One for each case
  - Use case name, suspect name, report number, etc.

- Case subfolders
  - Evidence cache
    - Segregation and control of temporary files
  - Local evidence
    - EnCase evidence files, logical evidence files, digital images, etc.

Another aspect of case management is the folder structure used during the examination. To begin, a case folder for each individual case should be created. This may be named in accordance with departmental or company policy and procedure. Typically, the case folders are named by report number or suspect name and are automatically created when the case is created. Afterwards, the important evidence cache subfolder is automatically created by digital forensic tools and another suggested folder is created by the examiner. The evidence cache subfolder is used for segregation and control of cache files created during the examination. The next optional subfolder is named local evidence. This is where the examiners will place evidence files. The required data will reside here for organizational purposes.

Insert your USB into your workstation.
Click Start => EnCase Imager
Click Tools => Wipe Drive and select "Next" to collect local devices

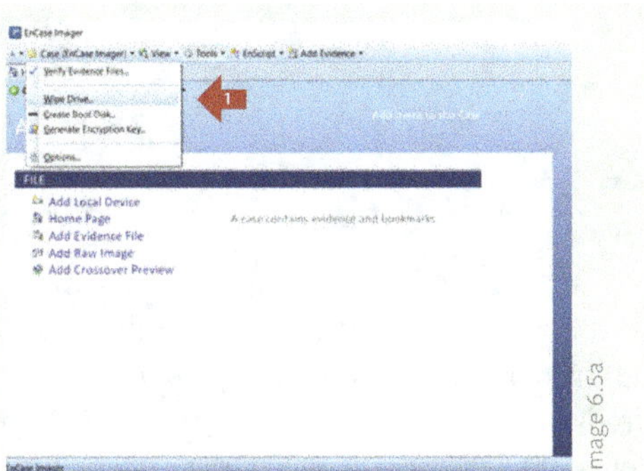

Image 6.5a

Select your physical USB and click Next

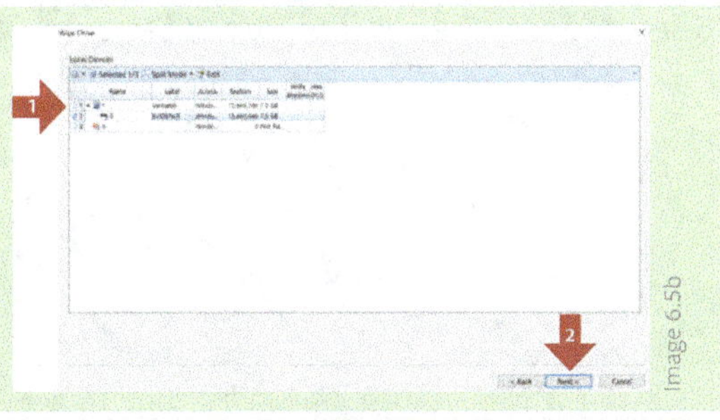

Image 6.5b

Click Finish => Type **Yes** => click OK

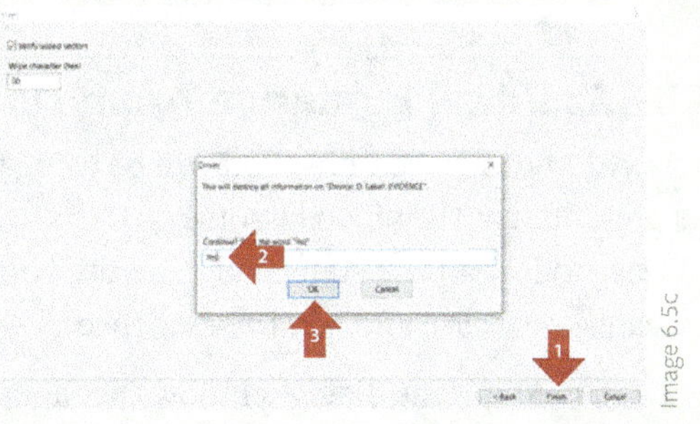

Image 6.5c

Notice the progress bar and close when finished.
Please make sure you use the right physical disk. If you use one GB of USB, it will show around 5 minutes. To cancel the wipe, double-click on the progress bar, and select Cancel.

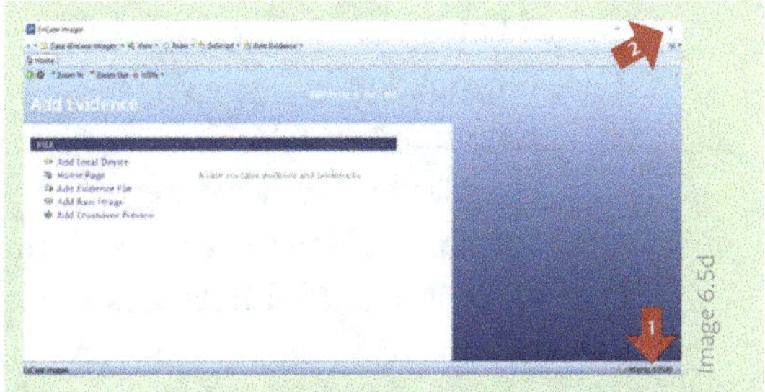

Image 6.5d

(continued)

File Explore => Select => Your USB
Click Format Disk => Select => FAT32 =>
Label = Evidence => click Start

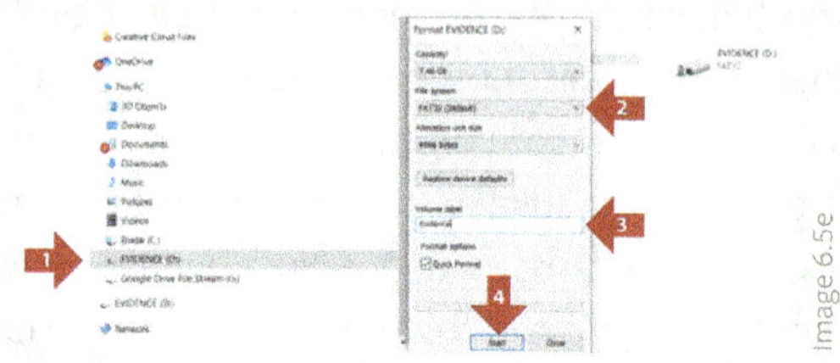

Image 6.5e

1. Format => click OK
2. Complete => click OK

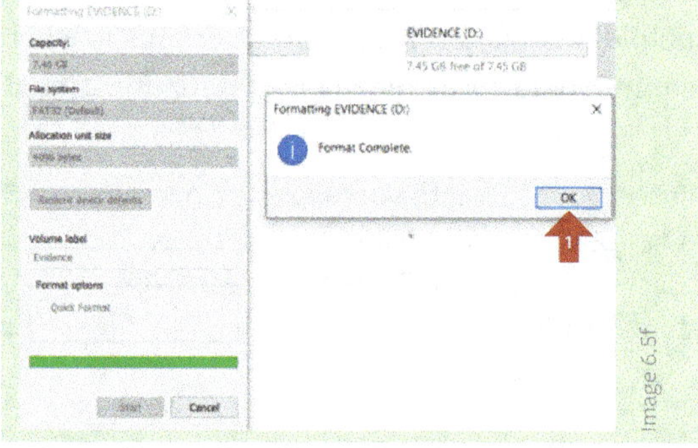

Image 6.5f

1. Download evidence file (4Dell Latitude CPi) from NIST website: https://www.cfreds.nist.gov/
2. Notice the EnCase evidence file

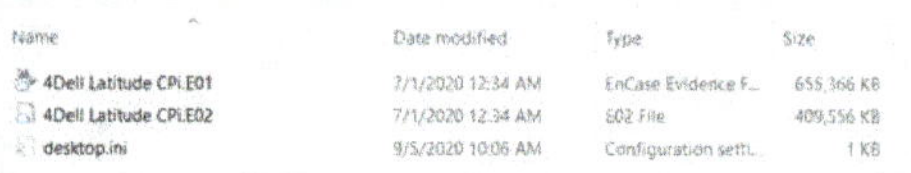

Select and copy the evidence file and paste

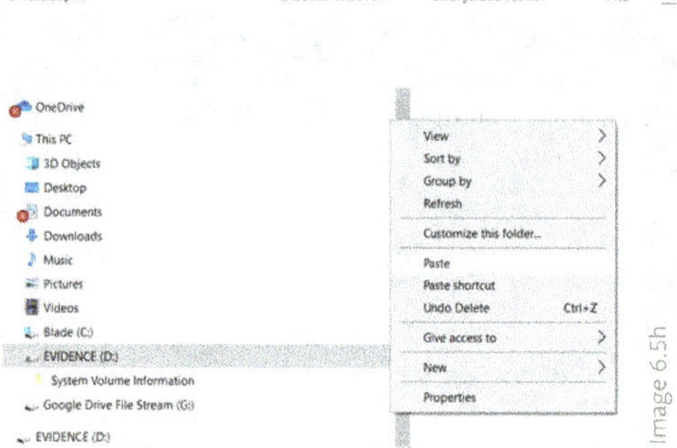

Image 6.5g

Image 6.5h

Lab 6.2: Email Evidence Analysis

Review all the video lessons. You will be using the same evidence file (4Dell Latitude CPi) that was presented in the video lessons.

Please follow all the steps and generate your report. In order to follow the steps, you need to create a new case and add the evidence.

1. In the Options dialog window that appears, type or select the following:

- Name: **Lab 6**
- Case information
    - Case number: **006**
    - Case Date: **Today's date**
    - Examiner Name: **Your Last Name**
    - Description: **Email Analysis**
- Click OK to create the new Preview case
- Please recheck your case folder entitled Lab 6
- Please recheck C:\EnCases\Cases for your case folder entitled Lab 6

2. Add evidence
   - You can find the Evidence File to process in your flash drive
3. Once you fully acquire and process the evidence, take your screenshot (MD5 and SHA1 hash verification)
4. Please answer the following questions:
   a. What time zone is present on the suspect's computer system? Manually locate the evidence.
   b. Search for the main user's web-based email address. What is it? Locate the evidence to support your answer.
   c. Yahoo! Mail, a popular web-based email service, saves copies of the email under what file name? Locate the evidence to support your answer.
   d. Search emails that are relevant to hacking information.
5. Save your report along with your answers as **a Word document** and name it as YOUR LAST NAME_LAB 6
6. Submit your lab report; explain (**written explanations**) and demonstrate (**screenshots and bookmarks**) all the steps and clearly answer the questions.

References

Brown, A. D. (2004). Authoritative sensemaking in a public inquiry report. *Organization Studies*, 25(1), 95–112.

CCIPS documents and reports. (2021, March 08). Retrieved April 30, 2021, from https://www.justice.gov/criminal-ccips/ccips-documents-and-reports

Daniel, L., & Daniel, L. (2012). *Digital forensics for legal professionals*. Syngress Book Co.

Rosenblatt, K. S. (1995). *High-technology crime: Investigating cases involving computers*. KSK Publications.

Volonino, L., Anzaldua, R., & Godwin, J. (2007). *Computer forensics: Principles and practices*. Pearson Prentice Hall.

Credits

Electronic Vandalism: Criminal Patterns and Countermeasures

Kyung-Shick Choi

Introduction

Within this chapter, electronic vandalism is mainly discussed. Students will gain an understanding of the definition of electronic vandalism, the risks of electronic vandalism, criminal patterns, and electronic vandalism investigation techniques.

Let's briefly discuss the history of the term vandalism before diving into the discussion of electronic vandalism. "Vandalism" is derived from the Vandals who famously sacked Rome in the 5th century, ruthlessly destroying aqueducts and other valuable property. Some examples of vandalism are breaking windows, shooting street signs, or defacing public property or the property of others.

Traditionally, vandalism is committed on tangible materials, like buildings or public entities. However, with the development of current technology, targets of electronic vandalism are intangible materials, including homepages and digital information. Electronic vandalism, therefore, can be defined as willful or malicious destruction of public or private intangible digital properties, including deleting or damaging data and thereby causing the proper services of a private or public website to cease. This definition of electronic vandalism includes an extensive range of patterns including viruses, worms, and internal data destruction.

In this lesson, we will explore cybercriminal profiles to understand why cybercriminals commit electronic vandalism.

Twelve Cybercriminal Profiles

Profiling delineates distinctive behavior patterns to narrow the range of suspects in a certain crime and is commonly used in the criminal justice field. Although cybercriminal profiling may not be widely available, cybercriminologists have gradually built profiles.

Shoemaker and Kennedy (2009) lists 12 profiles of the types of cybercriminals regularly encountered by law enforcement agencies and information assurance corporations.

TABLE 7.1

TYPES OF CRIMINAL GROUPS	INTENT	MOTIVATION	OFFENDER DESCRIPTION	ATTACK TOOLS	LEVEL OF DESTRUCTION
Kiddies	Trespass	Ego	• Any age • Technologically inept • New to crime	Preprogrammed tool kits	*
Cyberpunk hackers	Trespass or invasion	Ego or exposure	• Young • Technically proficient • Outsider	Virus, application layer, and DoS attacks	**
Old-timer hackers	Proving their art by trespassing	Ego	• Middle aged • The most technically proficient • Professional history	Website defacement	—
Code warriors	Theft or sabotage	Ego, revenge, or monetary gain	• Age range between 30 and 50 • Technically superb/A degree in Tech, but unemployed • Socially inept	Application layer and Trojan horse	***
Cyber-thieves	Illegal possession of valuable information or outright theft	Monetary gain	• Any age (usually younger than code warriors) • Most are organizational insiders • Use social engineering techniques	Surreptitious network attacks via sniffing or spoofing/ Simple programming exploits such as Trojans and malware	N/A

(continued)

TABLE 7.1 *(Continued)*

TYPES OF CRIMINAL GROUPS	INTENT	MOTIVATION	OFFENDER DESCRIPTION	ATTACK TOOLS	LEVEL OF DESTRUCTION
Cyber-hucksters	Commercialization	Monetary gain	• Older (business types) • Use social engineering techniques	Tracking cookies, spyware, and legal data mining to find victims	N/A
Unhappy insider	Theft, sabotage, or harm items of value to the company	Revenge or monetary gain	• Any age • Employed • Unhappy with company	Extortion or exposure of company secrets via destructive logic bombs or malicious applications	*****
Ex-insider	Theft, sabotage, or harm items of value to the company	Extortion, revenge, sabotage, or disinformation	• Any age • Terminated former employee • Unhappy with company	Extortion or exposure of company secrets via destructive logic bombs or malicious applications	****
Cyberstalker	Invasion of privacy	Ego and deviance	• Any age • Psychological issue	Keylogger, Trojan horse, or sniffers	N/A
Con man	Theft or illicit commercialization	Monetary gain	• Any age • Difficult to catch due to anonymity	Spoofing, Nigerian scam, or phishing	N/A
The mafia soldier	Theft, extortion, and invasion of privacy for the purposes of blackmail	Monetary gain	• Any age • Organized crime group member • Active in the Far East and Eastern Europe	All types with the best technology	N/A

(continued)

TABLE 7.1 *(Continued)*

TYPES OF CRIMINAL GROUPS	INTENT	MOTIVATION	OFFENDER DESCRIPTION	ATTACK TOOLS	LEVEL OF DESTRUCTION
Warfighter/cyber-solider	Protection for friends and harm to the enemy	Infowar	• Any age • Technically superb • Non-criminal type • If on the other side against law, violence can increase • Members of elite government agency	Application layer, logic bombs, DoS attacks	***** or -

Motives of Electronic Vandalism

The suggested 12 cybercriminal profiles are considered good general descriptions of the types of suspects who commit electronic vandalism. The six main motivations for electronic vandalism are summarized below:

- Exposure: An individual or a group shows off skills
- Revenge: Anger (destructive desire)
- Hacktivism (political purposes)
- Ego: Challenges and thrill
- Monetary gain
- Entertainment, boredom, etc.

The first reason is anger—people attempt a destructive behavior to reveal anger towards another person or group in order to achieve their own satisfaction. Unhappy insider and ex-insider would be classified within this category.

One of the most dangerous profiles is the unhappy insider. These attackers are inside most organization defenses. This type of attacker is often motivated by revenge or monetary gain and uses extortion or exposure of company secrets for the purpose of theft or sabotage. Overall, their intent is to steal or harm items of value to the company. They may steal information, set destructive logic bombs, or perform other malicious acts on the computer system. One distinctive characteristic of this perpetrator is unhappiness with the organization as a whole. The unhappy insider can be of any age and employed at any level. The only protection an organization has against this type of criminal is to identify early signs of unhappiness and closely monitor further actions.

The terminated former employee motivated by extortion, revenge, sabotage, or disinformation is referred to as the ex-insider. Among this group are those focused on harming the organization that dismissed them. If these employees can see their dismissal coming, they are likely to set logic bombs or perform other destructive acts to harm the organization. Otherwise, they will make use of insider information to damage or discredit the company from the outside. Comparable to the unhappy insider, the ex-insider can be of any age and work at any level. The only protection a company has from this type of exploit is to plan a dismissal to ensure a clean break. Attacks on company vulnerabilities that were not public knowledge are a sure sign of an ex-insider hacker.

Some individuals or groups, especially hacker groups, commit cybercrime to show off their expertise. An individual or group motivated by ego will invade well-secured companies or government organization systems and leave a signature. Kiddies, cyberpunks, and old-timer hackers can fall into this category.

Kiddies are considered technologically inept. They use preprogrammed tool kits, usually with the intent to trespass, and their motivation is ego. The more advanced kiddies will engage in invasion of privacy exploits. Kiddies can be of any age, but they are always outsiders and not technologically advanced. Kiddies are usually new to crime and can be tracked by matching a crime to the attacker who has downloaded a suitable tool set from hacker websites.

Cyberpunks are usually ego-driven, with the intent to always trespass or invade. If their purpose is invasion, the motive is exposure. They will engage in theft and sabotage but only of what they perceive as legitimate targets. Cyberpunks are responsible for many viruses, application layers, and DoS attacks targeted on well-established organizations, companies, and products. A cyberpunk is likely to be young, technologically proficient, and an outsider.

Old-timer hackers are one of the most technologically proficient members of the hacker community. They are ego driven and the last of the old guard, whose only intent was to prove their art by trespassing. They are relatively harmless because they know what they are doing and their motives are relatively nonthreatening. Where their actions cause harm, they generally specialize in website defacement. An old-timer is likely to be middle aged or older with a long personal and/or professional history in technology and possibly hacking.

"Hacktivism" is a compound of "hack" and "activism." Warfighters are usually considered hacktivists. When warfighters or cyber-soldiers are fighting on your side, they are not typically considered to be criminals. However, when the warfighter is on the other side of law, their actions would be viewed as destructive.

EXAMPLE

North Korea has attacked systems for a variety of reasons. In 2014, Sony was hacked to prevent reputational harm, the Bank of Bangladesh heist in February 2016 was for financial gain, and the WannaCry attack in May 2017 was a global ransomware cyberattack motivated by a desire to cause economic chaos.

The warfighters or cyber-soldiers tend to be motivated by infowar or politics. This group attempts to provide strategic benefit for friends and damage to the enemy. They are technologically superb and can be extremely dangerous. They can wreak havoc on the physical infrastructure of a country by attacking the electronic underpinnings. Warfighters can be any age, they are highly organized, and are frequently the best and brightest that the country has to offer. Since this profile really characterizes the members of any elite government agency, the best defense against a warfighter is another friendly set of warfighters.

Last, but surely not least, is electronic vandalism, which can be used for blackmail for money and/or goods. After committing electronic vandalism such as defacing a business's web server, stealing essential data, or denial of service, an offender will blackmail the victim by offering to halt the attack or return data in exchange for something of value. These types of crimes have significantly increased since the rise of e-business.

The mafia soldiers group is most frequently associated with seeking monetary gain via various blackmail techniques such as theft, extortion, and invasion of privacy. This group works within a highly organized group and paired with the best technological support. We can assume that most organized crime groups in the world will eventually be engaging in this business, considering that the development of technology offers ease and profitability.

Cyber-thieves are also motivated by monetary gain, but they tend to use network tools and simple programming exploits such as Trojans and malware rather than sophisticated targeted code. Similar to a con man who runs a traditional con game like the Nigerian scam and phishing, cyber-thieves adapt social engineering and spoofing. Cyber-thieves can be of any age, but the profile does not require a long history in technology. Unlike con man members, who are difficult to catch due to the nature of online anonymity, most cyber-thieves are organizational insiders; therefore, structured internal control in the organization is be the best defense.

Types of Cybercriminal Exploits and Profiling

Major cybercriminal exploits fall into two general categories: (1) forms of malicious code injection or (2) technological exploit for a specific target.

Malicious Code Injection

Virus

The term "virus" was first used to refer to any unwanted computer code; now the term generally refers to a segment of machine code that will copy itself into one or more host programs when it is activated. When infected programs are running, the viral code is executed and then the virus spreads further. Computer crimes of this type can be included in the system infection programs category.

Image 7.1

Computer virus malware tries to execute and replicate itself on the infected host. Viruses usually target data files or executables. Some of them target boot sectors. Virus are almost always intended to be harmful. They delete and encrypt data, damage the OS and sometimes the hardware, and can steal information.

Worms

Worms are computer programs designed to make copies of themselves automatically. A worm is self-executing, largely invisible to the computer users, and rapidly spreads from computer to computer over a network without any user action. They are payload (the part of transmitted data that is the actual intended message) free, which means they will not do any harm to the computer they infect, but

Image 7.2

in all cases, they harm the network by using its resources. As mentioned above, it is payload free, but hackers supply it with another piece of software, like a Trojan or virus, to spread the virus along the route of the worm. Most of the time, a botnet software is sent with it to create zombies or bot armies which are used for DDoS. Worms take advantage of weak security and outdated systems to breach and infect them.

For example, the Slammer worm spread during the weekend of January 25, 2003, and attacked a known flaw in systems of several organizations, including American Express and the Seattle police and fire 911 center. This attack is believed to be the fastest-spreading internet worm on record, infecting 90 percent of vulnerable computers nationwide. It challenged popular opinions that vital services were largely immune to such attacks (Wilson, 2005).

As another example, Stuxnet was discovered in June 2010 and was used to attack Iran's nuclear program in 2007. It was designed to attack industrial PLCs (programmable logic controllers) that allow the automation of electromechanical processes such as those used to control machinery on factory assembly lines, amusement rides, or centrifuges for separating nuclear material. Stuxnet exploited four zeroday[1] flaws in Microsoft Windows and then targeted Siemens Step 7 software on those machines. Stuxnet's design and architecture are not domain specific and it could be tailored as a platform for attacking modern SCADA (supervisory control and data acquisition, a computer system for gathering and analyzing real-time data) and PLC systems.

Increasingly, worms are being used for purposes of extortion and sabotage. The motivation for worm attacks is typically aggressive destruction. Thus, an organized type of perpetrator should be considered as the investigation begins.

Botnet

Botnet is malicious software that computer criminals can use to send out viruses or worms to online users via email. This malicious application is typically referred to as a "bot." The computer owners tend to be unaware that their systems are being controlled by computer criminals. These computers are known as zombies.

The bot on the infected PC logs into a particular IRC (Internet Relay Chat) server or web server. A spammer purchases access to the botnet from the computer criminals. The spammer sends instructions via the IRC server to the infected PCs, causing them to send out spam messages to mail servers.

An email is sent to online users telling them to call a telephone number to verify their personal information. This is a new form of identity theft called "voice phishing." The message normally says that your bank account will be closed due to a recent hacking incident. The scammers then ask callers for their security information.

Time Bombs/Logic Bombs (a.k.a. "Slag Code")

Time bombs are known as pieces of illicit software which are activated by computer clocks to initiate a fraud, a disruption, or some other sort of vicious activity. Similar to time bombs are logic bombs; logic bombs are activated by a combination of events rather than by the computer clock.

These attacks are only activated based on parameters because the programs are set in a host machine. Hands-on access is usually required to set a logic bomb, unless the perpetrator uses

---

1   Zero-day is a flaw in software, hardware, or firmware that is unknown to the party or parties responsible for patching or otherwise fixing the flaw. The term zero-day may refer to the vulnerability itself or an attack that has zero days between the time the vulnerability is discovered and the first attack.

a Trojan horse. In other words, the main suspect would be a person who is physically able to access the devices and set logic bombs. In the event of using a Trojan horse, investigators must track down a person who writes Trojan horses by building a profile to identify a limited set of suspects in a narrow physical location.

Trojan Horses

The term "Trojan horse" refers to a useful program (like a game program, a survey question, or media files) which contains hidden code that either executes malicious acts when triggered by some external event or provides a trap door to allow an intruder to secretly access the system.

Image 7.5

Malware

When an online user visits a website embedded in malware, a malicious code is transferred to the visitor's computer. Hijackers and keyloggers can also deliver malware to the homepage and adware; this is an example of internet commercialization.

Image 7.6

Since the victim can be anyone who visits the site, the site tends to be associated with "fringe" commerce such as a pornography site. It is difficult for an investigator to profile because the victims are almost self-selected by visiting unknown websites.

Ransomware

Ransomware is a form of malware targeting both human and technical vulnerability in an effort to deny the availability of critical data and/or systems. Ransomware is frequently delivered through phishing and Remote Desktop Protocol (RDP). RDP allows computers to connect to each other across a network. In one scenario, spear phishing emails are sent to end users, resulting in the rapid encryption of sensitive files on a corporate network. When victims determine they are no longer able to access their data, the cybercriminal demands the payment of a ransom, typically in virtual currency such as Bitcoin.

Image 7.7

Targeted Attacks Category

Similar to organized criminal activity, targeted attacks hold the motivation associated with a specific victimology. According to Shoemaker and Kennedy (2009), there are eight generic types of targeted attacks: (1) insider, (2) password, (3) sniffing, (4) spoofing, (5) man-in-the-middle, (6) application layer, (7) denial of service, and (8) social engineering.

Insider Attacks

The insider attack has a reputation for being the most prevalent. An insider, likely with knowledge of the security system, can easily steal valuable information by loading it on a flash drive and walking out the door with it without raising suspicion. The characteristics of the

Image 7.8

perpetrator almost always fit the organized type. It is always best to ensure a potential attacker is monitored and to have a policy in place to prevent the acts before they occur.

Password Attacks

Password attacks can be completed in a few different manners, including guessing a password or obtaining it via social engineering. By accomplishing password attacks, access to the system can also lead to theft and sabotage. According to a 2020 phishing benchmarking report, more than 90 percent of successful hacks and data breaches are associated with a social engineering type, which manipulates people to trick them out of sensitive information by using personal contact information (Kron, 2021). The primary suspect in this case would be profiled as an insider or client who is in close physical proximity to where the attack originated.

Password attacks are usually driven from social engineering exploits; however, these types of attacks can also originate from Trojan horses, network sniffing, or dictionary attacks that use software programs such as Word Password Recovery 1.0. Dictionary attacks can randomly apply all letters of the alphabet to a targeted password until it figures out the correct password to open the application. Using strong passwords consisting of a combination of numbers, letters, and characters can help prevent a successful dictionary attack.

If this type of attack is completed and causes very little harm, the basic hacker profile includes kiddies, cyberpunks, or old-timers. If the attack aim is sabotage, extortion, or theft, code warriors, cyber-thieves, and mafia warriors fit the profile.

Sniffer-Based Attacks

A sniffing-based attack involves detecting all data traveling through a network, enabling hackers to search for passwords that will allow for information access such as acquiring account or social security information, which can support ID theft or sabotage.

This type of attack is a carefully planned and executed exploit that requires a certain technical proficiency. When this attack involves simply gaining access to the system, the suitable suspect profile includes cyberpunks, cyber-thieves, and cyberstalkers. If the main aim of sniffing is theft or sabotage, the code warrior, mafia soldier, or warfighter fits in the profile. Additionally, if the purpose of a sniffing exploit is for the invasion of privacy, the perpetrator almost always has some form of direct relationship with the victim or the victim holds the potential to fulfill some type of common fantasy.

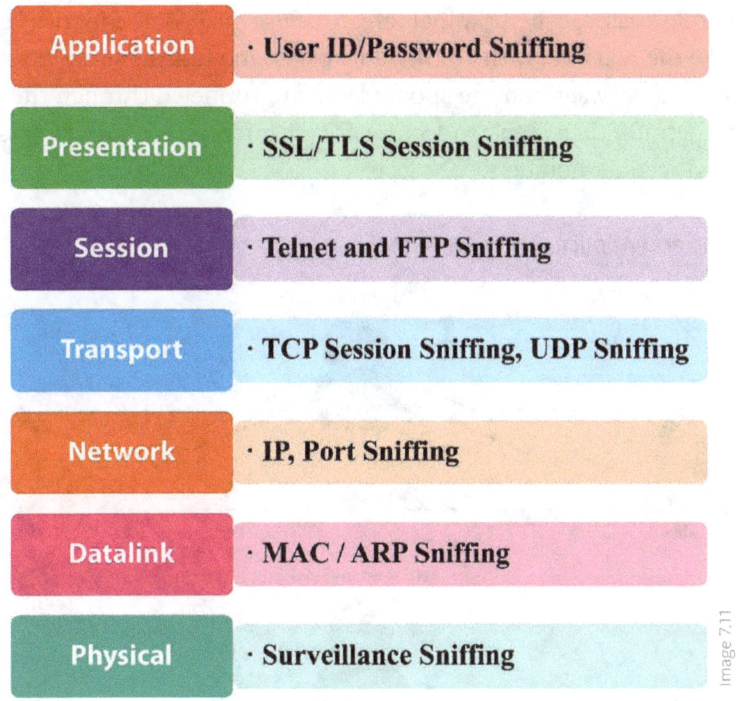

Spoofing-Based Attacks

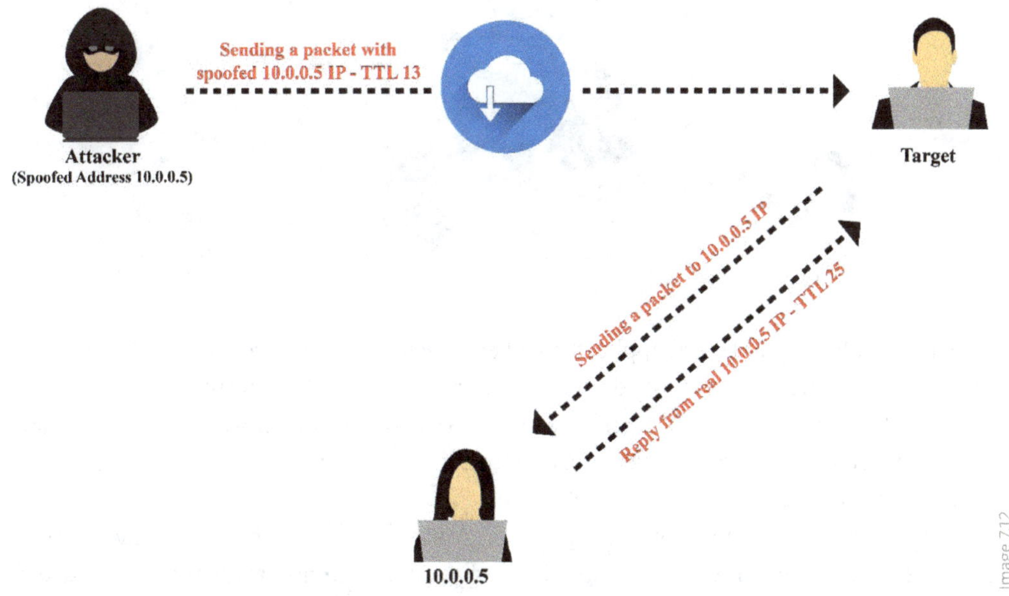

Spoofing is an organized type of crime. The attack is designed to convince others that the sender of an internet message is legitimate.

These attacks typically involve spoofing an IP address by changing the packet-header information, which can entail phishing scams or spamming using familiar email addresses. The Nigerian scam and the lottery scam would be typical examples of this exploit.

In advanced exploit cases, cybercriminals create a false or shadow copy of a legitimate website that looks like the real one, with all the same pages and links. The traffic of the network between the victim's browser and the spoofed site are funneled through the perpetrator's machine. The perpetrator then is able to acquire private information, such as passwords, credit card numbers, and account numbers.

Man-in-the-Middle and Application Layer Attacks

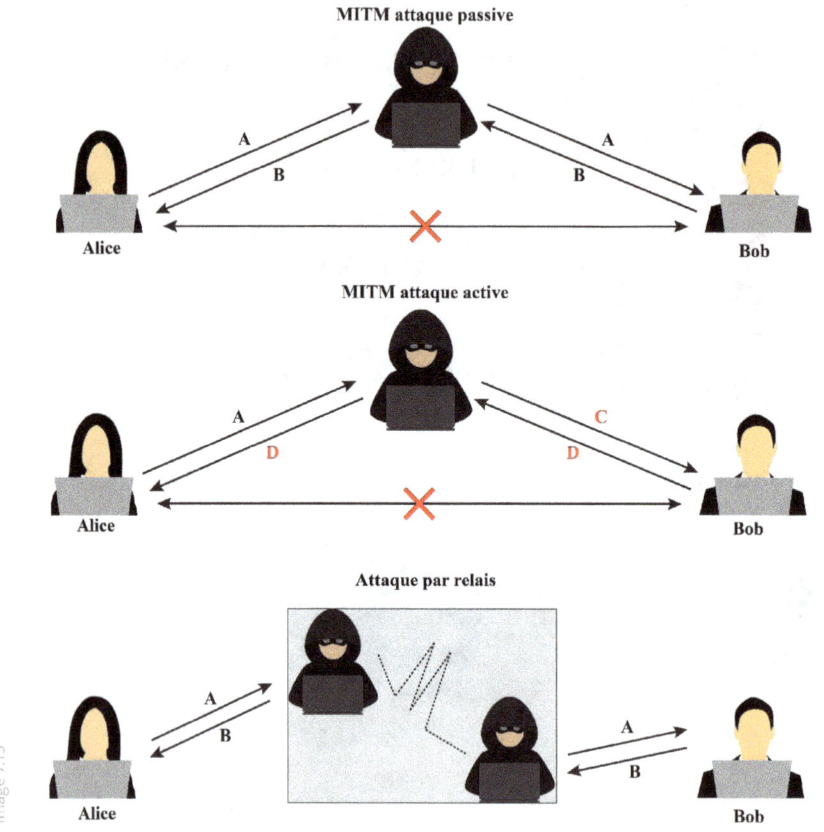

Man-in-the-Middle (MITM) and application layer attacks are a type of eavesdropping attack that intercepts, sends, and receives data never meant to be for the victims. MITM involves inserting malicious code like a Trojan horse into a two-party conversation as a third party. Application layer attacks come at a specific application through a defect or vulnerability in the code such as buffer overflow.

These types of exploits mainly target for theft, extortion, sabotage, and infowar (Das & Samdaria, 2014). The mafia soldier, cyber-thief, and the code warrior would fit into the suspect profile when the purpose of the attack is solely profit. Yet, if the target is a major corporation (like Microsoft), the cyberpunk can also fit into this typology. Law enforcement encounters extreme difficulty with these attacks since the perpetrator is almost always unknown.

Denial of Service (DoS)

DoS attacks ultimately shut down a server by flooding it with phony authentication methods to prevent people from accessing it.

The explicit attack can be in the simple form of vandalizing the server. However, worm-based DoS attacks can result in broad-spectrum damages. A single DoS exploit, such as MyDoom, has cost corporations upwards of $250 million in lost productivity. To offset costs and capture the attackers, the Microsoft company offered a $250,000 reward (Stein, 2004).

When the attacks are done for profit, a code warrior or a mafia soldier would be the main suspects. If the attack is strategic, however, warfighters should be considered. A cyberpunk-motivated DoS could also be considered if a major corporation is the intended victim.

Social Engineering
Social engineering scams can take the form of a dumpster diving approach in an effort to steal password or account information for profit. Social engineering can be also used for all other purposes including political, disinformation, infowar, and strategic purposes. However, when this technique is part of the art, the cyber-thief, the cyberpunk, and the cyberstalker should be considered as suspects.

Image 7.14

The Recent Trend of Attacks: Major Types of Criminal Exploits Against Corporations/Institutions

IT and communication devices are rapidly growing. Today, these technologies are utilized to create a new regime of e-business. Within the following section, four major types of criminal exploits against corporations will be discussed: (1) web page defacement, (2) domain redirection or hijacking, (3) DDoS attacks, and (4) ransomware attacks.

Today, most websites serve as the main electronic communication method of businesses and organizations advertising and selling their products. As electronic business expands, the everlasting risk of electronic vandalism has been dramatically increased. This major form of attack against businesses and organizations (like web page defacement) can be extremely damaging, ruining credibility and resulting in monetary loss.

Web Page Defacement

Image 7.15

The goal of web page defacement is to either change contents on private or public websites or block the service of websites. Attackers first test and diagnose weaknesses among a targeted active website and then intrude into the internal system of the website. At the final stage, attackers alter an index page and replace it with a hacker's signature for advertising successful attacks. The signature message often leaves a traceable hacker's nickname or indicates particular political slogans, exhibiting the characteristics of a suspect. If the purpose of the exploit is revenge for a certain opponent, a message containing malicious criticism can guide investigators toward the perpetrator. Kiddie attacks using tool kits can easily deface multiple websites within a short window of time if the website security is vulnerable.

- Zone-h.org is a database of defaced websites.
- Attackers who successfully deface certain websites post accomplishments to boost their fame.

Home   News   Events   Archive   Archive ★   Onhold   Notify   Stats   Register   Login

search…

Ads

1. **Edmunds® - Official Site** Begin Your Car Search With Edmunds. Research, Reviews,   www.Edmunds.com
2. **New Policy in MA** If you drive in MA you better read this…   Insurance Comparisons.org
3. **The Latest on Cell Phones** See what's trending on Yahoo & stay up to date with search.yahoo.com/CellPhone

[ENABLE FILTERS]

Total notifications: **174,669** of which **74,694** single ip and **99,975** mass defacements

Legend:
H - Homepage defacement
M - Mass defacement (click to view all defacements of this IP)
R - Redefacement (click to view all defacements of this site)
L - IP address location
★ - Special defacement (special defacements are important websites)

Date	Notifier	H	M	R	L	★	Domain	OS	View
2014/10/26	d3b~X	H		R		★	erzincanadsm.gov.tr	Linux	mirror
2014/10/26	Whoami	H	M			★	www.servioriente.gov.co	Linux	mirror
2014/10/25	HACKED BY LIBERO	H	M	R		★	www.proargex.gov.ar	Win 2008	mirror
2014/10/25	HACKED BY LIBERO	H	M			★	www.siia.gov.ar	Win 2008	mirror
2014/10/25	HACKED BY LIBERO	H		R		★	www.pronatur.gov.ar	Win 2008	mirror
2014/10/25	Zombie-Root	H	M	R		★	mysurat.kejora.gov.my	Linux	mirror
2014/10/25	Zombie-Root	H		R		★	e-lesen.lbjt.gov.my	Linux	mirror
2014/10/25	Drac-101code					★	www.lhsfj.gov.cn/coli.txt	Win 2003	mirror
2014/10/25	d3b~X	H		R		★	www.pamth.gov.gr	Linux	mirror
2014/10/25	medhack	H	M	R		★	sehutkamildh.gov.tr	Linux	mirror
2014/10/25	pk_Robot	H				★	www.governmentassurances.gov.pk	Linux	mirror
2014/10/25	Index Php					★	www.scholarships.gov.tt/zxcvbn…	Linux	mirror
2014/10/25	Mr.MaurgeulisX196			R		★	dishubkominfo.pariamankota.go….	Linux	mirror
2014/10/24	Yheya Alshami	H		R		★	cem.gov.vn	Win 2008	mirror
2014/10/24	VOTR3X	H		R		★	agrofea.fearp.usp.br	Linux	mirror
2014/10/24	gstaad & aymez	H	M			★	www.mairie-laturballe.fr	Linux	mirror
2014/10/24	sahrawihacker	H		R		★	prueba.tornquist.gov.ar	Linux	mirror
2014/10/23	i3r_cod3			R		★	www.bireuenkab.go.id/q.html	Linux	mirror
2014/10/23	KkK1337					★	xzzfdd.xwzf.gov.cn/tech.gif	Win 2003	mirror
2014/10/23	dEnny_Attacker	H	M			★	keu.dprd-manadokota.go.id	Linux	mirror
2014/10/23	dEnny_Attacker	H	M			★	deputi-dprd-manadokota.go.id	Linux	mirror
2014/10/23	dEnny_Attacker	H				★	ekonomi.dprd-manadokota.go.id	Linux	mirror
2014/10/23	by_yskr		M			★	www.thantong.go.th/php/	Linux	mirror
2014/10/23	by_yskr		M			★	www.maepao.go.th/php/	Linux	mirror
2014/10/23	by_yskr		M			★	www.mflhospital.go.th/php/	Linux	mirror

1 2 3 4 5 6 7 8 9 10 11 12 13 14 15 16 17 18 19 20 21 22 23 24 25 26 27 28 29 30

Image 7.16

When investigating website defacement cases, investigators should be aware of a website called Zone-h.com. This website is an encyclopedia of worldwide information, including previous examples of defaced websites, relevant statistics, and messages from the attackers who wish to glorify their accomplishments and share expertise and a blueprint of crime. More importantly, valuable information such as the crime date, damaged websites, and attackers' profiles can be easily attained. This website can be essential to investigators for building a profile sufficient to identify a set of suspects within a narrow group of website defacement cases.

According to Zone-h.com's statistics, web page defacement postings are decreasing, gradually. Reasons behind this may be due to hackers realizing that law enforcement agencies are constantly checking the list. However, a reduction in web page defacement implies the criminals may be switching gears to criminal activities involving monetary gains. In fact, the recent IC3 report indicated monetary loss due to cybercrime has been significantly increasing over the decade.

Domain Redirection or Hijacking

The second major category of cyberattacks against corporations and institutions shares a similar concept of web defacement. However, domain redirection or hijacking is not an alteration of website contents but an alteration of DNS. By entering through a URL, online users access a homepage. However, attacks of this type reroute the existing URL to a different website. This pattern is called domain redirection or hijacking. The main methods of attack are social engineering DNS attacks and cache poisoning DNS attacks.

How Does Domain Redirection Work?

Next is a review of DNS cache poisoning. In the event a user types the URL address in the website's browser, the user's request is sent to the DNS server and then the user receives a response from the DNS server. A local DNS server enables this faster because it has a cache of previous responses. If the local DNS has a cache of the user's requests, it will send a cache to the user instead of forwarding the user's request to the internet DNS.

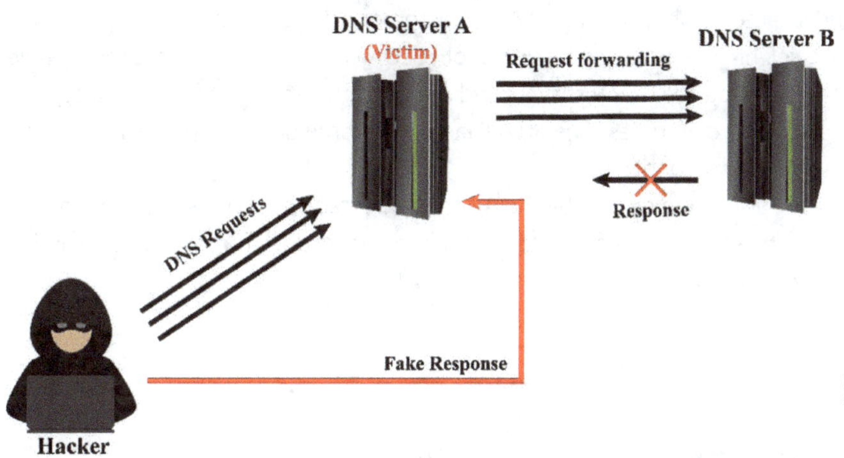

When a cache poisoning attack occurs, the scenario is as follows:

First, attackers send a request which is not cached in the local DNS, and then the query is forwarded to the internet DNS. Prior to the internet DNS response reaching the local DNS, fake responses flood the local DNS from hackers. If hackers succeed in this form of attack, the DNS cache in the local DNS is modulated to the hacker's response. Upon the successful attack, whenever a user types the normal URL to connect to the website, the user will reach the abnormal site, which was altered by hackers.

Denial-of-Service Attacks

As previously mentioned, denial-of-service attacks are formulated by hackers who intentionally attack a particular system with the goal of causing an overload or malfunction of regular service. Attackers typically formulate numerous connecting trails at a specific server to overload regular services or run out the TCP session.

After putting Yahoo! and eBay websites out of order in early 2000, this attacking pattern became very popular. Since 2007, the number of attacks has been increasing and its pattern, target, and motive are much more sophisticated than traditional DoS cases.

In previous cases involving web page alteration, electronic vandal's motives were frequently ego or exposure. In the past, the main targets used to be Microsoft or major corporations; however, current electronic vandalism cases tend to be targeting smaller companies. Smaller companies tend to lack the ability to defend their assets and the attacks become more aggressive—designed for theft and extortion with the purpose of monetary gain.

In short, a good way to describe DoS attacks is low tech yet highly impactful. In other words, attackers do not always require a high level of technological proficiency in order to commit a crime, and attackers can easily purchase premade DoS programs and structured botnets from online black markets. Presently, even with various DoS resources, crime fighters do not have sufficient solutions to combat this matter.

DDoS attacks are an advanced form of DoS attack combined with a botnet.

AWS DDOS ATTACK

Amazon Web Services, the 800-pound gorilla of everything cloud computing, was hit by a gigantic DDoS attack in February 2020. This was the most extreme recent DDoS attack ever and it targeted an unidentified AWS customer using a technique called Connection-less Lightweight Directory Access Protocol (CLDAP) reflection. This technique relies on vulnerable third-party CLDAP servers and amplifies the amount of data sent to the victim's IP address by 56 to 70 times. The attack lasted for three days and peaked at an astounding 2.3 terabytes per second.

While the disruption caused by the AWS DDoS attack was far less severe than it could have been, the sheer scale of the attack and the implications for lost revenue and brand damage were significant.

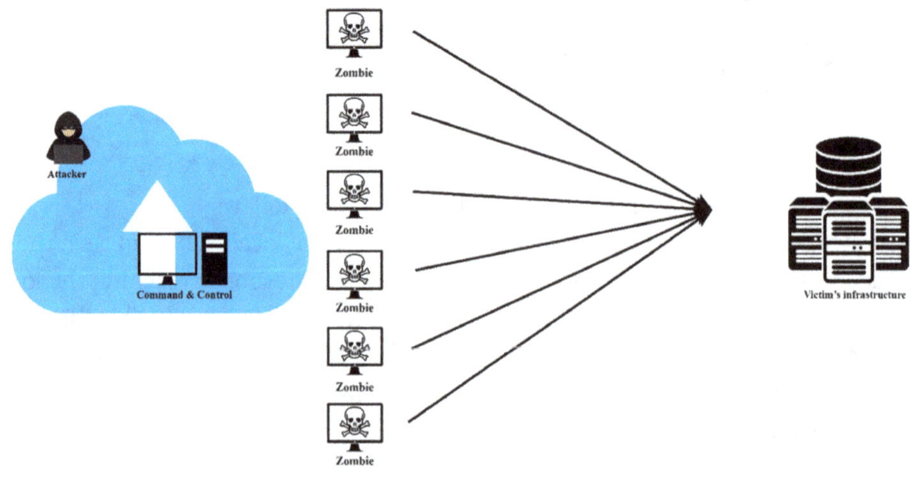

Image 7.18

DDoS Defined in Detail

A command and control server (C&C server) exists between the hacker and zombie computers and controls and delivers a command from the hacker to the zombie computer. It is very important to locate this server when a crime happens.

A zombie computer executes an actual attack to a victim's computer under the hacker's command, and a group of zombie computers is called a "botnet."

Botnet (from a combination of robot and network) is networks of computers infected by malware which computer criminals use to send out viruses or worms to online users via email. This malicious application is typically referred to as a "bot." The computer owner tends to be unaware that their system is being controlled by computer criminals and these computers are known as zombies.

The bot on the infected PC logs into a particular IRC (Internet Relay Chat) server or a web server. A spammer purchases access to the botnet from the computer criminals. The spammer sends instructions via the IRC server to the infected PCs, causing them to send out spam messages to mail servers.

There are an estimated 10,000 to 100,000 zombie computers for each DDoS crime committed. If the hacker achieves the botnet system on a specific target and orders a command, then the attack will have significant impacts on delivery of regular services, even on a well-secured and structured system.

Mobile Botnet Is the New Trend

This new type of botnet targets mobile devices (such as tablets and smartphones), attempting to gain complete access to the device and its contents while also providing control to the botnet creator. To provide attackers with root permissions over the compromised mobile device, mobile botnets take advantage of unpatched exploits, enabling hackers to send email or text messages, make phone calls, access contacts and photos, and more. Typical mobile botnets are likely to be undetected thus able to spread by sending copies of themselves from compromised devices to other devices via text messages, email messages, or free downloads.

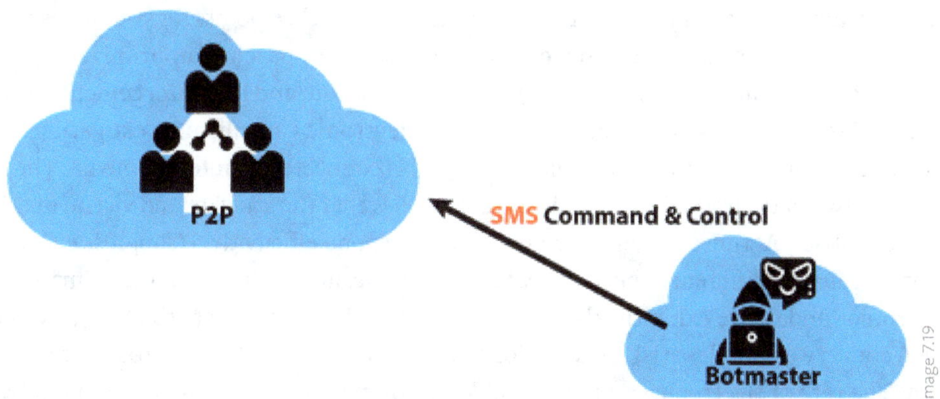

Examples of mobile botnets include the iPhone SMS attack that compromised iPhone and iPad devices; DreamDroid malware that affected Google Android devices; the ZeuS variant aimed at BlackBerry users; and Commwarrior and Sexy Space, both of which compromised Symbian Series mobile devices.

Ransomware Attacks

Ransomware is a form of malware that encrypts all of the data on a user's computer device and network, then it denies the user access to the device until a ransom has been fully paid. Recent ransomware is capable of encrypting files on a shared network system, which can essentially disable an entire organization's data.

Once the files are fully encrypted, the hacker usually displays instructions explaining how to pay to unlock data along a 48- to 72-hour deadline. If you pass the deadline, the ransom is likely to be increased. Most ransoms start around $100–$600 and, once the deadline has passed, will likely increase to over $1,000.

Most hackers request the payment of Bitcoin and ask victims to pay the ransom through a darknet site. To gain access to darknet sites, victims need to use TOR (The Onion Relay), which is a special network encryption browser for hiding all the origination and ending destination of the traffic.

According to Johnson (2016), in August 2015, a ransomware attack under the username of Albanian Hacker demanded payment of two Bitcoin, which accounted for $500 at the time, from an Illinois internet retailer in exchange for removing bugs from their computer. While seizing control of computers, Albanian Hacker generated a kill list used as one of the first kill lists issued by the Islamic State of Iraq and Syria (ISIS). The WannaCry attack and its relationship with the Lazarus Group, sponsored by the North Korean intelligence agency, is an additional bona fide example of terrorist activities against another nation utilizing a new technology. The WannaCry outbreak shut down computers in more than 80 National Health Service organizations in England alone, resulting in 600 general practices having to return to pen and paper and five hospitals simply diverting ambulances to handle more emergency cases (Hern, 2017). The WannaCry 2.0 global ransomware attack also resulted in damage to massive amounts of computer hardware and extensive loss of data, money, and other resources (U.S. Department of Justice, 2018).

Prevention Efforts Against Electronic Vandalism

So far, electronic vandalism's definition and patterns have been discussed. At last, various methods of electronic vandalism prevention, countermeasures, and key points for your investigation will be outlined—namely, **international cooperation, security patches, appropriate security equipment, backup and encryption for sensitive information, and law enforcement training.**

Cooperation among nations is very important to electronic vandalism investigations. With the growth of communication technologies, the world is connected by internet media. The main difference between traditional crimes and cybercrimes is that the attacker and victim may be in different countries. At times, law enforcement fails to respond efficiently to suppress cybercrime incidents because of differences between each country's legal system and physical jurisdiction.

Electronic vandals are exhibiting bolder movement with the idea that attacking a system in a foreign country carries less risk of detection because of the difficulty of a cross-nation investigation. The rate of electronic vandalism incidents will not be decreased or prevented by one country's effort. Every investigator must commit to close cooperation with other countries to succeed in the arrest of electronic vandals.

In an effort to prevent electronic vandalism, regular secured patch inspections should be conducted to analyze any potential weakness of services that are exposed to the public. In addition to a regular security inspection, it is strongly recommended to have adequate security equipment including IPS (intrusion-prevention systems) and firewalls (Choi, 2015).

When an incident happens, an investigation of the security equipment can provide strong evidence and resources to assist in tracking the attacker. However, security equipment alone is not sufficient to prevent or secure one's intangible assets. Ideally, it is good practice to have a technology expert team within the organization to minimize the potential risks and strengthen internal control.

With the growth of digital and information technology, issues of cybercrime have become more prevalent, as seen in numerous recent cybercrime cases. Even during the COVID-19 pandemic, cybercriminals are leveraging fear and panic to take advantage of many vulnerable victims for bogus cures, coronavirus phishing, federal stimulus check scams, and fake coronavirus charities. In addition, the COVID-19 shelter-in-place order required millions of people working at home to access corporate and government computer networks via their home Wi-Fi—home networks which are often shared by kids and family members and are more vulnerable than their workplaces. Furthermore, this type of cybercrime will ultimately result in poor financial, physical, and mental health outcomes for the general public (Finklea, 2020).

Unfortunately, the current capabilities of many law enforcement agencies are very limited despite such a heightened level of awareness and concern for the role recent technology has in facilitating cybercrime and instances of online victimization.

Since most local and state law enforcement officers lack knowledge concerning processing computer data and related evidence (Choi, 2015), more specialized computer forensics and digital evidence training programs are warranted.

Summary

Ultimately, there is no protection from electronic vandalism; yet, collective efforts from the whole organization or institution can minimize the incidents. Capacity-building efforts are crucial to strengthen international support. To prepare for any possibilities of electronic vandalism, individuals or public organizations must back up important information regularly and create an encrypted security program for valuable information (e.g., credit card transactions or identity information) to minimize damages from wrongful use. Law enforcement personnel must be trained to handle the digital forensics involved within these crimes.

Lab 7.1: Wireless Network Scanners

In February 2019, Mr. Evil Noodle hacked into the computer system of a man named Curry. Conducting his crimes behind his personal computer screen, Mr. Evil Noodle utilized the tool Vistumbler to **scan nearby wireless networks** where he could then find potential victims. Mr. Evil Noodle believed that it would be a good idea to target a place where users frequently utilized **public Wi-Fi networks**. He then decided to create **a fake Wi-Fi** name that was almost identical to the real Wi-Fi network that belonged to American University, where Curry was employed. Unfortunately, Curry fell victim to the crime because he inadvertently accepted the fake public Wi-Fi sign-in agreement which gave Mr. Evil Noodle access to Curry's computer system. Mr. Evil Noodle's ultimate goal was to hack into a victim's computer system where he could acquire personal information. Curry happened to work as the employment office director where he had access to every employee's paychecks and tax returns. Gaining access to Curry's computer system gave Mr. Evil Noodle complete control over every employee's highly sensitive information. Mr. Evil Noodle had no other motive than to utilize this information maliciously.

Below is a useful tool that is utilized to map and visualize the access points around a user based on their wireless and GPS data:

Vistumbler is a wireless network scanner written in AutoIt for Windows. It is a free tool that scans for nearby wireless networks within range of a user's Wi-Fi adapter. Once Vistumbler finds a wireless network it will display **the network's SSID, signal strength, encryption, MAC address, network channel, access point manufacturer, etc.**

Please answer the following questions (2–3 pages):

1. How did Mr. Evil Noodle gain access to Curry's computer system containing highly sensitive information?
2. How could Mr. Evil Noodle share the acquired information with other criminals or cybercriminals?
3. If Mr. CIC and Mr. Hero were to investigate this case, how would they go about doing so?
4. Prevention: How would one stop Mr. Evil Noodle from acquiring personal information from a hacked Wi-Fi network?
5. Discussion: What are your thoughts on this type of case? What would your reaction be as a victim?

Lab Instruction

In this lab, you will have to check the Wi-Fi that surrounds you. First, download Vistumbler and launch it. (If you want to scan geo data, you need an extra tool.) Once you open Vistumbler, click Scan APs and it will scan all Wi-Fi around you.

1. Go to a public place near you (such as Starbucks, your campus, or public library).
2. Scan APs. Take a screenshot and upload it. After you scan your area's Wi-Fi, you will see all the data for each Wi-Fi. Specifically, keep an eye out for the MAC addresses.

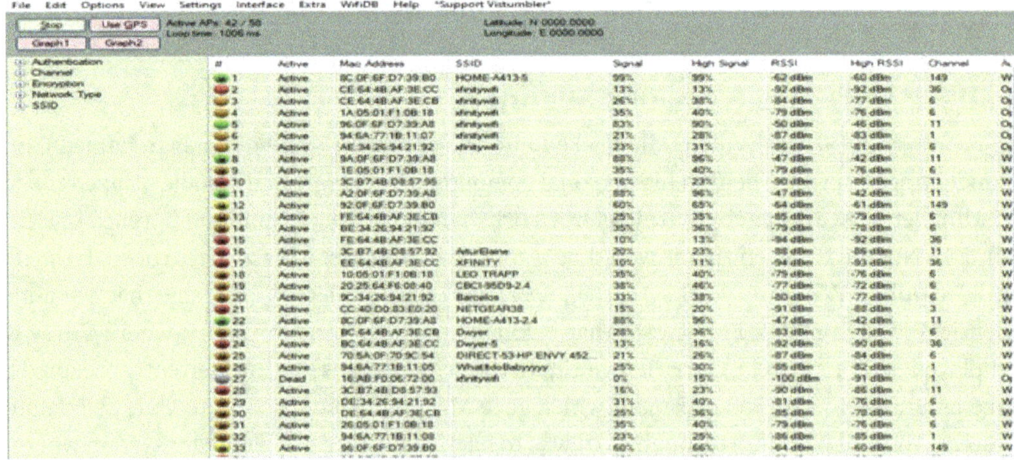

#	Active	Mac Address	SSID	Signal	High Signal	RSSI	High RSSI	Channel	A
1	Active	8C:0F:6F:D7:39:B0	HOME-A413-5	99%	99%	-62 dBm	-60 dBm	149	W
2	Active	CE:64:4B:AF:3E:CC	xfinitywifi	13%	13%	-92 dBm	-92 dBm	36	O
3	Active	CE:64:4B:AF:3E:CB	xfinitywifi	26%	38%	-84 dBm	-77 dBm	6	O
4	Active	1A:05:01:F1:0B:18	xfinitywifi	35%	40%	-79 dBm	-76 dBm	6	O
5	Active	96:0F:6F:D7:39:A8	xfinitywifi	83%	90%	-50 dBm	-46 dBm	11	O
6	Active	94:6A:77:1B:11:07	xfinitywifi	21%	28%	-87 dBm	-83 dBm	1	O
7	Active	AE:34:26:94:21:92	xfinitywifi	23%	31%	-86 dBm	-81 dBm	6	O
8	Active	9A:0F:6F:D7:39:A8		88%	96%	-47 dBm	-42 dBm	11	W
9	Active	1E:05:01:F1:0B:18		35%	40%	-79 dBm	-76 dBm	6	W
10	Active	3C:87:4B:D8:57:95		16%	23%	-90 dBm	-86 dBm	11	W
11	Active	A2:0F:6F:D7:39:A8		88%	96%	-47 dBm	-42 dBm	11	W
12	Active	92:0F:6F:D7:39:B0		60%	65%	-64 dBm	-61 dBm	149	W
13	Active	FE:64:4B:AF:3E:CB		25%	35%	-85 dBm	-79 dBm	6	W
14	Active	BE:34:26:94:21:92		35%	36%	-79 dBm	-78 dBm	6	W
15	Active	FE:64:4B:AF:3E:CC		10%	13%	-94 dBm	-92 dBm	36	W
16	Active	3C:87:4B:D8:57:92	AturBane	20%	23%	-88 dBm	-86 dBm	11	W
17	Active	EE:64:4B:AF:3E:CC	XFINITY	10%	11%	-94 dBm	-93 dBm	36	W
18	Active	10:05:01:F1:0B:18	LEO TRAPP	35%	40%	-79 dBm	-76 dBm	6	W
19	Active	20:25:64:F6:08:40	CBCI-95D9-2.4	38%	46%	-77 dBm	-72 dBm	6	W
20	Active	9C:34:26:94:21:92	Barcelos	33%	38%	-80 dBm	-77 dBm	6	W
21	Active	CC:40:D0:83:E0:20	NETGEAR38	15%	20%	-91 dBm	-88 dBm	3	W
22	Active	8C:0F:6F:D7:39:A8	HOME-A413-2.4	88%	96%	-47 dBm	-42 dBm	11	W
23	Active	BC:64:4B:AF:3E:CB	Dwyer	28%	36%	-83 dBm	-78 dBm	6	W
24	Active	BC:64:4B:AF:3E:CC	Dwyer-5	13%	16%	-92 dBm	-90 dBm	36	W
25	Active	70:5A:0F:70:9C:54	DIRECT-53-HP ENVY 452...	21%	26%	-87 dBm	-84 dBm	6	W
26	Active	94:6A:77:1B:11:05	WhatIdoBabyyyyy	25%	30%	-85 dBm	-82 dBm	1	W
27	Dead	16:AB:F0:06:72:00	xfinitywifi	0%	15%	-100 dBm	-91 dBm	1	O
28	Active	3C:87:4B:D8:57:93		16%	23%	-90 dBm	-86 dBm	11	W
29	Active	DE:34:26:94:21:92		31%	40%	-81 dBm	-76 dBm	6	W
30	Active	DE:64:4B:AF:3E:CB		25%	36%	-85 dBm	-78 dBm	6	W
31	Active	26:05:01:F1:0B:18		35%	40%	-79 dBm	-76 dBm	6	W
32	Active	94:6A:77:1B:11:08		23%	25%	-86 dBm	-85 dBm	1	W
33	Active	96:0F:6F:D7:39:B0		60%	66%	-64 dBm	-60 dBm	149	W

Image 7.20

A MAC address (media access control address) is a unique identifier assigned to a NIC (network interface controller) for use as a network address in communications within a network segment. MAC addresses are primarily assigned by device manufacturers and are therefore often referred to as the burned-in address, Ethernet hardware address, hardware address, or physical address. Each address can be stored in hardware, such as the card's read-only memory, or by a firmware mechanism.

3. Now, it is time to search a MAC address. Go to MA:CV:en:do:rs, enter a MAC address, and the results will show up below.

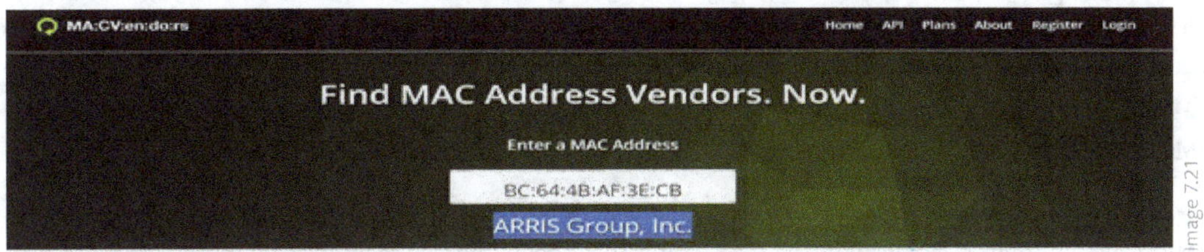

Image 7.21

Lab 7.2: Keywords and Searching

- You will be using the same evidence file (4Dell Latitude CPi) that was presented in the previous chapter.
- Answer the following questions:
    1. A search for the name "Greg Schardt" reveals multiple hits. One of these proves that Greg Schardt is Mr. Evil and is also the administrator of this computer. What file is it? What software program (email account) does this file relate to?
    2. Find six installed programs that may be used for hacking.
    3. List five hacking-related newsgroups that Mr. Evil has subscribed to.
    4. We want to search any media file that contains more than 1 megabyte created in August 2004. Please list files that meet the requirements.

- Add your screenshots/bookmarks and briefly describe to support your answer.

References

Choi, K. S. (2015). *Cybercriminology and digital investigation*. LFB Scholarly Publishing.

Das, M. L., & Samdaria, N. (2014). On the security of SSL/TLS-enabled applications. *Applied Computing and informatics*, 10(1-2), 68–81.

Finklea, K. (2020). *COVID-19: Cybercrime opportunities and law enforcement response*. Congressional Research Service.

Hern, A. (2017, December 30). *WannaCry, Petya, NotPetya: How ransomware hit the big time in 2017*. The Guardian. https://www.theguardian.com/technology/2017/dec/30/wannacry-petya-notpetya-ransom-warehttps://www.theguardian.com/technology/2017/dec/30/wannacry-petya-notpetya-ransomware

Johnson, T. (2016, July 20). *Computer hack helped feed an Islamic State death list. McClatchy DC Bureau*. https://www.mcclatchydc.com/news/nation-world/national/article90782637.html.

Kron, E. (2021). The human problem behind credential theft and reuse. *Cyber Security: A Peer-Reviewed Journal*, 4(3), 223–231.

Nicholson, P. (2020, August 12). *Five most famous DDoS attacks and then some*. A10 Blog. https://www.a10networks.com/blog/5-most-famous-ddos-attacks/

Shoemaker, D., & Kennedy, D. B. (2009). Criminal profiling and cyber criminal investigations. *Crimes of the Internet*, 439–455.

Stein, A. (January 30, 2004). Microsoft offers MyDoom reward. CnnMoney. http://www.moneymag.com/2004/01/28/technology/mydoom_costs/

U.S. Department of Justice. (2018, September). *North Korean regime-backed programmer charged with conspiracy to conduct multiple cyber-attacks and intrusions*. https://www.justice.gov/opa/pr/north-korean-regime-backed-programmer-charged-conspiracy-conduct-multiple-cyber-attacks-and

Wilson, C. (October 17, 2003). Computer Attack and Cyber terrorism: Vulnerabilities and Policy Issues for Congress, CRS Report for Congress. http://www. fas. org/irp/crs/RL32114. pdf.

Credits

Phishing and Online Child Sexual Exploitation

Kyung-Shick Choi and Hannarae Lee

Introduction

This chapter discusses cybercrimes that focus on social engineering: phishing and online child exploitation. In the first part of the chapter, the authors demonstrate the history of phishing, the different types of phishing, their delivery mechanisms, and investigative efforts and tactics. By exploring the variety of phishing techniques and their vulnerabilities, the phishing section goes over crucial areas for investigators to collect and analyze forensic evidence while providing the vital procedural steps to conduct the investigation.

The remaining portion of the chapter illustrates a prime example of cybercrime: online child sexual abuse. In this section, the authors provide the criminal patterns and countermeasures of online sexual predators, and mobile forensics basics for SNS (social network service) environments to locate suspects and collect the evidence. After reading this chapter, readers will be able to acknowledge the prevalence and seriousness of each crime, along with investigative and prevention mechanisms.

Phishing

What Is Phishing?

Exactly what is phishing? The term has more than one origin. For some, the term is a compound of "fishing" and "private data." Others claim that the term is a play on the word "phreaking," which is the practice of hacking telecommunication services and public telephone networks. Phishing is a name given to security attacks that employ social engineering methods. Someone manipulates others into revealing information that can then be used to steal data or access systems, cellular phones, money, or even identity. Such attacks can be either quite simple or highly complex.

In gaining access to information over the phone or through websites, a new dimension to the role of the social engineer has been added. Social engineering is the acquisition of sensitive information or inappropriate access privileges by an outsider based upon the building of inappropriate trust relationships among insiders. The overall goal of social engineering is to trick a victim into providing valuable information or access to that information.

The History and Current Status of Phishing

Over the past decade, phishing attacks were used to gain internet service provider account information, with America Online being a popular target. In 1996, the term was coined by hackers who were stealing America Online (AOL) accounts by scamming passwords from unsuspecting AOL users (Moore, 2011).

The first mention of phishing was made famous on the internet in the alt.2600 hacker newsgroup in January 1996. It is quite possible the term may have been used even earlier in the popular hacker newsletter 2600. By 1996, hacked accounts were called "phish." In 1997, phish were actively being traded between hackers as a form of electronic currency.

There are instances whereby phishers would routinely trade 10 working AOL phish for a piece of hacking software or warez. Warez refer to stolen copyrighted applications and games. Over time, the definition of what constitutes a phishing attack has blurred, while also expanding.

The term phishing covers much more than merely obtaining user account details; it now also includes access to all personal and financial data.

Originally, what was entailed in the meaning of phishing was tricking users into replying to emails or passwords and credit card details. However, the method of phishing has now expanded into using fake websites; installing Trojan horses; and using keyloggers, screen loggers, and man-in-the-middle data proxies, which are delivered through any electronic communication channel.

Phishing Message Delivery: Email and Spam

Phishing attacks initiated by email are the most common. Phishers use spam to deliver specially crafted emails to millions of legitimate "live" email addresses within a few hours or minutes.

Techniques of phishing by email:

Image 8.1

- Official looking and sounding emails
- Copies of legitimate corporate emails with minor URL changes
- HTML-based email used to obfuscate target URL information
- Fake postings to popular message boards and mailing lists

Phishers are then able to create emails with fake mail from headers and impersonate any organization they choose. In some cases, the phishers may also set the RCPT command to tell the mail server an email address of their choice. You can send more than one RCPT command for multiple recipients. The server will respond with a code of 250 to each command. The syntax for the RCPT is RCPT TO:<forward-path>.[1] Thereby customer replies to the phishing email will be sent to the phishers.

The growing press coverage of phishing attacks demonstrates most customers are very wary of sending confidential information like passwords and PIN information via email. However, phishers still have much success in many cases.

Techniques used within phishing emails will include official-looking emails, copies of legitimate corporate emails with minor URL changes, HTML-based email used to obfuscate target URL information, and often fake postings to popular message boards and mailing lists.

Phishing Message Delivery: Instant Messenger and Website

Instant messengers became a famous phishing ground As many instant messenger clients allow embedded dynamic content to be sent by channel participants, many phishing techniques have been used in standard web-based attacks.

An increasingly popular method of conducting phishing attacks is utilizing malicious website content. Web-based delivery techniques include:

- The use of disguised HTML links;
- The use of fake banners on advertising graphics to lure customers to the phisher's website;
- The use of pop-up or frameless windows to disguise the true source of the phisher's messages; and
- New on the phishing radar, instant messenger forums are likely to become popular phishing grounds.

Instant Messenger

While various communication channels become more popular with home users, more functionality is included within the software and phishing attacks will increase. Many instant messenger clients allow embedded dynamic content, such as graphics, URLs, and multimedia, which can be sent by channel participants.

The common usage of bots in many popular channels also means that it is straightforward for a phisher to send semi-relevant links and fake information to would-be victims anonymously.

Image 8.2

---

1   If the recipient is not known to the mail server, the response code will be 550. You might also get a response code indicating that the recipient is not local to the server. If that is the case, you will get one of two responses back from the server: (1) 251 User not local; will forward to <forward-path>—this reply means that the server will forward the message; the correct mail address is returned so that you can store it for future use; (2) 551 User not local; please try <forward-path>—this reply means that the server won't forward the message; you need to issue another RCPT command with the new address.

Website

Conducting phishing attacks by using malicious website content is an increasingly popular method. This method may be included within a website operated by the phisher or a third-party site hosting embedded content.

Web-based delivery techniques include, but are not limited to, the use of disguised HTML links within popular websites and/or message boards, the use of fake banners on advertising graphics to lure customers to the phisher's website, and the use of pop-up or frameless windows to disguise the trustworthy source of the phisher's message.

Types of Phishing Attacks

For a successful phishing attack, phishers utilize several methods to lure the online customer to their server or disguise the page with legitimate-looking content. There is an ever-increasing number of ways to do this; the four most common methods are explained in detail below:

Man-in-the-Middle Attacks

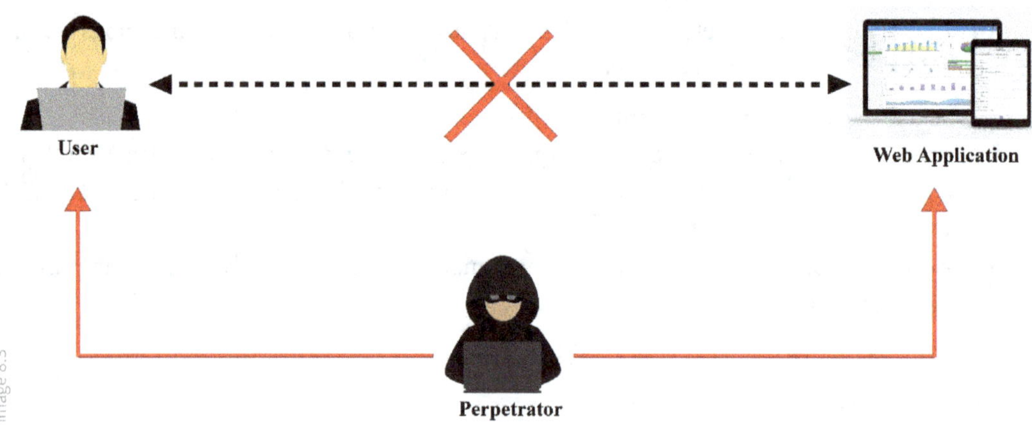

Image 8.3

As previously discussed in Chapter 7, one of the most effective methods for gaining control of customer information and resources is the utilization of man-in-the-middle attacks. During this type of attack, the attackers situate themselves between the customer and the real web-based application, and the attacker proxies all communications between the systems. From this vantage point, the attacker can observe and record all transactions. This form of attack is successful for both HTTP and HTTPS communications. The customer connects to the attacker's server as if it were the real site; meanwhile, the attacker's server makes a simultaneous connection to the actual site. Then, the attacker's server proxies all communications between the customer and the real web-based application server in real time.

In the case of secure HTTPS communications, an SSL connection is established between the customer and the attacker's proxy. In contrast, the attacker's proxy creates its own SSL connection between itself and the real server.

For man-in-the-middle attacks to be successful, the attacker must be able to direct the customer to their proxy server instead of the real server. This attack may be carried out through a number of methods such as transparent proxies, DNS cache poisoning, URL obfuscation, or browser proxy configuration.

The types of man-in-the-middle attacks include the following.

Transparent Proxies

These are situated on the same network segment or located on route to the real server. A transparent proxy service can intercept all data by forcing all outbound HTTP and HTTPS traffic through itself. In this transparent operation, no configuration changes are required at the customer end.

DNS Cache Poisoning

This may be used to disrupt normal traffic routing by injecting false IP addresses into key domain names. For example, the attacker poisons the DNS cache of a network firewall so that all traffic destined for the bank IP address now resolves the attacker's proxy server IP address.

URL Obfuscation

The attacker is able to trick the customer into connecting to their (the attacker's) proxy server instead of the real server using URL obfuscation techniques. For example, the customer may follow a link to mybank.com.ch instead of mybank.com

Altering Browser Proxy Configuration

By overriding the customer's web browser set up and setting proxy configuration options, an attacker can force all web traffic to their nominated proxy server. This method is not transparent to the customer, but the customer may easily review their web browser settings to identify an offending proxy server. In many cases, phishers may have already used a malware attack to infect a victim's computer and alter browser proxy configurations.

Malware-Based Phishing

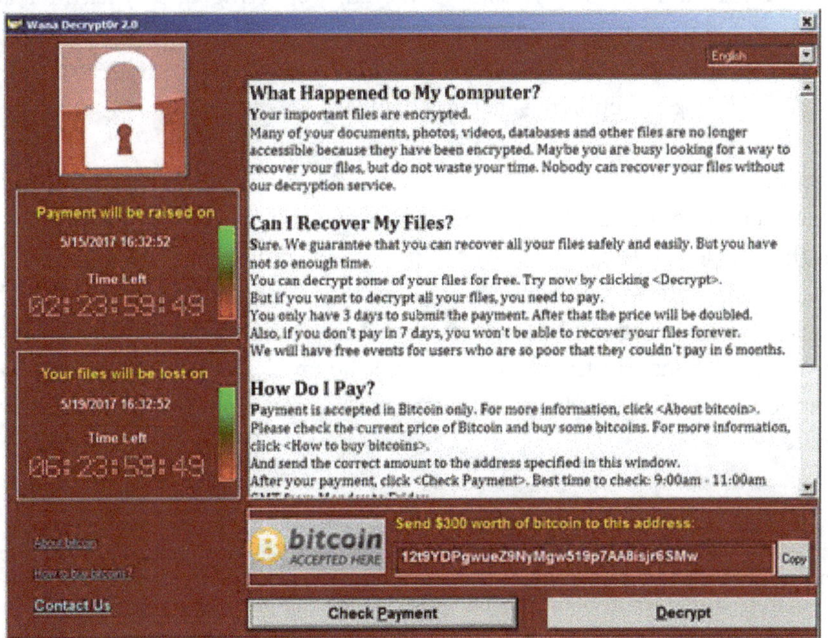

Image 8.4

Generally, malware-based phishing refers to any phishing that involves running malicious software on the user's machine.

Malware is spread in one of two ways: either by social engineering or by exploiting security vulnerabilities. A typical social engineering attack is to convince a user to **open an email attachment or download a file from a website** (**e.g.,** an email with a link to celebrity gossip). Additionally, some **downloadable software** can contain malware. Another way malware is spread is by propagating a worm or virus that takes advantage of a security vulnerability to install the malware or makes the malware available on a website that exploits security vulnerability. Traffic may be driven to a malicious website via social engineering. This can be in the form of spam messages promising some appealing content at the site or injecting malicious content into a legitimate website by exploiting a security weakness, such as a cross-site scripting vulnerability.

Now, let's break down the forms of malware-based phishing.

Keyloggers and Screen-loggers

Programs that install themselves either into a web browser or as a device driver. Data is then monitored as it is being input and relevant data is sent to a phishing server. Keyloggers use a number of different technologies and can be implemented in various ways.

For example, a browser helper object detects changes to the URL and logs information when the URL is at a designated credential collection site. A device driver monitors keyboard and mouse inputs in conjunction with monitoring the user's activities.

A screen-logger monitors both the user's inputs and the display to thwart alternate onscreen input security measures. Keyloggers may collect credentials for a wide variety of sites; often keyloggers monitor the user's location and only transmit credentials for particular sites. Often, hundreds of such sites are targeted, including financial institutions, information portals, and corporate VPNs.[2] Various secondary damage can be caused after a keylogger compromise.

In one real-world example, the inclusion of a credit reporting agency within a keylogger spread via pornography spam once led to the compromise of over 50 accounts with access to the agency. In turn, this has been ultimately used to compromise over 310,000 sets of personal information from the credit reporting agency's database.

Session Hijacking

An attack of a user's activities as they are monitored, typically by a malicious browser component. When the user logs into their account or initiates a transaction, the malicious software hijacks the session to perform malicious actions once the user has legitimately established their credentials. Session hijacking can be performed on a user's local computer by either malware or remotely as part of a man-in-the-middle attack. When malware is activated locally, session hijacking can look exactly like a legitimate user interaction being initiated from the user's home computer.

---

2   A virtual private network (VPN) extends a private network across a public network, such as the internet. It enables a computer or network-enabled device to send and receive data across shared or public networks as if it were directly connected to the private network while benefiting from the private network's functionality, security, and management policies.

Host File Poisoning

If a user types "www.company.com" into the URL bar or uses a bookmark, the user's computer needs to translate that address into a numeric address prior to visiting the site. Many operating systems, such as Windows, have shortcut host files for retrieving host names before a DNS (domain name system) lookup is performed. If this file is modified, then www.company.com can be made to refer to a malicious address. In the event the user goes there, they will see a site that appears to be legitimate and subsequently enter confidential information which actually goes to the phisher instead of the intended legitimate site.

DNS-Based Phishing Attacks ("Pharming")

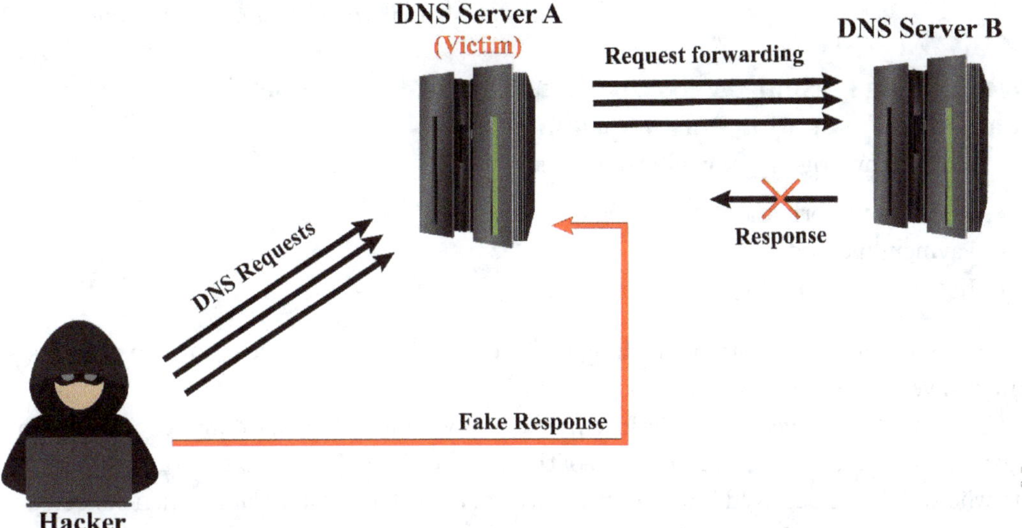

Image 8.5

DNS-based phishing is used here to generally refer to any form of phishing that interferes with the integrity of the lookup process for a domain name. This includes host file poisoning, even though the host file is not properly part of the domain name system. Within the malware section, host file poisoning is discussed since it involves changing a file on the user's computer.

Another form of DNS-based phishing involves polluting the user's DNS cache with incorrect information. This information will be used to direct the user to an incorrect location. If the user has a misconfigured DNS cache, it can direct the user to an incorrect location. By hacking a legitimate DNS server, attaching a system reconfiguration that changes the user's DNS server to a malicious server, or by polluting the cache of a misconfigured legitimate DNS server, personal credentials and other access credentials can be obtained.

Search Engine Phishing

Another approach taken by phishers is creating web pages for fake products, getting the pages indexed by search engines, and waiting for users to enter their confidential information as part of an order, sign up, or balance transfer. These pages usually offer prices slightly too good to be true (because they are).

Image 8.6

Scams involving fraudulent banks have been particularly successful. A phisher is able to create a page advertising an interest rate slightly lower?? than any real bank. Victims discover the online bank via a search engine and enter their bank account credentials for a balance transfer to the new account. Sometimes judgment is clouded as greed can be a powerful motivator. In fact, some victims even provide their full bank account details to the fraudulent phishing websites.

Investigating Phishing Websites: The Organization of Phishing and Key Points of Investigating Phishing Domains

Key points of investigating phishing domains:

- Registrant information
- Payment method
- IP address

The definition and history of phishing have been reviewed. Now let's move on and look at how to investigate when a victim reports a phishing website.

First, an email arrives in the victim's inbox. The victim is then lured into connecting to a phishing website, and upon the first click of the link the victim is connected to a phishing website where they are deceived into inputting private data. The private data is then collected in the suspect's email or database. In short, phishing is organized as a phishing domain, phishing website, and email or database used to collect victims' private data.

As mentioned, the suspect uses URL obfuscation attacks to avoid raising suspicion. For example, the URL of Citibank is www.citi.com. If a suspect were to use www.sitibank.com, everyone would be suspicious and would not access that website. However, if a suspect were to employ www.citi-bank.com or www.citybank.com, people may connect to that phishing website without any question and willingly use the online banking.

Suspects use URL obfuscation to attack the victims. Because of this strategy, the starting point of the investigation will be searching phishing domains. Luckily, we are able to investigate registrant information from the domain registration company or payment solution from the phishing domain—for example, an investigator can trace a credit card payment or wire transfer that occurred when the suspect purchased the domain.

Investigation of Domain Registrant Information

Investigators are able to acquire domain registrant information of the phishing website from the Whois service. The Whois service is a domain name registry officially offered by the network information center (NIC) of each country. For example, the URL and NIC homepage of the United States is www.arin.net.

Please check: http://www.whois.com/whois/

Investigators are also able to acquire domain registrant information by using the homepage of the domain registrant company. If the phishing website's domain is not registered to the NIC or a particular domain registry company, the investigator theoretically would need to visit the NIC homepage of every country or countless homepages of domain registrant companies. Fortunately, however, there are many useful domain lookup sites on the internet which can prevent the trouble of going through so much work just like the Whois service.

For example, visit and review www.domaintools.com. Besides simply offering domain registrant information, this site provides information such as additional domains of the domain owner (the phisher) or other assigned domains of the phishing server.

Investigation of Payment Solution

To inquire about the payment methods, investigators will need an official document or a warrant. If a phisher used a credit card to purchase a phishing domain, the investigator will be able to obtain the credit card number. Through further investigation, the credit card holder's personal information, payment history, and other information can be obtained. If a phisher paid by wire transfer, the investigator would be able to obtain the account number.

> **INVESTIGATING IP ADDRESS OF PHISHING DOMAINS AND ANALYZING PHISHING WEB SERVERS**
>
> - Investigating IP address of phishing domain
>   - Confirming phishing web server
>   - Investigating the user of IP address and web server (name, SSN, mobile phone number, email, payment history)
>   - Confirming the physical location of phishing web server
> - Analyzing phishing web server
>   - Analyzing phishing web page files
>   - Finding an email account or database server which is used to store victims' private data and financial data
>   - Analyzing access logs to confirm where the suspect connected

Investigating the IP Address of a Phishing Domain

In order to convert a domain name into an IP address, use the Nslookup command in the Windows operating system. Click the **start button**, then click **run**, and type **cmd** to use the **NS lookup** command. You will then see the black screen terminal command. Type **nslookup** and then type **the phishing domain next to it with a space**.

For example, if you want to look up an IP of www.google.com, you need to type nslookup www.google.com (notice the space between nslookup and www.google.com). Upon confirming the IP address of the phishing domain, a user of the IP address and a web server (which is allocated in the IP address) should be investigated. User information of the IP address, the physical location of phishing web servers, and payment history from the ISP (internet service provider) or IDC can be investigated.

After verifying the physical location of a phishing web server, confiscate it and analyze the phishing web page files to find the email account or database server. These files store private and financial data. Analyze access logs to verify where the suspect is connected.

Online Child Sexual Exploitation

Policymakers, law enforcement, educators, and parents in the United States and elsewhere have expressed concern about the victimization of children, particularly online. While traditional personal victimization presents its own challenges, the police usually have a single event, a perpetrator, and a victim in a unique time and space—discrete evidentiary clues that

traditional law enforcement strategies can effectively investigate. However, sexual perpetrators "hiding" in cyberspace with their identity disguised can be located anywhere in the world and can victimize hundreds of youths simultaneously (Wolak et al., 2006), and the "cyber clues" represent a different kind of evidence for which traditional law enforcement does not have training or experience. In addition, many children, even preteen youth, are now using the internet unsupervised. The advance of mobile phone technology has expanded the scope of online child sexual exploitation investigations beyond the laptop or desktop devices, and this, coupled with the difficulties of hidden perpetrators and murky evidentiary clues, makes internet crimes against children (ICAC) a challenge for modern law enforcement investigation.

According to the National Internet Safety Survey (NISS-3) there are four types of online child sexual exploitation (Mitchell et al., 2014, p. 1):

- **Sexual solicitations.** Requests to engage in sexual activities or sexual talk or give personal sexual information that was unwanted or, whether wanted or not, made by an adult.
- **Aggressive sexual solicitations.** Sexual solicitations involving offline contact with the solicitor through the mail, telephone, or in person or attempts or requests for offline contact.
- **Harassment.** Threats or other offensive behavior (not a sexual solicitation) sent online to the youth or posted online about the youth for others to see.
- **Unwanted exposure to sexual material.** Without seeking or expecting sexual material, being exposed to pictures of naked people or people having sex when doing online searches, surfing the web, or opening an email or links in an email.

Mitchell and her colleagues (2014) report that approximately 9% of all youth in a nationally representative sample report being sexually solicited online by an adult, 2% self-report aggressive sexual solicitation by an adult (in which the adult offered to meet offline), and 23% self-report receiving unwanted naked sexual images. The only behavior that has increased over time is non-sexual harassment, in which 11% of youth self-reported being victimized non-sexually in cyberbullying or other online harassment (Mitchell et al., 2014).

Online Sexual Predator Typology

According to Hartman, Burgess, and Lanning (1984), four different types of individuals collect sexually abusive images of children:

- **Closet collectors** are offenders who secretly collect abusive images of children without contact.
- **Isolated collectors** collect abusive images in conjunction with contact offenses.
- **Cottage collectors** share their collection and experiences with like-minded others, primarily for validation.
- **Commercial collectors** are those who seek financial benefits from their collection.

Although the typology proposed by Hartman et al. (1984) predates the internet era, similar behavior and motivational drives have been found within the typologies that describe the offenders who commit technology-facilitated crimes against children.

For example, based on the results from a collaborative program between the Australian Institute of Criminology and the Australian High-Tech Crime Centre, Krone (2004) utilized information on the seriousness of three online behavioral factors to identify offender subtypes. The three factors are: (1) the nature of the abuse (either indirect or direct); (2) the level of networking by the offenders; and (3) the level of security offenders employ to avoid detection. Using these factors regarding offenders' progression from abusive online images through the crossover and the commission of offenses, Krone (2004) classified nine groups of offenders: browser, private fantasy, trawler, non-secure collector, secure collector, groomer, physical abuser, producer, and distributor.

Briggs, Simon, and Simonsen (2011) also classified internet-initiated sex offenders who used an internet chat room to either attempt or to entice an adolescent into a sexual relationship. Within internet chat room sex offending, they discovered two subgroups: a contact-driven group and a fantasy-driven group. As the name suggests, a contact-driven group is motivated to engage in offline sexual behavior with minors. In contrast, a fantasy-driven group engages in online cybersex without an expressed intent to meet offline.

Recently, utilizing case files involving offender chat logs, email threads, and social networking posts gathered from state and local ICAC task forces resulting in a sample of 200 offenders, DeHart and colleagues (2017) identified key elements in these cases and proposed a typology of online solicitation offenders. The study categorized the offenders into four groups: cybersex-only offenders, schedulers, cybersex/schedulers, and buyers. The offender characteristics of each category were also introduced.

Even though scholars have proposed numerous ways to classify and understand the characteristics of online child sex offenders, many studies contain, either in part of or in its entirety, the following four criteria: (1) curiosity and impulse-based without physical contact; (2) fulfillment of sexual fantasies without contact; (3) the grooming and facilitation of the contact; and (4) production and distribution of abusive images for financial gain.

Smartphone and Digital Evidence Within the SNS Environment

Image 8.8

Social Networking Sites and Smartphones

The mobile phone is a powerful and vital tool in the daily lives of children and young people, providing easy and continuous access to perform a wide range of activities. Typical uses include but are not limited to accessing the internet, downloading and forwarding pictures or film clips, checking email, listening to music, and playing games.In addition to storing music, taking photos and videos, and sending these to other phones, children can also share this content with other phones via wireless connections. To share information, wireless personal area network technology uses radio waves, providing a freeway for enabled devices (such as phones, computers, and handheld game consoles) in close range of each other. Content can then be posted on social networking sites, sent from phone to phone, or shared using a short-range wireless connection between devices bypassing the phone network altogether.

Popular social networking sites such as Facebook, Instagram, Twitter, and Snap Chat allow users to create personalized homepages, set up blogs and add friends. Typically, social network sites allow users to set up a profile page, list their interests and other details, and enable contact with other users.

Hosted by major social networking service providers, there are many online chatting sites; some smaller independent websites also offer chat rooms. It is common practice for chat rooms to be categorized by interest, age, or location of the users. In real time, chat rooms allow groups of people from across the world to hold text conversations.

Since accounts usually only require an email address to verify a user's identity, public chat rooms can be populated by anyone. More often than not, chat rooms do not carry age verification; thus, children are able to visit chat rooms disguised as adults or an adult disguising as a child. Because chat rooms are not necessarily moderated or monitored, there have been cases of adults using public chat rooms to begin relationships with children and young people to abuse them sexually. Children and young people may be persuaded to give out private information or enter into apparent friendships with people who are lying to them about who they are in order to develop a friendship. Later these unsuspecting children are maneuvered and exploited, or worse.

Identifying a Sexual Predator on Mobile Devices

If a victim receives SMS messages, it can be determined whether the perpetrator has used a mobile phone or SMS service website. It is then possible to check through a mobile network operator as to whether messages were sent via mobile device or website.

Image 8.9

When a mobile phone is used, the investigator can easily identify the cell phone owner. In the case of website usage, however, the investigator must identify the SMS service company used to send the sextortion messages through the mobile network operator. The investigator must then provide the SMS service company with the victim's phone number and the exact time the obscene messages were sent to the victim in order to specify a suspicious account.

Email, Instant Messenger, Chat Rooms, and Social Network Services (SNS)

SNSs are used to initiate sexual relationships, offer a means of communication between victim and offender, access information about the victim, disseminate information or pictures about the victim, and get in touch with the victim's friends (Mitchell et al., 2010). Mitchell and her colleagues (2010) indicated that an estimated 2,322 arrests for internet sex crimes against children involved SNSs in some way, including an estimated 503 arrests in cases involving identified victims and the use of SNSs by offenders (the majority of arrests involved undercover operations undertaken by police).

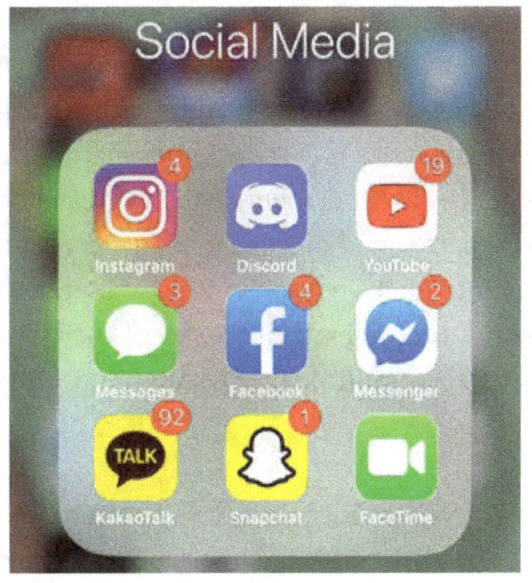

Image 8.10

Upon beginning the investigation, first verify if the suspect used their own account. Confirming who requested child sexual materials is not a simple task. The suspect is likely not going to use their own account. If the suspect uses another person's account, the best course of action to track the suspect is by **analyzing the access log**. Following the execution of a warrant on the email or instant messenger service provider to acquire the access log, you should check the suspect's locations through the internet service provider. Upon obtaining the results, it is not difficult to pinpoint the suspect if the access place is usually the home.

In instances of access occurring in a public location, such as an internet café, the investigator must analyze the PC's web browser histories to find whether the suspect utilized a different account.

When the investigator traces log-in evidence placing a person on a particular website before and after requesting sexual materials, there is reason to believe the suspect likely used the account. The investigator can then cross-reference the account details with the server information to build a case against the suspect.

Forensics on Smartphones

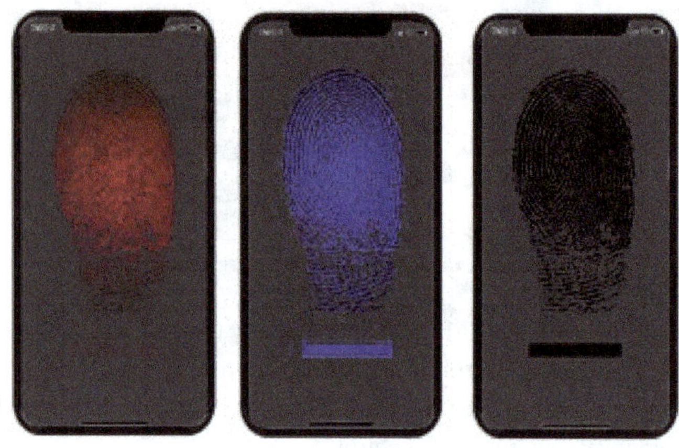

Image 8.11

Forensics on smartphones is relatively new and rapidly changing due to the development of mobile technology. Because of its capability to carry valuable digital or electronic evidence, a smartphone has become a significant part of digital forensics. Like other computers, smartphones are considered a small computer, and information stored in smartphones is admitted. Smartphone evidence is easily manipulated and vulnerable to being changed by the external environment; thus, an investigator must pay special attention when managing digital evidence. The process of digital evidence is quite similar to other digital forensics and keeping the integrity of the forensics is critical for validating the source. As previously emphasized, securing digital evidence is a fundamental step to treat any digital evidence in a court of law. All procedures need to be performed by considering the admissibility of evidence in the court. The mathematical hashing function is an important method used to compare the values between the original and a copy and prove the integrity of the evidence.

As discussed in previous chapters, practical digital forensic investigation follows the steps below:

1. Hash values using a calculated one-way encryption algorithm need to be produced from a disk or file in the process of collecting digital evidence.
2. After producing the hash value, the investigator needs to sign and date to confirm the original hash value's identity.
3. The investigator makes a copy; the copy must then be reconfirmed with the same analysis for proof of authenticity.
4. Finally, the validated copy can be used for performing the digital evidence analysis.

The hash value determines the authenticity of the source. However, this method for securing general integrity cannot be applied to smartphone investigation. Smartphones have become daily necessities as the rapid development of information technology enables mobile phones to store a significant amount of information. Thus, mobile device forensics must be performed to extract digital evidence or recover deleted information from the SIM card and memory card.

Additionally, mobile forensics indicates any data changes in a mobile device. A mobile device has a function to produce a different output when entered data is changed by even one slight difference. Therefore, the integrity of data can be verified as long as the hash value of the data collected from the original target of the investigation matches with the hash value of the copy submitted to the court. PDAs, laptops, and tablet PCs are also considered to be mobile devices, so the same procedure is applied to prove the integrity of digital evidence.

Overview of Smartphone Forensic Procedures

Although smartphone forensic procedures are similar to computer forensics, smartphone communication functions must be carefully considered.

Let's examine "Guidelines on Cell Phone Forensics" (Jansen & Ayers, 2007) listed below:

Site Preservation. If a smartphone is found at the crime scene, the investigator should take pictures to preserve the scene. If a smartphone screen is turned on, the investigator should capture the screen of the smartphone.

Securing Evidence. When the smartphone is secured to be investigated, the network should be cut off to avoid others changing the state of the smartphone over the network. The investigator can use **airplane mode. If the smartphone is on, there needs to be an adequate power supply for the smartphone.**

Data Collection and Analysis. When collecting data from the smartphone, the following legal issues must be considered:

- Is there a risk of evidence destruction?
- Does the investigator need to modify the data on the smartphone?
- Is the smartphone connected to the network?
- What is the scope of the investigation?
- Does the investigator have jurisdiction?

After considering all the above requirements, the investigator can begin data collection. It is also essential to select appropriate methods for assessing the operating system and the smartphone's memory characteristics as the investigator extracts and analyzes data from a smartphone.

Investigation Report. Included in the report should be incident information, the process of obtaining evidence, and the analysis results. Research is underway in terms of search and seizure of smartphones, disconnecting a communication function, and securing the integrity of the electronic information within the smartphone; however, unified procedures have not yet been established.

Smartphone General Information

When the investigator checks the program list or the network connection, the methods of saving the executed command and the results as a file or taking pictures of contents shown in the smartphone for attachment to the report can be used. Since the investigator has attempted to improve the proof of evidence, the suggested methods can be considered a legitimate verification process.

Collecting the Smartphone Load Data

In some cases, it is necessary to collect the loaded data or the temporary memory data processed by the information processing system. It may also be necessary to store the data in the main memory or temporarily store the data on a hard disk while running a specific program. A search and seizure process to collect such data can be completed if the data from the digital medium is temporarily stored for executing certain commands. The collected data should be evaluated and the investigator will then determine if the digital evidence can be considered after reviewing the nature of the collected data.

For example, if the evidence of the existing document is found in the temporary work file, the evidence can serve as the original source.

The 'Smart Revolution' has been facilitated by the increased use of smart devices, social networking services, and machine-to-machine (M2M) communication. The problems associated with limited data storage and management capabilities have increased exponentially.

In the current era of big data, collecting quality evidence is not easy. Depending on the types of digital evidence confiscated, different standards and considerations need to be applied to rebuild the case and reiterate evidence when necessary. Therefore, it is essential to develop a digital forensic method to gather concrete digital evidence from a large amount of information.

However, privacy infringement can be of great concern because of the high-risk of a third party exposing information through the search of big data. Therefore, special care must be taken to prevent invasion of privacy.

The management of data integrity for SNS data, particularly real-time management of the synchronization service, is one of the most challenging issues. Suspects or others can easily create the same account to access, change, and remove SNS data from one device to other devices. Digital forensic techniques need to invest extra consideration on a method of minimizing the privacy issue, since ensuring the integrity of digital evidence on SNS is very difficult.

Interventions for Online Child Sexual Exploitation

According to the United Nations (UN) (2019), the issue of online child sexual exploitation and the sale of a child as a commodity both within and across national borders has been growing at an alarming rate. Rapid technological advancement in communication and information sharing online enables offenders to connect and share encrypted information with one another. Essentially, these technological developments help to increase efficiency while lowering the potential risks. By utilizing the dark web, these offenders also create more burdens and challenges for law enforcement. In efforts to prevent further technology-facilitated crimes against children and to identify offenders while fostering a deeper understanding of the issue, several worldwide intervention approaches have taken place along with national, civil, and private approaches.

International Approaches

Historically, laws did not consider children as the bearers of rights or requiring legal protection. Over the course of the last century, however, many countries have recognized children as either a deserving party or a special protection party in legal codes. Furthermore, several international instruments require countries to implement measures to protect children from abuse and exploitation while engaging in international cooperation in the investigation and prosecution of crimes against children (UNODC, 2015). For example, the UN's Convention on the Rights of the Child (CRC) sets the minimum standard for the protection of children from harmful influences, abuse, and exploitation (UN, 1989). The minimum standard includes the right to an education based on equal opportunity.

While the CRC sets the standard, in 2002 the Optional Protocol to the CRC on the Sale of Children, Child Prostitution, and Child Pornography (OPSC) emphasized the importance of adopting and implementing criminal legislation on child sexual abuse and exploitation. This instrument requires countries who have ratified or acceded the OPSC to implement not only criminal code for crimes but also establish criminal, civil, or administrative accountability of legal persons concerning child sexual abuse and exploitation (UN, 2002).

In addition to the CRC and the OPSC, United Nations Convention Against Transnational Organized Crime (UNODC, 2003) considers penal matters and a range of provisions concerning international cooperation against transnational organized crime, including information and communication technology. The United Nations Economic and Social Council also adopted the Guidelines on Justice in Matters Involving Child Victims and Witnesses of Crime (2005) to ensure full respect for the rights of child victims and witnesses of crime. The ramifications of these international approaches to national legal implementations, however, vary among countries. For example, the United Nations Office on Drugs and Crime's (2013) study on cyber-crime indicated that while 80 percent of countries in Europe report sufficient criminalization of cybercrime acts, 60 percent of countries in other regions of the world report insufficiency of criminal code regarding cybercrime.

The issue is prevailing in the case of the production of child sexual abuse material. Many countries have criminalized the production of child sexual abuse material; however, there is no gold standard on the definition of the term child, and the elements of the crime are not concrete (UNODC, 2013). Furthermore, even though many technology-facilitated crimes against children overlap with existing crimes (i.e., trafficking, cybergrooming, cyber-solicitation, cyberstalking, cyber-harassment, and exposure to harmful content), many countries still apply general crim-inal offenses rather than implementing cyber-specific criminal laws.

There is also an international task force comprising a select cadre of 53 online child sexual exploitation investigators and law enforcement experts representing 48 countries. This task force is known as the Violent Crimes Against Children International Task Force (VCACITF), formerly known as the Innocent Images International Task Force (Halpern, 2019). The VCACITF serves as the largest task force of its kind in the world to formulate and deliver a dynamic global response to online child exploitation. This coalition approach enables the task force to coordinate and operate strategically through the extensive use of liaison and operational support (Halpern, 2019).

U.S. Approaches

Within the United States, the Internet Crimes Against Children (ICAC) Task Force, the Federal Bureau of Investigation (FBI), and the National Center for Missing and Exploited Children (NCMEC) jointly collaborate to identify, investigate, prosecute, and prevent crimes against children online. In general, the 61 ICAC task forces, located in all 50 states, and their affiliates, initiate their investigation when the task forces receive information regarding technology-facilitated crimes against children through CyberTipline. The CyberTipline was created by the

NCMEC. The NCMEC is a private, non-profit organization designated by Congress to serve as a national clearinghouse on issues related to missing and exploited children and works in cooperation with the United States Department of Justice (USDOJ) and other federal, state, and local law enforcement agencies, as well as education and social service agencies, families, and the public (USDOJ, 2016).

Image 8.12a

Image 8.12b

According to the USDOJ's 2016 report to Congress regarding child exploitation prevention and interdiction, 344,801 complaints were filed between the fiscal year of 2010 and 2015 across all 61 ICAC task forces. Among the lead agencies for the 61 task forces, ten agencies filed more than 10,000 complaints during the six years examined. The highest number of complaints was filed in the California-Los Angeles Police Department (15,208), followed by Ohio-Cuyahoga County Prosecutor's office (13,772), Pennsylvania-Delaware County District Attorney's Office (13,613), Michigan-Michigan State Police (11,423), Hawaii-Hawaii Department of Attorney General (11,176), New York-State Police (11,102), North Carolina-North Carolina State Bureau of Investigation (10,815), New York-New York City Police Department (10,392), Texas-Dallas Police Department (10,325), and Texas-Office of Attorney General of Texas (10,201).

Once a victim files a complaint through the CyberTipline, NCMEC staff will review the case, analyze the content, add relevant publicly available information, and make the report available to law enforcement agencies for independent review and possible investigation (USDOJ, 2016). Generally, ICAC task forces and their affiliates receive reports from the CyberTipline and initiate investigations. According to Mitchell and Boyd (2014), who surveyed 144 investigators from ICAC task forces and affiliate agencies, most of their respondents initiate investigations using manual and human-centric technological tools. For instance, investigators start their investigation by tracking known websites such as online classifieds, niche websites, and search engines. They also monitor specific characteristics of websites or advertisements that fit targeted crime characteristics such as photos or buzz words.

Since the perpetrators of technology-facilitated child sexual exploitation do not solely rely on one type of technology, investigators also utilize various technologies to identify a case or a perpetrator. Investigators inspect a wide range of everyday communication platforms such as social networking sites, instant messaging, email, SMS or text messaging, and underground communication channels like the darknet. Even though it was not the case for the majority, some investigators initiate their examination using more sophisticated technologies like face recognition programs (Mitchell & Boyd, 2014).

On top of the initial investigation, ICAC task forces and their affiliates also perform forensic examinations. Between 2010 and 2014, ICAC investigators in all 50 states performed forensic examinations on 253,778 cases. Investigators also conducted in-person interviews with victims when it was necessary. According to Roby and Vincent (2017), the majority of children and adolescents who have been exploited in commercial sex acts have typically experienced prior abuse, neglect, or other forms of trauma. Therefore, investigators are also trained to work with the victims through various training opportunities often provided by non-profit organizations.

After thorough investigations by each agency, a total of 41,851 arrests were made between 2010 and 2015. Once an arrest is made, ICAC task forces can refer these cases for further investigation in their state legal system or refer these cases to the United States Attorneys' Offices for prosecution. More recently, the coordinated operation known as Broken Heart arrested more than 2,300 suspected online child sex offenders between March and May in 2018, and more than 1,700 suspected online child sex offenders between April and May in 2019. According to the USDOJ (2018; 2019), the operation targeted suspects who:

- Produce, distribute, receive, and possess child pornography;
- Engage in online enticement of children for sexual purposes;
- Engage in the sex trafficking of children; and/or
- Travel across state lines or to foreign countries and sexually abuse children.

During the course of these two operations, the ICAC task forces investigated more than 25,200 and 18,500 complaints, respectively. Considering the highest number of filed complaints between 2010 and 2015 were 86,390 complaints in 2015, the Broken Heart operation also demonstrates a drastic increase in the CyberTipline complaints and a high number of arrests within the short time frame of its operations.

Private Sector Approaches

Since regulating the flow of content on the internet is impractical to any government, assistance from the private sectors is critical to combat technology-facilitated child sexual exploitation. Perpetrators require internet access to commit these crimes involving a child. Therefore, the participation of internet service providers (ISPs) in the prevention and regulation approaches are crucial. Though several countries, including the United States, mandate ISPs to report any identified sites containing child sexual abuse materials to the police within a reasonable period, many countries do not have policies that help regulate ISP behavior (UNODC, 2015). The UNODC (2015) also revealed that the ISP industry itself is moving toward drafting a formal code of conduct that requires members to refrain from "knowingly accepting illegal content on their sites and to expediently remove content when they are alerted to its existence" (p. 50).

Other private sector organizations also willingly install specific programs that aid in disrupting the spread of known child sexual abuse images online, such as PhotoDNA technology. Microsoft donated PhotoDNA technology to help online service providers and others to disrupt the spread of known child sexual abuse images online in 2009 (USDOJ, 2016). The technology works by creating a unique signature, like a fingerprint, for a digital photograph. Once a signature, also known as a hash, is created, the signature will contain the essential characteristics of the image. Generally, a hash can be altered once a simple change like resizing of the file occurs. However, the hash created by PhotoDNA can consistently match the signature across extensive data sets regardless of alterations. Currently, NCMEC has the legal rights to sublicense the technology for free to domestic and foreign email service providers (ESPs) who are interested in taking proactive steps to identify and eliminate child pornography from their servers.

Additionally, there are several international coalitions seeking to disrupt and dismantle technology-facilitated child sexual exploitation. The International Center for Missing and Exploited Children's (ICMEC) Technology Collections is a voluntary collaboration of nine major internet companies to develop and execute plans for technology-based solutions to identify, disrupt, and possibly dismantle child exploitation criminal enterprises (UNODC, 2015). Similarly, ICMEC's Financial Coalitions against Child Sexual Exploitation, which consists of numerous leading banks, credit card companies, electronic payment networks, third-party payment companies, and internet service companies, assists in identifying, disrupting, and imposing curbs on the economics of commercial child sexual abuse materials.

Furthermore, according to the UNODC (2015), many private sector organizations implemented a wide range of internal policies and external obligations concerning domestic and foreign law enforcement data requests. For instance, to protect the privacy of consumers while cooperating with investigations, most of the major social media companies enable public access to law enforcement guidelines, which specify procedures to request information regarding a target of investigation (Choi, 2015).

Summary

The increasing trends of reporting suggest the severity and prevalence of technology-facilitated crimes against children. To understand the trends and the characteristics of offenders, this chapter provided an overview of technology-facilitated crimes against children, offender typology, and various intervention and prevention approaches. The shifting nature of online child exploitation cases and the challenge they pose to investigative resources make it particularly important to monitor trends and developments in offenders' nature and methods. According to the NCMEC, the CyberTipline has received more than 57 million reports, including more than 18.4 million in 2018 alone (NCMEC, 2020). Most reported incidents are related to abusive images of children, online enticement, child sex trafficking, and child sexual molestation. Due to an exponential rate of evolution in technology and the rise of technological availability to children, however, efforts to effectively and comprehensively combat technology-facilitated crimes against children will require a collective approach. As Merdian, Perkins, Webster, and McCashin (2019) suggest, it is also critical to invest in resources that can help address the knowledge gaps that currently limit individuals' abilities to investigate online or transnational child sexual abuse. Lastly, the process of combating such crimes requires the active involvement of children, families, communities, governments, members of civil society, and private sectors.

Lab 8: Social Media

Mr. Evil Noodle decided to utilize the social networking site Twitterfall to scope out a potential victim. Mr. Evil Noodle was tracking tweets in real-time and watching what people around him were talking about. After some time passed, Mr. Evil Noodle spotted a user named Curry talking to a friend. At that point, he chose to monitor Curry's account. While stalking Curry's account, Mr. Evil Noodle decided to make the first move by tapping 'Like' and posting comments on one of Curry's tweets. Curry quickly responded to Mr. Evil Noodle's comments and decided to 'follow' him. Over the next few months, the two got close and shared a lot of personal information. Specifically, Curry shared sexually explicit photographs with Mr. Evil Noodle. Not long after this, Curry's photographs and personal information (i.e., name, telephone number, address, and email address) were exposed on different social media. During this same time period, many of Curry's friends, family members, and associates received emails containing semi-nude photos of her. In addition, various Twitterfall accounts sent similar seminude photos to some of Curry's contacts and several fake dating profiles associated with Curry were created. Curry reported the matter to authorities and unfollowed Mr. Evil Noodle on Twitterfall. As a result, Mr. Evil Noodle threatened repercussions for Curry and his family if he did not respond, especially through social media. Mr. Evil Noodle then proposed a deal that if Curry wired him $100,000, all communications between them would cease.

Below is a useful tool to track tweets posted by others in real time. Users can search keywords and they can also search by geolocation.

Twitterfall is a UK-based website designed to allow users of the social networking site Twitter to view upcoming trends and patterns posted by users in the form of tweets. Twitterfall takes advantage of Twitter's search trends (listed on the Twitter search page), which reveal the topics that are currently most popular and most discussed at that time.

Please answer the following questions (2–3 pages):

1. For the purposes of this exercise, how did Mr. Evil Noodle use Twitterfall to acquire all of Curry's personal information?
2. How could Mr. Evil Noodle share the acquired information with other criminals or cybercriminals?
3. If Mr. CIC and Mr. Hero were to investigate this case, how would they go about doing so?
4. Prevention: How would one stop Mr. Evil Noodle from acquiring personal information from Twitterfall?
5. Discussion: What are your thoughts on this type of case? What would your reaction be as a victim?

Lab Instruction

Search your address on Twitterfall and you will see tweets by people nearby.

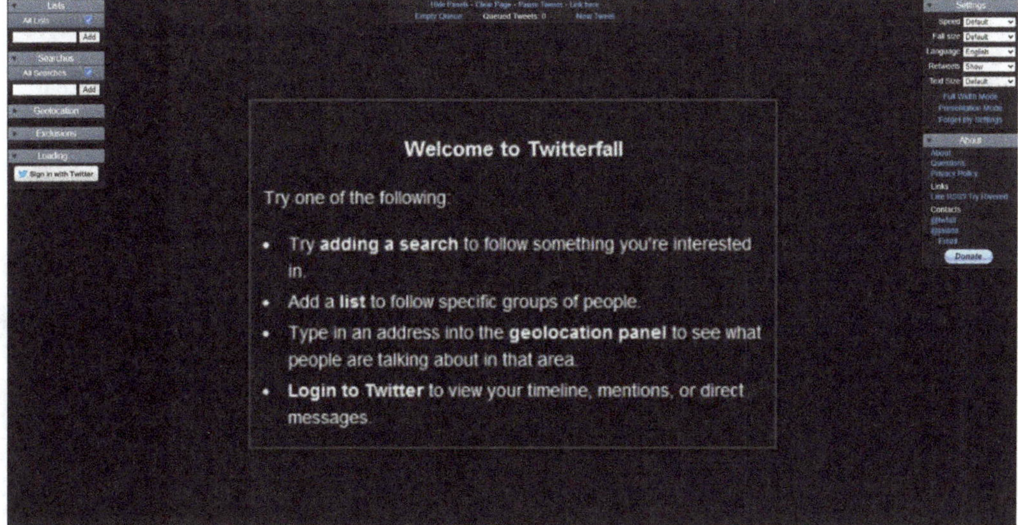

Image 8.13

References

Briggs, P., Simon, W. T., & Simonsen, S. (2011). An exploratory study of internet-initiated sexual offenses and the chat room sex offender: Has the internet enabled a new typology of sex offender? *Sexual Abuse: A Journal of Research and Treatment, 23*(1), 72–91.

Choi, K. S. (2015). *Cybercriminology and digital investigation.* LFB Scholarly Publishing.

DeHart, D., Dwyer, G., Seto, M. C., Moran, R., Letourneau, E., & Schwarz-Watts, D. (2017). Internet sexual solicitation of children: A proposed typology of offenders based on their chats, e-mails, and social network posts. *Journal of Sexual Aggression, 23*(1), 77–89.

Halpern, M. (Host). (2019, September 3). Violent crimes against children international task force expands. [Audio podcast episode]. In *Inside the FBI.* Federal Bureau of Investigation. https://www.fbi.gov/audio-repository/inside-podcast-vcac-international-task-force-090319.mp3/view

Hartman, C. R., Burgess, A. W., & Lanning, K. V. (1984). Typology of collectors. In A. W. Burgess & M. L. Clark (Eds.), *Child pornography and sex rings* (pp. 90–109). Lexington Books.

Jansen, W., & Ayers, R. P. (2007, May 30). *Guidelines on cell phone forensics.* National Institute of Standards and Technology, Special Publication. https://www.nist.gov/publications/guidelines-cell-phone-forensics

Krone, T. (2004). *A typology of online child pornography offending. Trends and Issues in Crime and Criminal Justice, 279,* 1–6.

Merdian, H. L., Perkins, D. E., Webster, S. D., & McCashin, D. (2019). Transnational child sexual abuse: Outcomes from a roundtable discussion. *International Journal of Environmental Research and Public Health, 16*(2), 1–14.

Mitchell, K. J., & Boyd, D. (2014). *Understanding the role of technology in the commercial sexual exploitation of children: The perspective of law enforcement.* Crimes Against Children Research Center. https://scholars.unh.edu/cgi/viewcontent.cgi?article=1036&context=ccrc

Mitchell, K. J., Finkelhor, D., Jones, L. M., & Wolak, J. (2010). Use of social networking sites in online sex crimes against minors: An examination of national incidence and means of utilization. *Journal of Adolescent Health, 47*(2), 183–190.

Mitchell, K. J., Jones, L. M., Finkelhor, D., & Wolak, J. (2014). Trends in unwanted online experiences and sexting: Final report. Crimes Against Children Research Center.

Moore, R. (2011). *Cybercrime: Investigating high-technology computer crime* (2nd ed.). Anderson Publishing.

National Center for Missing & Exploited Children. (2020). *Exploited children statistics.* http://www.missingkids.com/footer/media/keyfacts

Roby, J. L., & Vincent, M. (2017). Federal and state responses to domestic minor sex trafficking: The evolution of policy. *Social Work, 62*(3), 201–210.

United Nations. (1989). *Convention on the rights of the child adopted by the general assembly of the United Nations.* https://www.ohchr.org/en/professionalinterest/pages/crc.aspx

United Nations. (2002). *Optional protocol to the CRC on the sale of children, child prostitution, and child pornography.* https://www.ohchr.org/EN/ProfessionalInterest/Pages/OPSCCRC.aspx

United Nations Convention on the Rights of the Child. (2019). *Guidelines regarding the implementation of the optional protocol to the convention on the rights of the child on the sale of children, child prostitution and child pornography.* https://www.ohchr.org/Documents/HRBodies/CRC/CRC.C.156_OPSC%20Guidelines.pdf

United Nations Economic and Social Council. (2005, July 22). *Guidelines on justice in matters involving child victims and witnesses of crime.* ECOSOC Resolution 2005/20.

United Nations Office on Drugs and Crime. (2003). *Convention against transnational organized crime.* https://www.unodc.org/unodc/en/organized-crime/intro/UNTOC.html

United Nations Office on Drugs and Crime. (2013). *Comprehensive study on cybercrime.* https://www.unodc.org/documents/organized-crime/cybercrime/CYBERCRIME_STUDY_210213.pdf

United Nations Office on Drugs and Crime. (2015). *Study on the effects of new information technologies on the abuse and exploitation of children.* https://www.unodc.org/documents/Cybercrime/Study_on_the_Effects.pdf

United States Department of Justice. (2016). *National strategy for child exploitation prevention and interdiction: A report to Congress.* https://www.justice.gov/psc/file/842411/download

United States Department of Justice. (2018, June 12). *More than 2,300 suspected online child sex offenders arrested during operation "Broken Heart"* [Press release]. https://www.justice.gov/opa/pr/more-2300-suspected-online-child-sex-offenders-arrested-during-operation-broken-heart

United States Department of Justice. (2019, June 11). *Nearly 1,700 suspected child sex predators arrested during operation "Broken Heart"* [Press release]. https://www.justice.gov/opa/pr/nearly-1700-suspected-child-sex-predators-arrested-during-operation-broken-heart

Wolak, J., Mitchell, K., & Finkelhor, D. (2006). Online victimization of youth: Five years later. Report from Crimes Against Children Research Center, University of New Hampshire. Retrieved from http://www.unh.edu/ccrc/pdf/CV138.pdf

Credits

Investigative Techniques on Interpersonal Crimes Based on Youth Population

Marlon Mike Toro-Alvarez

Introduction

Interpersonal crimes include the concept of aggression. Aggression is a human's thoughtless reaction to an attack. More importantly, aggressiveness can put coexistence at risk. Aggressive behavior usually generates a violent reaction in the person being subjected to the aggression, making them feel violated in their personal environment and their way of interacting in society (Gallego, 2010). This interaction includes all the environments in which the human being relates, such as cyberspace.

Some forms of violence are related to the intimacy of the victim, others against thought, against social recognition, and against different moments in the life of that human being. Online interactions today include all the above scenarios and break traditional statements about human relationships such as those by Brantingham and Brantingham (1984) who stated that human interactions and distance between people have an inversely proportional relationship (i.e., if there is more distance there is less interrelation).

With the increase in human interrelationships supported by information technologies, there are signs of the aggression and violence which shape many different types of online interpersonal victimization (Choi & Toro-Alvarez, 2017). This victimization carried out by interpersonal crimes in cyberspace can harm a victim's honor, reputation, and good name. Likewise, it can affect the victim's life in the physical world. According to Lipton (2011), victims can confuse their fantasies with their physical life and be stalked at work because of the interactions they have on social networks. There are even records of sexual abuse investigations where offenders allege that they were motivated by private online encounters with their victims. The concern of this kind of victimization scenario increases when the victim has suffered from personality disorders or their maturity as an individual has been affected by traumatic events or physical limitations. However, there is a large population that meets these two characteristics of vulnerability to interpersonal crime in cyberspace (Roberts, 2008).

According to Harter (1993) children and adolescents have not fully developed their self-esteem, and their degree of maturity to make decisions individually is still in the process of strengthening. These conditions lead us to approach interpersonal crimes in cyberspace from a more specific investigation perspective. Therefore, this chapter presents the most common forms of aggressors and interpersonal cyber-victimization among young people along with lessons learned in the management of this class of victims.

New Cyber-Offenders

Interpersonal crimes in cyberspace are not committed only by adults against minors. Minors also perpetuate cycles of offense where there are permanent and sometimes fatal consequences for victims (Choi & Lee, 2017). Among these offensive behaviors in cyberspace are happy slapping, cyberbaiting, and motivation towards suicide through high-risk challenges.

Happy Slapping

Happy slapping is a deviant behavior that centers around violence. Here young offenders record on cell phones incidents of pushing, hitting, or street fighting. Some episodes of happy slapping have included rape, with the young offenders recording sexual abuse. This type of victimization has been common among closed communities and has claimed the lives of young people and adults (Mann, 2009).

Some useful characteristics for identifying happy slapping are:

- Presence of physical aggression;
- Group practice with a minimum of one offender who plays the role of aggressor and another who films the aggression;
- The victims in most cases are not known to the assailants;
- A random process to choose the victims;
- Potential victims are those of the same age as the aggressors, beggars, or persons with cognitive disabilities;
- Permanent digital storage of the victimization;
- Facilitated by costly technological devices (smartphones, tablets, etc.); and
- Social media usage and instant messaging applications.

Cyberbaiting

Cyberbaiting is a kind of deviant behavior also known as cyber-harassment directed at teachers. According to Garrett (2014), cyberbaiting consists of recording teachers to ridicule them. The offenders engage in verbal aggression and insults to provoke a negative reaction from their teachers. Once the provoked teacher reacts, the situation is recorded. Then the cyber-offenders publish this to victimize the teachers.

Some useful characteristics for identifying cyberbaiting are:

- The offenders are youth and are usually students of victimized teachers;
- Frequently, there is a violation of a teacher's personal data;
- Cyberbaiting takes place in academic-related environments; and
- The offender's anonymity is difficult to maintain in this kind of victimization.

Online High-Risk Challenges

Online high-risk challenges are another kind of deviant behavior. Young offenders seek to influence others to engage in high-risk activities. The participants use social networks to publish multimedia evidence of having carried out the risky activity on social networks (Meza, 2008). There are different forms of victimization motivating these challenges, but they can result in wounds, permanent physical sequelae, or disease. Unfortunately, some challenges can also lead to suicide (Prinstein, 2008).

Below are some examples of online high-risk challenges.

Blue whale challenge. It was an online closed game directed to adolescents. Gamers were invited through social networks. The game motivates players to complete for 50 days with 50 challenges assigned by the game administrator. The administrators used false profiles on social networks. The challenges were physically dangerous, with the tasks frequently involving self-harm which culminated in the player committing suicide by jumping from a building at the end of the game. For each test completed, the player had to record multimedia evidence showing that they had completed the challenge (Sumner et al., 2019).

Coronavirus challenge. It motivates the young population to lick the surfaces of public toilets, record it on video, and upload it to social networks such as TikTok (Toro-Alvarez, 2020).

Momo's challenge. This challenge starts with seeing a statue of a girl with a frightening appearance. Then, players are encouraged to hurt themselves, generating negative psychological impact and, in some cases, leading to suicide. The players receive a link that provides access to the information on the cell phone. According to some reports, the risks could be mainly related to the security of the users' cell phone data. Like the blue whale challenge, the participants are mostly adolescents who are more easily influenced by trends and social media content (Singh et al., 2020).

Suffocation/fainting/loss of consciousness challenges. This type of challenge seeks to generate a strong emotion from the participant or cause fainting. Often, children suffocate other children, forcefully press their chest, or hyperventilate. This is obviously very dangerous and could cause death (Madge et al., 2008).

Soap or Tide Pods challenge. This challenge involves children biting down on a detergent capsule and posting videos of the challenge on social media. The outer layer of the capsules is designed to dissolve and release the contents very quickly, which causes chemical burns and kidney and lung problems when ingested (McCarthy, 2018).

Online Young Victims

Along with the expansion of social networking sites, the use of smartphones has also increased. This increase has allowed more children to be connected with one another, making them more vulnerable to different forms of victimization in cyberspace. This risk of victimization becomes even more worrisome when parents, guardians, and custodians are unaware of the risks (Rezgui & Marks, 2008). Another factor to consider is the care provided by investigators and authorities when an interpersonal crime occurs in cyberspace. In some cases, this attention is scarce or not useful. For this reason, some forms of victimization are presented below in order to help tutors and researchers to identify the main characteristics of some forms of cyber-victimization.

Gossip

Gossip is also called crime against honor, and it involves offenders stating false imputations or wrong information about a victim on internet sites and/or social networks. This kind of cyber-victimization falls under traditional crime and punishment standards in most countries. Not only is the slander/perpetrator punished, but the one who publishes and who reproduces by any means the insults or slander expressed by another also faces punishment (Kenyon, 2019).

This is one of the most common crimes on the web because it is very easy to create blogs and fake accounts on social networks, WordPress pages, or anonymous web accounts in order to publish information against a victim's reputation. Some relevant characteristics in this kind of victimization are:

- There are multiple fake or new accounts on social networks;
- There is some relationship between perpetrators and victims; and
- Revenge may be the offender's main motivation.

Cybergrooming

Online grooming is defined as an intentional approach or harassment of a boy, girl, or adolescent, which is carried out by an adult for sexual purposes or for indirect economic victimization of a child's family. For example, stealing the credit card numbers of parents through the internet, social networks, or video game consoles. Cybergrooming potentially represents an imminent danger to the young population.

According to Anderson et al. (2012), cyber-perpetrators create fake profiles in some social networks, chat rooms, forums, or video game platforms where they pose as young people. The mechanism they usually use is sending photographs, initially of their faces. They then escalate from innocent pictures to images of a sexual nature. The cybercriminals then begin asking children for photos of their private parts. Once cyber-perpetrators have the first photos or videos, they start blackmailing. The types of blackmail range from public photography on the internet to talking to parents to asking the children for money if they do not agree to send more photos or have a personal meeting.

This type of criminal modality triggers other types of traditional crimes such as child pornography, human trafficking, and extortion (Choi & Toro-Alvarez, 2017).

Cyberstalking

Cyberstalking occurs when an individual harasses another individual on the internet using various modes of victimization such as email, chat rooms, newsgroups, and the World Wide Web (Choi, 2015).

Cyberstalkers can also obtain personal information about their victims from the internet and use this information to meet their victims in person. Cyberstalking takes different forms, such as an email containing a threatening message, spamming, live chat harassment (verbal abuse online), sending electronic viruses, and tracking someone else's computer and internet activity (Ogilvie, 2000).

Cyberstalkers are usually men, and cyberstalking victims are more likely to be women and children (LeClair, 2015). According to Choi (2016) there are four different types of stalkers:

- **Distorted paranoid.** Despite the fact that harassment does not exist, victims say they are being harassed. Victims distort the situation and point to someone as a stalker.
- **Affective obsession.** There is no previous relationship between the victim and the offender. The victim is generally well known to the media. This type of perpetrator has a problem in their social life and usually has paranoid schizophrenia.
- **Simple intensive.** The victim and the perpetrator are familiar with each other, and the intention of the stalking tends to be revenge.
- **Affection delirium.** Criminals have fantasies about the victim's love for them. In order to make the visions real, violators could use violence. There are few cases of such violence resulting in actual physical harm. (Choi, 2016)

A common example of this kind of victimization is an event in Kings County, Washington. Over a period of less than four months, a 23-year-old suspect sent 269 emails and voicemails to a 17-year-old girl asking her to send him nude photos and threatening to kill her and her family if she did not comply (KIRO News Seattle, 2010).

According to the police, the suspect stalked the girl on MySpace, where the victim had disclosed a large amount of personal information. The harassment continued for several months before police involvement (Hinduja & Patchin, 2008a).

Cyberbullying

Bullying at school has a long history. Many generations can tell stories of how they or someone they grew up with was bullied at school. Familiarity with this topic can cloud the ability to effectively deal with the growing problem of cyberbullying. Many adults' response to cyberbullying is very nonchalant and unflappable; they feel that bullying is a part of growing up. The problem with this approach is that bullying has changed significantly since the turn of the century and has increasingly negative effects throughout a victim's life. These events can motivate young people to vandalism, for example (Motta & Toro-Alvarez, 2017).

Bullying in schools may not be a new concept, but the difference today is that children cannot escape their bullies (Smith et al., 1999). If a child is being bullied at school, it is very likely that the torment will continue through social networks, emails, and even text messages when the child returns home. This type of unrelenting bullying is what has led some children to kill themselves.

While bullying has been around for generations, cyberbullying is a problem that has been developing in recent decades, as the use of information and communication technologies has been increasing. The term is said to have originated with either Canadian politician Bill Belsey or U.S. lawyer Nancy Willard (2003). Whatever its origin, online bullying and careful study of that behavior date back to at least the 1980s (Holfeld & Grabe, 2012).

Today, the biggest problem with cyberbullying is not about its treatment, but how to clearly define it. As with any crime, you need to have a clear definition so that there is no confusion when it comes to prosecution. This, however, has been quite difficult. For example, Hinduja and Patchin define cyberbullying as, "deliberate and repeated harm inflicted through the use of computers, cell phones, and other electronic devices" (2010, p. 208).

According to Choi et al. (2019), a definition accepted by some academics describes cyberbullying as an aggressive or intentional act that is repeatedly and overtly carried out by a group or an individual against a victim who cannot easily defend themselves. Bullying is a form of abuse that is based on an imbalance of power. Bullying can also be defined as a systematic abuse of power. Using these definitions for bullying, we can extend them to define cyberbullying.

Cyberbullying, therefore, can be defined as an aggressive and intentional act carried out by a group or an individual through repeated electronic forms of contact and over time against a victim who cannot easily defend themselves. Cyberbullying has become more prevalent in recent years due to the increased use of electronic devices, such as computers and smartphones. Bullying is no longer about the strong singling out the weak in the schoolyard; the physical assault has been replaced by an online beating 24 hours a day, 7 days a week. Students can use instant messaging, emails, chat rooms, and websites they have created to humiliate a selected victim. Parents can no longer count on the telltale physical signs of bullying (black eye, bloody lips, or ripped clothes), but the damage caused by cyberbullies is no less real and can be infinitely more painful (United Nations News, 2014).

The other problem that comes along with cyberbullying involves limiting the freedoms and obligations of citizens. Many may argue that what someone writes on an online social media site is protected by the right to freedom of expression: "The judiciary has long struggled to balance freedom of expression with the darker side of digital communication" (Holladay, 2010).

Types of Cyberbullying

According to Toro-Alvarez (2018), cyberbullying comes in many different forms, some of which are more difficult to detect than others. Below are the top five types of cyberbullying.

Harassment. Repeated spam messages, regardless of whether they are explicitly offensive or not, are a form of harassment. Online stalking, where a person's online activities are constantly monitored, can cause psychological harm and fear. Bullying and harassment can occur in various forms online and may or may not be a continuation of offline harassment or lead to physical harassment and stalking.

Threats and intimidation. Serious threats can be sent to both staff and students via mobile phones, email, and comments on websites, social networking sites, or message boards.

Vilification or defamation. Cyberbullying can include posting disturbing or defamatory comments or insults about an individual online. This may also include general insults. For example, students can use their mobile phones or email accounts to send sexist and racist messages.

Rejection or exclusion of peers. Online exclusion may be more difficult to detect than when exclusion occurs in an enclosed space, such as a classroom where adults are present. Social networking sites like Facebook and Twitter offer a platform for young people to establish an online presence and talk to other members of the network, which is configured as an important extension of a young person's space and social activity. Most social networking sites operate as closed communities, allowing only contact between members, so it is common for only a small number of social networking sites to be popular with students at a school. It is also possible for a group of students to establish a closed group that can protect them from unwanted contact. However, this also means that excluding someone is easily accomplished by simply not answering messages, removing select contacts from friends lists, or using ignore functions (against the person to be excluded). This situation can have negative impacts on young victims and can lead to extremely damaging behaviors and emotions.

Publishing or sending personal or private information or images. Videos or images can be transmitted between mobile phones, either over a local wireless connection, sent via text to other phones, uploaded to websites, or posted on public video hosting sites. Most young people are aware of the phenomenon known as happy slapping, a term that has been used to refer to physical attacks that are generally recorded and circulated through mobile phones. People who log the attacks can actively participate in cyberbullying. Circulation of images of the attacks can also be a form of intimidation and certainly complement the victimization caused by the initial attack.

Characteristics of Cyberbullying

Impact

The perpetrators of cyberbullying can be multiple and easily accessible to the victim. This means that the degree and severity, as well as the possible risks and repercussions, must be evaluated differently in each case. If the cyberbullying is shared on mobile phones or posted online, it is difficult to control who can see it or have copies of it. The victim will maintain a permanent concern about more cyberbullying events, even if there has been a successful intervention by relatives or authorities. This is a particularly significant feature of cyberbullying which differentiates it from other forms of bullying (Huus, 2011).

Furthermore, a single incident can be experienced as multiple attacks. For example, a humiliating video posted on the web can be copied to many different sites. A single instance of bullying, creating a nasty website or sending a personal email, can have long-term and repeated consequences, since content removed from the internet may reappear or be recirculated (Choi & Toro-Alvarez, 2017).

Victims and Perpetrators

According to Petrosino et al. (2010), children and youth are not the only ones who can be subject to cyberbullying. School administrative officials have also been victims of cyberbullies and have suffered from bullying. Victims do not need to have special physical characteristics, which means that it is not necessary to be stronger, taller, or greater than the target to be a cyberbullying perpetrator. A victim doesn't even need to share the same physical space as the bullying person.

Bystanders of cyberbullying can easily become perpetrators by passing or showing others an image designed to humiliate another child or official or by recording an assault or bullying on a mobile phone and circulating it. As with other forms of bullying, it is important for the entire school community to understand the responsibility of reporting cyberbullying and supporting the person being bullied. It is advisable that anti-bullying policies also address bystanders who actively support cyberbullying incidents and establish penalties for this kind of conduct (Hoff & Mitchell, 2009).

Location

Cyberbullying can take place at any time and can intrude on spaces that could have previously been considered safe or personal. The victim is left feeling that there is no place to hide and that an attack is possible at any time. Sending abusive text messages, for example, makes cyberbullying possible at any time of the day or night, and the goal of cyberbullying can be achieved in a victim's own home (Patchin & Hinduja, 2013).

Anonymity

According to Armstrong and Forde (2003), perpetrators can remain anonymous and this is extremely worrying for those who are being bullied. Although the person being bullied may know that their cyberbully is within their circle of friends (e.g., at their school), they may not know the real identity of the perpetrator (cyberbully). This can upset victims, making them suspicious of all their relationships.

Motivation

Some cyberbullying incidents can be clearly intentional and aggressive. However, some cases of cyberbullying have been unintentional, and the perpetrators did not think or were not fully aware of the consequences. Online behaviors are generally less inhibited than offline behavior, and some children may say things to others online that they would not have said or done in the physical world (Dooley et al., 2009).

Evidence

Unlike other forms of offline bullying, many cyberbullying incidents can, by themselves, act as evidence, such as in the form of text messages or computer screenshots. It should be noted that these records, in addition to alerting about the possible materialization of an intimidation incident, may also help identify the possible perpetrator. An unpleasant text message, for example, will contain the message, the date and time it was sent, and information about the offender's phone (Law et al., 2012). The last part of this chapter presents a guideline for addressing this kind of investigation.

Applied Lessons for Investigators

According to Bhat (2008), anti-cyberbullying measures can be taken at the policy level and the level of case investigators. In the policy arena, the investigator should be familiar with the social and legal infrastructure against cyberbullying. If there is no such cyberstalking legislation, the possibility of criminalizing such an act should be seriously considered. At the investigator level, victims need to be understood and have confidence that law enforcement will help capture the cybercriminal and restore a sense of cybersecurity. In addition, specialized knowledge of digital evidence must be acquired. Here, it is important to consider previous learning paths which can help to co-create applicable knowledge against interpersonal cyber-victimization.

Behavioral Analysis of Cyberstalking

It is important to examine the behavioral analysis of cyberstalking. The reason the investigator should consider a behavioral analysis is to establish the direction of the case by judging the types of harassment from previous cases and, eventually, to effectively protect victims. The four main types of cyberstalking behavior are presented below (Reyns et al., 2011).

Collecting information from the victim. The offender, in order to broaden the range of harassment, uses the internet to gather information about the victim. To obtain more personal information from the victim, the offender could make hacking attempts against the victim and the victim's acquaintances.

Defamation. Defamation is an effective intimidation tool that can have a significant impact on the victim. False information spread by the media, such as the internet or emails to the victim's acquaintances, will lead to despair. Meanwhile, cyber-offenders will enjoy the victim's suffering.

Disguising oneself as the victim. The offender will disguise themselves to torture the victim and use a bulletin board, post a threatening message, or send pornography as if it came from the victim so that the victim suffers in the community.

Threatening. This is accomplished through the use of telecommunication devices, constantly contacting and threatening the victim or their family members. This includes sending pornography or calling regardless of the time of day to induce stressful situations.

In many legal systems, understanding what types of behaviors constitute a crime is a very basic element in establishing the correct direction of the investigation of the case. The laws of each country may differ, but, in general, there are three behaviors that together constitute the crime of cyberstalking (Choi & Toro-Alvarez, 2017).

Intentional threat carried out through a communication network. The threat must be delivered through the internet, telephone, fax, or other telecommunications device in any form, such as sound, text, image, or video. Any form of a file that is transferable over the internet is included. The threat is not limited to threatening the life or physical integrity of the victim but includes a threat to property, liberty, fame, or credit. The threat does not have to be feasible nor must it be carried out by the cybercriminal. The threat may also be about the intention of a third party to harm the victim.

Messages causing fear or anxiety. The contents of the message are threatening in some way (e.g., a scary sound, acoustics, strange illustration, or image of a murder scene). If such communication is repeatedly transferred to the victim and, as a result, the recipient/victim feels fear or anxiety, then this behavior constitutes cyberstalking. If such content does not reach the victim and the victim feels no fear or anxiety at all, then the content itself will not be punishable.

Repetitive and continuous. The behavior must be frequent and permanent—this is one of the most distinctive characteristics of cyberstalking. A one-time threat is not considered punishable as cyberstalking. At least two or more successful deliveries of a threat are required for them to apply. The criteria for judging repeatability depend on the condition of each person, taking into account the need to protect the recipient.

Detecting the Means of Cyber-Victimization

Cyber-victimization can take many forms, among which are six main categories: mobile phone, email, instant messaging, chat rooms and message boards, social networking sites, and gaming sites (Kowalski et al., 2014).

Mobile Phone

Today, children and young people use their mobile phones for more activities than just talking and sending text messages. The most common uses, apart from talking and texting, including accessing the internet, downloading and forwarding images or movie clips, checking email, listening to music, and playing games. In fact, the wide range of activities carried out by phones, together with the management of different social networks by young people, makes the phone a powerful and important tool. In addition to being able to store music, take photos and videos, and send files to other phones, children can also share this content with other phones through wireless connections. In this way, cyber-intimidators abuse these devices by making unpleasant calls, sending malicious text messages, recording and sharing humiliating images, and filming and sharing acts of intimidation and assault through mobile phone cameras (Toro-Alvarez, 2018).

Email

This is an essential part of the working life of most people. Cyberbullies can send threatening messages by email or repeatedly send spam messages. Inappropriate images or video clips may be transmitted. Personal emails may be sent inappropriately. Most computer viruses are forwarded by email (Casey, 2000).

Instant Messaging (IM)

Instant messaging applications allow the user to see which contacts are online while using the computer. They also offer chat services using text. Like social networking sites, instant messaging services work among a network of people who have signed up for the same service and authorize mutual permission to see and talk to each other when they are online.

Unlike chat rooms, which are generally public and open to anyone who has subscribed to the chat service, instant messaging services are more private and are generally one-on-one conversations. According to Hinduja & Patchin (2008b) cyberbullies can use instant messaging to send unpleasant messages or content to other users. People can also hack instant messaging accounts and send bullying messages to contacts.

Chat Rooms and Message Boards

There are many online chat sites that are hosted by major telecommunication service providers as well as smaller independent websites. Chat rooms are generally organized by user interest, age, or location. Chat rooms allow groups of people from all over the world to have text conversations in real time. These public chat rooms can be occupied by anyone since accounts only require an email address to associate the possible identity of a user. Most chat rooms do not carry age verification; therefore, children can visit chat rooms disguised as adults. Chat room exchanges tend to be less inhibited than when people first meet in the real world. Disgusting or threatening messages can be sent without the recipient of these messages necessarily knowing about the authors of the messages.

On the other hand, chat rooms are not necessarily moderated or monitored, so there have been cases of adults using public chat rooms to initiate relationships with children and young people and subsequently sexually abusing them. Children and young people may be persuaded to give out private information or enter into friendships with people who are lying to them about who they are; the cyber-offender then uses the friendship to exploit the victim (Choi, 2016).

Social Networks

The most popular social networking sites, such as Facebook, allow users to create their own homepages, set up blogs, and add friends. Social networking sites often allow users to set up a profile page, list their interests and other details, and allow contact with other users. Many social networks focus on interests and services. They can also provide blogs or website creation tools.

Social networking sites are designed to help people find and make friends and make it easy to stay connected. However, social networking sites can be abused in various ways. Most of those sites allow nasty comments to be posted; however, other sites allow users to review or approve content before it is displayed.

Similarly, users can use their own sites to spread rumors or make unpleasant comments about others or post humiliating images or videos of them. Fraudulent profiles are also quite common, and these could be used to pretend to be someone else in order to intimidate or harass (Solove, 2007).

Games on the Web

Young people spend a significant amount of time using technology to play a wide variety of online video games. Computer games can be accessed through online gaming sites where conversation is facilitated between players around the world, or on handheld consoles that use a wireless connection to allow people in the same location to play against each other. As with other programs that allow people to communicate with each other, there have been instances of intimidating calls and abusive and derogatory comments. Additionally, the players can choose weaker or less experienced users and repeatedly kill their character. Activated wireless consoles can be used to forward spam messages to other compatible devices (Choi, 2015).

Tracking Cyberbullies

It is important to recognize cyberbullying as a serious crime. In the case of cybercrimes, by default, the offender has no physical contact with the victims and therefore tends not to be considered as serious as sexual assault, physical intimidation, and other similar acts of traditional violent crime.

Investigators can only have a superficial understanding of their seriousness because they tend to neglect the pain and suffering of victims, just as in other investigations. The cyber investigator must have the trust of the victims so that they can feel safe and a level of cooperation can be successfully achieved.

On the other hand, it is difficult to trace the cyberbully because the behavior is often initiated by an anonymous figure through contact via text message or chat. In recent years, this prosecution has become increasingly difficult as suspects use more complex techniques to circumvent police investigation.

The first step in a successful case is information provided by the victim. Investigators should understand the detailed situation and take note of the victim's exposure to internet use and the use of other means of connectivity. Like traditional research, cyberspace researchers also need to review acquaintances and potential enemies. Based on the information collected, it is crucial to develop a primary range of suspects, such as the number of offenders and the approximate relationships with the victim.

In the second step, efforts should focus on the types of media used in the situation. This could lead to contacting the suspect. Examples of these potentially useful means include internet sites, blogs, emails, bulletin boards, instant messages, mobile phones, telephones, and video games.

In the third step, identify the type of harassment and thoroughly check whether it has been done by the victim's acquaintances. Harassment can be done through theft of personal information, defamatory content, threats, delivery of sexually explicit material, SMS, or repeated calls. If the investigator has reason to assume that it is the act of an acquaintance, they may need to focus on the victim's use of the internet and see if the suspect is available. However, if no acquaintances or few related acquaintances are found in the investigation, the victim or witness statements would not help much in the search for the suspect. In this case, the investigator will be limited to tracking the communication records.

When investigators first meet the victim, it is important to verify whether any erased or missing evidence can be restored. In many cases of cyberbullying, the initial stage of the victim's contact largely does not indicate that the act may be offensive. Therefore, the information at this stage can be easily neglected. For example, a single man who sends a love letter, email, text message, gift, or flowers would not be considered offensive. One of the difficulties in this initial phase is that victims often eliminate that communication out of fear or embarrassment. When the victim finally reaches out to the police for help, there may not be enough information that can be helpful in charging the offender. A cyberbully tends to repeatedly deliver threats to the victim; therefore, even if the investigator has a lack of focal evidence, they need to prepare for the next offense.

The investigator must secure any information related to the case and use it as the basis for tracking the suspect. Quickly securing communication records is the key to the investigation because internet service providers and mobile communication providers will not keep records for an extended period of time. Slow response time can result in lost evidence. In addition, the investigator must exhaustively collect existing data and produce a copy for analysis (Toro-Alvarez, 2018).

From the Victim's Side

According to Lumsden and Morgan (2017), before starting the investigation, investigators should advise students and school personnel to try and keep a record of the abuse, particularly the date and time, the content of the message, and, where possible, the sender's ID (e.g., name of the user, email, mobile phone number, the web address of the profile or content). For example, it is recommended to take an exact copy or recording of the entire website address as it will help the service provider locate relevant content. The evidence provided by the service provider will assist in any investigation of cyberbullying. It may also be helpful to show what has happened to those who may need to know (e.g., parents, teachers, and after-school care staff). Also, it is always helpful to keep a written record, but it is best to keep bullying evidence, if any, on the device itself.

Mobile Devices

On mobiles, you must ensure that the person being bullied records and saves any message, be it audio, image, or text. Unfortunately, forwarding messages, for example to a staff member's phone, will cause loss of information from the original message, such as the sender's phone number.

Emails

The person who is being cyberbullied should be asked to print and send any email messages to the investigative team. The victim should be encouraged to continue sending and saving subsequent messages. Preserving the whole message and not only the body is more useful since it will contain headers and information where the message comes from.

Instant Messaging

Some instant messaging services allow users to record all conversations. The user can also copy and paste, save, and print these messages. When the user reports to the service provider or even the police, copied and pasted conversations are less useful as evidence, as this can be easily edited. Conversations recorded or archived by the instant messaging service are best for evidence. Conversations can also be printed on paper or certain sections can be saved with a screenshot.

Chat Rooms

In chat rooms, you must print the page or take screenshots of it. To take a copy of what appears on the screen, press the Control and Print Screen keys simultaneously on the computer keyboard, and then paste it into a word processing document or graphics display program. The screenshot can also be done from mobile devices (the easiest way to execute these records should be consulted in the manuals of each manufacturer).

Social Networks

On social media sites, video hosting sites, or other websites, the user or the investigator must maintain the site link, print the page, or create a screenshot of the page and save it.

Identifying the Bully

Although technology allows anonymity, there are ways to find information on where cyberbullying originated (LeClair, 2015). However, it is important to note that this may not necessarily lead to an identifiable individual. For example, if the offender has used the phone or school network account of another person, the location where the information was originally sent will not itself determine who the offender is. There have been cases of people using another individual's phone, or "hacking" their email account or school messaging, to send unpleasant messages.

On Mobiles

When a victim has received text or multimedia messages, you can check whether the offender has used a mobile phone or messaging service website. It is possible to verify through a mobile network operator the source of the intimidating messages. In the case that the source is a mobile phone, the investigator can only track the owner of the mobile phone. However, if the source is a website, the investigator must first identify the messaging service company that was used through the network operator. The investigator can then specify suspicious accounts, providing the messaging company with the message delivery time and the mobile phone number of the victim or recipients.

In this process, the investigator must first verify whether the offender used their own account. Sometimes, it is a very simple job to confirm who sent bullying messages. However, the perpetrator may recognize cyberbullying as an illegal act and therefore may avoid using their own account.

If the perpetrator uses someone else's account, the easiest way to track the cyber-offender is by analyzing the access log. After executing an order in the email or instant messaging service provider to acquire the access log, the locations of the cyberbully can be verified through the internet service provider.

If the place of access is a residence, it is not difficult to specify the offender; however, if the access point is a public place, such as an internet café, the histories of the PC's web browsers should be analyzed to determine if the cyber-intimidator has used another account.

If the investigator finds evidence that someone was connected to another website before and after the bullying message was sent, it is likely that the account was used by the perpetrator. Therefore, the investigator may be able to point to the cyber-offender by referring to who used the account and where the account was connected.

Finally, it is necessary to remember that the investigation of cyber offenses depends on the amount of information that can be collected as digital evidence. Victims are the primary source of this information; hence, the investigators must earn their trust and communicate clear instructions on the relevance of their collaboration. The procedures carried out by the investigator depend on the kind of technology that cybercriminals use; however, the digital devices used by service providers permanently record the behavior of their users, which ultimately allows successful digital investigations (Toro-Alvarez, 2018).

Means to Trace

This section provides some tips for tracking down a suspected cyberbully. Cyberbullies can use multiple means of communication, so it is essential to gain knowledge and experience in dealing with each network of subscribers. Five subscriber networks are usually linked to cyberbullying: (1) email; (2) text messaging; (3) blogs, websites, or newsletters; (4) telephony; and (5) fax networks (Choi, 2016).

Various types of information can be found from email services. Investigators can obtain concrete evidence to support the charges and trace the physical location through access records. Sometimes, personal information about the subscriber can be found. Even if the offender used someone else's account, the access records and transmission details may still belong to the offender. Therefore, such records may be the optimal test for moving forward with the investigation.

Whenever possible, the investigator should obtain subscriber details and access records from the email service provider. However, in many cases, criminals will try to circumvent the police investigation and try to disguise their true identity or try to erase their tracks. Some email service providers include the sender's information within the email itself. In such cases, the investigator should check the email header that has such information; it could be very helpful in tracking down the suspect.

Text Messages

In recent years, crimes using mobile phones have been on the rise. Mobile phone features have advanced rapidly, so harassers can easily send pornographic photos or video files or multimedia text messages. Mobile phone network providers often support CID (caller identification) on the recipient's mobile phone; however, criminals may also be able to manipulate the caller's ID in text messages or calls.

If a stalker speaks on the phone, the victim can recognize the offender or the investigator can track the stalker by following the phone records. However, criminals can manipulate the caller ID in the text message or calls, making it more difficult to track them down.

Also, the offender does not have to use mobile phones to send those messages to others, because many internet services provide mobile text messages. In this case, tracing the offender could be a difficult task. Either way, the investigator should obtain as much information as possible when stalkers are using text messaging.

Blogging, Websites, or Newsletters

Newsletters or blogging sites are primarily featured in criminal defamation cases. In such events, the investigator should be aware that the investigation could cause further harm to the victim's privacy. In general, tracking articles in newsletters and blogs is similar to tracking email; however, the investigator should keep in mind that articles published in blogs can be created by a different person than the one who publishes the article. In many cases, articles are linked from somewhere else or copied and pasted from somewhere else. Therefore, the investigator should collect the post as evidence, keep it intact, and work with the blog service operator if necessary.

Telephony

Cell phones, home landlines, public phones, and the internet can be used as stalking media. It is highly recommended for the investigator to keep a list of such service providers or approach them to quickly find out who is the provider of that kind of service in each area.

In countries where the caller pays all communication fees, such as South Korea, voice operators on the caller side will keep the number of the receiver, but the carrier on the receiver side cannot keep the record of the caller's number. When caller identification (CID) service is available, the caller can easily be found. However, in cases where the CID service does not apply or the CID itself is tampered with, the carrier will have to be found on the caller's side. For example, if the criminal used VoIP (Voice over Internet Protocol) or calls over the internet, the cell phone carrier or landline will have no records about the caller.

Fax

The internet can also be used to send faxes. Many investigators already understand such possibilities, but still seem to forget that faxes can also have an online origin. One more thing to consider is scheduled faxing: it should be noted that the offender does not have to be in front of the fax machine at the time of transmission.

In addition, the victims must be trained in preserving the content of the communication. In many cases, the investigator may have to describe, in detail, how to capture the screen in the web browser and what to do when receiving the email or mobile phone messages from the cyberbully. If the investigator does not take such steps, deletion can erase the content, and restoring the data can be extremely difficult. The investigator may even have to take steps to examine a postal mail or package to the victim.

According to Choi and Lee (2017), teenagers who engaged in risky social networking activities (such as providing personal information to SNS strangers and accepting strangers as friends) were at a higher risk of being victimized by traditional bullying and cyberbullying.

Teenagers who engage in risky social networking site activities are also likely to commit cyber-interpersonal violence.

It is important to examine cybercrime victims' social media activities as a criminal investigator. As we all know, you can view an individual's pictures, videos, and conversations through social media such as YouTube, Facebook, Twitter, and Instagram.

Those postings offer timestamps, locations, friends, their relationships, and much more. Advanced technology such as AI is already assisting cybercrime investigation tasks using machine learning to narrow massive amounts of digital evidence such as pictures, message contents, and media. The current digital forensic technology can assist a crime investigator in finding a place to start connecting all the potential dots from finding evidence to conducting interviews and arrest proceedings.

It would be helpful for the investigator to provide the victim with contact information for the investigation team 24 hours a day, 7 days a week so that the victim can notify the investigator when additional harm is attempted. Providing this information is a good way to give confidence to the victim and also allows a faster response against the cybercriminal's actions (LeClair, 2015).

Summary

New cyber-offenders: The chapter introduces the most common forms of interpersonal cyber-victimization among young people as well as presents useful information for a formal investigation. From conceptual description to behavioral analysis, the reader finds applied knowledge to counteract interpersonal crime in cyberspace.

Happy slapping: Commonly young offenders use cell phones to record incidents of pushing, hitting, street fighting, or rape. It is important to be alert to this kind of behavior because its victimization can include serious injuries to the victims.

Cyberbaiting: This victimization is directed at teachers. Some characteristics of cyberbaiting were identified to assist investigators and profilers.

Online high-risk challenges: Young offenders can influence peers to engage in high-risk activities and publish them on social networks. There are diverse examples of this kind of challenge and it is highly advisable to be alert in order to prevent fatal victimization related to this deviant behavior.

Online young victims: Some types of victimization focus on the young. Attacks can be facilitated by risky interactions in social networking sites and the ignorance of this victimization vector. Hence, some concepts about these forms of victimization are introduced.

Gossip: Cyber-offenders can affect the reputation of the victim by stating false imputations on the internet. It is also known as a crime against honor and can be prosecuted under traditional law without requiring cybercrime special procedures.

Cybergrooming: Cyber-perpetrators create fake profiles in cyberspace and reach young victims in order to gain their trust, then other ways of victimization are perpetrated.

Cyberstalking: Investigators can identify four different types of cyberstalkers: (1) distorted paranoid, (2) affective obsession, (3) simple intensive, and (4) affection delirium.

Cyberbullying: This kind of cyber-victimization is described, including five types of cyberbullying and its main characteristics. It is important for investigators to consider impact, location, anonymity, motivation, and possible evidence.

Learned-lessons for investigators: Details in regards to the investigation of interpersonal crime in cyberspace are mentioned. From profiling to tracing cyber-perpetrators, this chapter introduces the reader to a deep understanding of online victimizations.

Behavioral analysis of cyberstalking: Four main types of cyberstalking behavior are analyzed: (1) collecting information from the victim, (2) defamation, (3) disguising oneself as the victim, and (4) threatening.

Detecting means of cyber-victimization: There are six main vehicles for cyber-victimization: (1) mobile phone, (2) email, (3) instant messaging, (4) chat rooms and message boards, (5) social networking sites, and (6) gaming sites.

Tracking cyberbullies: Investigators can deploy three main steps in order to trace cyber-bullies: (1) collecting information provided by the victim, (2) detecting types of media used in the situation, and (3) identifying the type of harassment.

This learning activity helps to consolidate knowledge about interpersonal crime in cyber-space. Five questions will be applied to a hypothetical school environment.

Lab 9: Mr. Evil Is a Bully—Cyberbullying Characterization Lab

Mr. CIC (our expert investigator) is receiving a lot of emails from local school parents and teachers. Because of the number of received emails, it is not possible for Mr. CIC to review all of the information. Fortunately, Mr. CIC is not working alone. You have been asked to help profile a possible pattern of cyber-victimization. Please review your knowledge about stalkers, bullies, and cyberbullying before analyzing the emails.

1. You are prepared to find victimization clues about online affective obsession. Indeed, it is possible to find some characteristics related to that obsession in the school emails. Could you please choose from one to three affective obsession characteristics from the following list? (You can choose more than one answer.)
   a. The victim is well known in the school.
   b. There is no previous relationship between the victim and the offender.
   c. The suspect perpetrator has a problem in their social life and usually has paranoid schizophrenia disorders.
   d. None of the above.
2. Maybe there is some information about online simple intensive victimization. Could you please identify some of those characteristics from the following list?
   a. The victim had some previous relationship with the perpetrator.
   b. The perpetrator shows revenge motivation.
   c. All of the above.
   d. None of the above.
3. You are also prepared to find victimization clues about cyberbullying. Indeed, there are emails with evidence of the common three types of cyberbullying. Could you please choose those types from the following list? (You can choose more than one answer.)
   a. Harassment and repeated spam messages
   b. Vilification or defamation

    c. Threats and intimidation

    d. Affection delirium

4. Some emails can be related to rejection or exclusion of peers. Is it a cyberbullying characteristic?

    a. No, it is not.

    b. Never, cyberbullying would not include that aspect.

    c. Maybe, when exclusion is related to university professors.

    d. Yes, it is and, unfortunately, online exclusion may be more difficult to detect than when exclusion occurs in an enclosed space, such as a classroom where children are present.

5. Some emails are related to adults who, in social networks, are pretending to be teenagers in order to get closer to the young victims. This cybercriminal technique is well known as

    a. Gossip

    b. Cyberbaiting

    c. Bugchasing

    d. Cybergrooming

References

Anderson, G., Ktoridou, D., Eteokleous, N., & Zahariadou, A. (2012). Exploring parents' and children's awareness on internet threats in relation to internet safety. *Campus-Wide Information Systems, 29*(3), 133–143.

Armstrong, H. L., & Forde, P. J. (2003). Internet anonymity practices in computer crime. *Information management & computer security, 11*(5), 209–215.

Brantingham, P. J., & Brantingham, P. L. (1984). *Patterns in crime.* Macmillan.

Casey, E. (2000). *Digital evidence and computer crime.* Academic Press.

Choi, K. (2015). *Cybercriminology and digital investigation.* LFB Scholarly Publishing LLC.

Choi, K. (2016). CJ 710 Applied digital forensic investigation [classroom material]. Study guide (Module 3). Boston University.

Choi, K. S., Earl, K., Lee, J. R., & Cho, S. (2019). Diagnosis of cyber and non-physical bullying victimization: A lifestyles and routine activities theory approach to constructing effective preventative measures. *Computers in Human Behavior, 92,* 11–19.

Choi, K. S., & Lee, J. R. (2017). Theoretical analysis of cyber-interpersonal violence victimization and offending using cyber-routine activities theory. *Computers in Human Behavior, 73,* 394–402.

Choi, K., & Toro-Alvarez, M. M. (2017). *Cibercriminología: Guía para la investigación del cibercrimen y mejores prácticas en seguridad digital.* UAN Fondo Editorial.

Dooley, J. J., Pyżalski, J., & Cross, D. (2009). Cyberbullying versus face-to-face bullying: A theoretical and conceptual review. *Zeitschrift für Psychologie/Journal of Psychology, 217*(4), 182–188.

Gallego, M. M. Á. (2010). Prácticas educativas parentales: Autoridad familiar, incidencia en el comportamiento agresivo infantil. *Revista Virtual Universidad Católica del Norte,* (31), 253–273.

Garrett, L. (2014). The student bullying of teachers: An exploration of the nature of the phenomenon and the ways in which it is experienced by teachers. *Aigne, 5*(1), 19–40.

Harter, S. (1993). Causes and consequences of low self-esteem in children and adolescents. In Roy F. Baumeister (Ed.), *Self-esteem* (pp. 87–116). Springer.

Hinduja, S., & Patchin, J. W. (2008a). Personal information of adolescents on the internet: A quantitative content analysis of MySpace. *Journal of Adolescence, 31*(1), 125–146.

Hinduja, S., & Patchin, J. W. (2008b). Cyberbullying: An exploratory analysis of factors related to offending and victimization. *Deviant Behavior, 29*(2), 129–156.

Hinduja, S. & Patchin, J. W. (2010). Bullying, cyberbullying, and suicide. *Archives of Suicide Research, 14*(3), 206–221.

Hoff, D. L., & Mitchell, S. N. (2009). Cyberbullying: Causes, effects, and remedies. *Journal of Educational Administration, 47*(5), 652–665.

Holfeld, B., & Grabe, M. (2012). An examination of the history, prevalence, characteristics, and reporting of cyberbullying in the United States. In Q. Li, D. Cross, & P. K. Smith (Eds.), *Cyberbullying in the global playground: Research from international perspectives* (pp. 117–142). Wiley Blackwell.

Holladay, J. (2010). Cyberbullying. *Teaching Tolerance.* http://www.tolerance.org

Huus, K. (2011, December 28). Bullied girl's suicide has ongoing impact. *NBC News.* http://usnews.nbcnews.com

Kenyon, A. T. (2019). Libel, slander, and defamation. In T. Vos & F. Hanusch (Eds.), *The International Encyclopedia of Journalism Studies* (pp. 1–8).

KIRO News Seattle. (2010, October 15). Man charged in prolific cyberstalking case; police search for more victims. http://www.kirotv.com

Kowalski, R., Giumetti, G., Schroeder, A., & Lattanner, M. (2014). Bullying in the digital age: A critical review and meta-analysis of cyberbullying research among youth. *Psychological Bulletin, 140*(4), 1073–1137. https://doi.org/10.1037/a0035618

Law, D. M., Shapka, J. D., Hymel, S., Olson, B. F., & Waterhouse, T. (2012). The changing face of bullying: An empirical comparison between traditional and internet bullying and victimization. *Computers in Human Behavior, 28*(1), 226–232.LeClair, D. (2015). The victim's contribution to the crime problem. [classroom material]. Study guide #3. Boston University.

Lipton, J. (2011). Combating cyber-victimization. *Berkeley Technology Law Journal, 26,* 1104–1155.

Lumsden, K., & Morgan, H. (2017). Media framing of trolling and online abuse: Silencing strategies, symbolic violence, and victim blaming. *Feminist Media Studies, 17*(6), 926–940.

Madge, N., Hewitt, A., Hawton, K., Wilde, E. J. D., Corcoran, P., Fekete, S., van Heeringen, K., De Leo, D., & Ystgaard, M. (2008). Deliberate self- harm within an international community sample of young people: Comparative findings from the Child & Adolescent Self-Harm in Europe (CASE) Study. *Journal of Child Psychology and Psychiatry, 49*(6), 667–677.

Mann, B. L. (2009). Social networking websites: A concatenation of impersonation, denigration, sexual aggressive solicitation, cyber-bullying or happy slapping videos. *International Journal of Law and Information Technology, 17*(3), 252–267.

McCarthy, C. (2018). Why teenagers eat Tide Pods. *Harvard Health Blog.* https://www.health.harvard.edu/blog/why-teenagers-eat-tide-pods-2018013013241

Meza, R. S. (2008). Violencia interpersonal en una sociedad tradicional. Formas de agresión y de control social en Chile: Siglo XIX. *Revista de Historia Social y de las Mentalidades, 12*(2).

Motta, D., & Toro-Alvarez, M. M. (2017). Social innovation articulators to counter threats to public safety. *Revista Logos Ciencia & Tecnología, 8*(2), 24–34. https://doi.org/10.22335/rlct.v8i2.315

Ogilvie, E. (2000). Stalking: Legislative, policing and prosecution patterns within Australia. *Research and Public Policy Series, 34.* https://www.aic.gov.au/publications/rpp/rpp34

Patchin, J., & Hinduja, S. (2013). *Words wound: Delete cyberbullying and make kindness go viral.* Free Spirit Publishing.

Petrosino, A., Guckenburg, S., DeVoe, J., & Hanson, T. (2010). What characteristics of bullying, bullying victims, and schools are associated with increased reporting of bullying to school officials? *Issues & Answers, 92.*

Prinstein, M. J. (2008). Introduction to the special section on suicide and nonsuicidal self-injury: A review of unique challenges and important directions for self-injury science. *Journal of Consulting and Clinical Psychology, 76*(1), 1.

Reyns, B. W., Henson, B., & Fisher, B. S. (2011). Being pursued online: Applying cyberlifestyle–routine activities theory to cyberstalking victimization. *Criminal Justice and Behavior, 38*(11), 1149–1169.

Rezgui, Y., & Marks, A. (2008). Information security awareness in higher education: An exploratory study. *Computers & Security, 27*(7–8), 241–253.

Roberts, L. (2008). Jurisdictional and definitional concerns with computer-mediated interpersonal crimes: An analysis on cyber stalking. *International Journal of Cyber Criminology, 2*(1).

Singh, S., Thapar, V., & Bagga, S. (2020). Exploring the hidden patterns of cyberbullying on social media. *Procedia Computer Science, 167,* 1636–1647.

Smith, P. K., Madsen, K. C., & Moody, J. C. (1999). What causes the age decline in reports of being bullied at school? Towards a developmental analysis of risks of being bullied. *Educational Research, 41*(3), 267–285.

Solove, D. J. (2007). *The future of reputation: Gossip, rumor, and privacy on the internet.* Yale University Press.

Sumner, S. A., Galik, S., Mathieu, J., Ward, M., Kiley, T., Bartholow, B., Dingwall, A., & Mork, P. (2019). Temporal and geographic patterns of social media posts about an emerging suicide game. *Journal of Adolescent Health, 65*(1), 94–100.

Toro-Alvarez, M. M. (2018). *Programa de entrenamiento integral de prevención y contención del cibercrimen contra niños, niñas y adolescentes.* Escuela de Postgrados de Policía (ESPOL).

Toro-Alvarez, M. M. (2020). *Stability and changes after pandemic.* Escuela de Postgrados de Policía (ESPOL).

United Nations News. (2014, March). Child trafficking, exploitation on the rise, warns UN expert. https://news.un.org/en/story/2014/03/463842-child-trafficking-exploitation-rise-warns-un-expert.

Willard, N. (2003). Off-campus, harmful online student speech. *Journal of School Violence, 2*(1), 65–93.

The History of Cybersecurity

Jennifer LaPrade and Sinchul Back

Introduction

With each technological advance, cybercriminals also advance their capabilities to invade computers, steal data, and create havoc in our society. Therefore, cybersecurity plays an increasingly vital role in the critical protection of our computers, our data, our infrastructure, and many growing aspects of our lives. Cybersecurity is a concept many people assume is relatively new; however, cybersecurity has a history than spans over more than 50 years. This chapter will provide an overview of the history of cybersecurity.

Early Stages of Cybersecurity
Early Days of Computing

In the early days of computing, computers consisted of gigantic machines that filled up entire rooms, were not connected to a network, and only a handful of carefully selected personnel had access to them or even knew how to operate them. Therefore, there was little to no threat of cyberattack or intrusion. In 1949, John von Neumann was the first to speculate that computer programs could reproduce, marking the first time computer viruses were mentioned as a theoretical possibility (Szor, 2005).

In the 1960s, computers became more widely used in businesses, government, and corporations to store data and perform complex calculations, but they were still huge mainframes locked behind closed doors with only a small number of employees with the technical knowledge necessary to operate the machines. There was still no network or internet that others could access from outside the computer; however, computer owners began to institute more physical measures to limit unauthorized access, such as passwords, as technical knowledge grew. Stricter fire-safety measures so the computers and data stored inside of them would be safe in the case of a blaze could also be considered early versions of cybersecurity.

The 1960s also brought the first "hacking" reference, but it was not related specifically to computer hacking. The term "hack" actually originated in 1961 when members of the Signals and Power Committee of Massachusetts Institute of Technology's (MIT) Tech Model Railroad Club "hacked" into their high-tech train sets in efforts to improve the trains' functionality (Levy, 1984). These same MIT students eventually moved on to "hacking" computers; however, their goal was again to improve the systems and make them more rapid and more efficient.

The first known network computer hacking took place in 1967 when IBM brought their new computer to students of a computer club at a Chicago high school. The computer was connected with a system through dial-up modems, which was revolutionary at the time. The students were given full access to the computers and in no time they had cracked the code and were able to gain access to restricted parts of the system. This hacking work by the students caused IBM to realize the importance of strengthening the security of the network, claiming gratitude for "a number of high school students for their compulsion to bomb the system" (Rosenzweig, 1988, p. 1532).

Increased Computer Vulnerability

As society learned more about computers in the 1970s and how they could be used to improve efficiency in many aspects of people's lives, more computer vulnerabilities were also being discovered. In 1967, researchers began to develop the groundbreaking ARPANET (Advanced Research Project's Agency Network) which served as the precursor to the internet and was the first time computers in different states across the country were connected through a packet-switching network (Cerf, 2009). By the 1970s, ARPANET was functional, but the computers did not speak the same language; therefore, they could not communicate with each other.

In 1971, ARPANET researcher Bob Thomas discovered it was possible to gain access to computers across the network and created the first ever computer worm which would move through the network and show up unexpectedly on computer screens (Guice, 1998). This worm was called "Creeper" because of the message it would display on any infected screen: "I'm the creeper: catch me if you can."

In response, Ray Tomlinson wrote a code called "Reaper" that chased the "Creeper" code and deleted it, marking the creation of antivirus software (Linden, 1976). Even though this may seem harmless, Thomas and Tomlinson's work revealed that despite how useful it can be to connect computers across a network, it also exposes the computers to additional vulnerabilities providing more access points and needs for additional security measures, perhaps those that have not yet been invented. Therefore, the cybersecurity era began.

The U.S. federal government was quick to realize how vulnerable open access computing could be to outside troublemakers, causing U.S. Representative Martin Russo to propose the first cybersecurity legislation called the Federal Computer Systems Protection Act on August 4, 1977. The bill sought to "make a crime the use, for fraudulent or other illegal purposes, of any computer owned or operated by the United States, certain financial institutions, and entities affecting interstate commerce." However, the bill failed to gain enough support to even go to a vote.

Nonetheless, the 1970s made cybersecurity an important part of the conversation in the fast-evolving technological era. Government officials, corporations, and scholars began to see the need to make sure cybersecurity developments kept up with technological advances, with one author stating, "Security has become an important and challenging goal in the design of computer systems" (Linden, 1976).

Developing Cybersecurity

If cybersecurity was born in the 1970s, it began to develop and mature in the 1980s. Unfortunately, so did the cyberattacks. Over the years that followed, computers became more and more connected, computer viruses became more advanced, and cybersecurity systems struggled to keep up with the continuous bombardment of new and innovative hacking techniques. But in the 1980s, there were many advancements in the world of cybersecurity.

First, the U.S. government became more focused on cybersecurity. In 1983, a movie called *War Games* was released. It brought national attention to hacking and the damage and havoc that could be created by any curious individual from the comfort of their own bedroom (Fuller, 2019). The movie starred two normal teenagers who hacked into the military's nuclear missile program and almost started World War III.

After viewing the film, then U.S. President Ronald Reagan was so alarmed by the prospect that the military's computer systems could be that vulnerable that shortly after watching the movie he asked his top military commanders if an intrusion like that was really plausible. A week later, after some research, the general responded, "Mr. President, it's worse than you think" (Fuller, 2019, p. 167). And the administration began extensive work on new and improved cybersecurity measures, especially in the Department of Defense.

The movie also sparked new legislation against cyberattacks. Previously, computer crimes were prosecuted as mail and wire fraud, but confusing applications of these laws made convictions difficult. In 1986, in groundbreaking legislation, the Computer Fraud and Abuse Act was passed by Congress making it a federal crime to access a computer without authorization (Griffith, 1990). Additionally, the Computer Security Act was passed in 1987 which established minimum security practices in federal computer systems.

Advancements for cybersecurity began to come not only in in the government realm, but also in the private sector. For example, in 1983, the first patent for cybersecurity was issued in the United States to Massachusetts Institute of Technology (MIT) for a cryptographic communications system, which is a method still used today in cybersecurity (Rivest et al., 1983). In 1987, multiple commercial antivirus programs came on the market including the first antivirus product for the Atari and John McAfee's popular and widely used McAfee antivirus and VirusScan software (Raymond, 1996). In 1988, the antivirus company Avast was also founded in Prague and is still used by millions today.

Even though the 1980s saw advancements in cybersecurity, the methods and sophistication of cyberattacks also increased. For example, the first denial-of-service (DoS) attack in history occurred in 1988 when Robert Morris created a computer worm that caused the internet to slow down considerably. Even though Morris only intended for the worm to highlight security flaws, the aggressive replication of the worm caused damages estimated to be up to $10 million, causing Morris to be the first person charged under the Computer Fraud and Abuse Act. He was convicted and received 3 years of probation, 400 hours of community service, and a fine of $10,500 (Raymond, 1996).

As a result of this new attack and the severity of the damage, the Computer Emergency Response Team (CERT) was formed to prevent such attacks in the future. CERT is still active in cybersecurity today with the goal of creating "a safer, stronger Internet for all Americans by responding to major incidents, analyzing threats, and exchanging critical cybersecurity information with trusted partners around the world" (Fuller, 2019).

Just a year later in 1989, the first ransomware attack in history occurred as Joseph Popp created malware called the AIDS Trojan, which was distributed via a floppy disk through the postal service. Popp maliciously intended to extort money out of victims, but fortunately, the ransomware was flawed and easily removable (Raymond, 1996). However, this did usher in a new way to use computers as an attack weapon that cybersecurity experts must add to their growing list.

Furthermore, foreign nations were also finding ways to use computer systems in international warfare. For example, in 1986 during the Cold War, a German hacker Martin Hess broke into over 400 U.S. military computers connected to ARPANET and gained access to sensitive files with plans to sell the data to Russian intelligence (Dixon, 2018). He was caught, tried, convicted, and given a 20-month suspended sentence.

In another technological milestone that would create even more cybersecurity vulnerabilities, during the 1980s, ARPANET transformed into the internet and became widely available to the public as the World Wide Web by the early 1990s (Gillies, Gillies, & Cailliau, 2000).

The Internet and the World Wide Web

The 1990s was a turning point in the history of cybersecurity. With millions of homes full of unsavvy computer users all across the United States and the world all connected using the World Wide Web, cyberspace was more vulnerable than ever and cybercriminals had easy targets. New virus and malware numbers went from tens of thousands in the early 1990s to 5 million each year by 2007 (Middleton, 2017). So many new viruses were created that cybersecurity, again, had a tough time keeping up with demand.

Not only did cybercriminals have millions of new connected targets, but they made their own capabilities more sophisticated. For example, in 1990, the first polymorphic viruses were created which can cause the virus to mutate to avoid detection. Additionally, as more and more antivirus scanners were installed on personal computers, attackers responded with the first anti-antivirus program in 1992. Attackers also developed macro viruses and viruses with stealth capabilities.

As each new virus was created, antivirus software companies scrambled to develop new detection and removal capabilities such as heuristic detection and generic signatures, which were both developed in the 1990s. An important tool in cybersecurity, the firewall, was also created in 1990 by a NASA researcher (Avolio, 1999). The firewall was modeled after actual physical firewall structures designed to prevent the spread of flames throughout buildings.

As more and more people accessed the internet and used internet-connected computers to store data, purchase items, do banking, and communicate, hackers only became more determined. So even though cybersecurity measures increased, the threats grew exponentially during the 1990s and cyberattacks were prevalent in households across the world.

Towards the end of the 1990s, email had also become widely used creating a new and relatively easy avenue for hackers to gain access to personal computers and data. For example, in 1999, the Melissa virus created over $80 million in damage by entering a user's computer as a Word document then immediately emailing copies of itself to email addresses located on the same computer (Garber, 1999).

In 1999, Microsoft Windows 98 was released, creating more sophisticated security for the everyday computer user. The Secure Sockets Layer (SSL) also came into use in 1995, which has served as a critical security protocol and was later further developed into Hyper Text Transfer Protocol Secure (HTTPS), which is used widely in cybersecurity today (Condron, 2007).

The 1990s began to see more international interest in cybersecurity with the establishment of the European Institute for Computer Antivirus Research (EICAR) in 1990 and the passage of the Computer Misuse Act in the United Kingdom in the same year, making it illegal to gain unauthorized access to a computer (Macewan, 2008).

And finally, in 1993, the first ever DEF CON conference was held with an attendance of approximately 100 people, marking another milestone in cybersecurity history (Young et al., 2007). Now that conference is attended by over 20,000 cybersecurity professionals each year.

New Era of Cybersecurity
New National Focus on Cybersecurity

In the late 1990s, there was widespread concern that computers and critical infrastructure could possibly shut down because of the inability of the machines to switch from the year 1999 to 2000. Referred to as Y2K, legislators responded to this issue by passing the Year 2000 Readiness and Responsibility Act, spending billions of dollars to prepare for the millennial switch to the year 2000 (Fuller, 2019). Because of the extensive preparation into cybersecurity issues before 2000, the government was able to make the switch without major issues. However, this event did highlight for many society's growing dependence on computers and the power they have on our critical infrastructure, making cybersecurity an even greater priority (Landwehr, 2010).

Shortly after the Y2K scare, the attacks of September 11, 2001, occurred in New York; Washington, D.C.; and Pennsylvania stunning the nation and the world. Weeks later, Osama bin Laden declared that he was working with scientists who would use their technological expertise against the United States in further attacks (Fuller, 2019).

After these events, attention shifted to defending against possible cyberterrorism and exploring our vulnerabilities to cyberterrorist attacks. U.S. President George W. Bush named Richard Clarke as the first ever National Cybersecurity Advisor to examine these issues and to lead the creation of a national strategy for cybersecurity (see Clarke, 2016). This event became a turning point in cybersecurity history and created a national priority for cybersecurity measures that continues today (Roesener et al., 2014).

One of the major initiatives to come out of this new national cybersecurity focus was the creation of a new U.S. executive department in 2002—the Department of Homeland Security (DHS). Although the National Security Agency takes the lead when it comes to protection of federal computers, DHS is responsible for the protection of civilian computers (Johnson & Hunter, 2017; Kemp, 2012). The six overarching DHS missions are:

- Counterterrorism and homeland security threats;
- Secure cyberspace and critical infrastructure;
- Strengthen preparedness and resilience;
- Secure U.S. borders and approaches;
- Preserve and uphold the nation's prosperity and economic security; and
- Champion the DHS workforce and strengthen the department.

And specifically, under the mission to "secure cyberspace and critical infrastructure", DHS said the department works to:

- Secure federal civilian networks;
- Strengthen the security and resilience of critical infrastructure;
- Assess and counter evolving cybersecurity risks; and
- Combat cybercrime.

One of the agencies created under the DHS umbrella is the Cybersecurity and Infrastructure Security Agency (CISA) whose mission is to "lead the national effort to understand and manage cyber and physical risk to our critical infrastructure" (Kemp, 2012). Since September 11, 2001, the United States has stepped up its cybersecurity efforts and uses the National Security Agency (NSA) primarily for intelligence interception, DHS primarily for assessing cyberthreats and improving national cybersecurity, and the Federal Bureau of Investigation (FBI) primarily for investigating and prosecuting cyberattacks after they happen.

DHS also leads the Cyberterrorism Defense Analysis Center (CDAC) which helps train state and local law enforcement officers and others at the state and local level on best cybersecurity and cyberterrorism prevention practices.

While the U.S. government reorganized itself to create a greater focus on cybersecurity issues after 9/11, hackers only increased their efforts in the new millennium with larger scale attacks than ever before. In 2000, the ILOVEYOU worm infected over 10 million Windows users starting as an email with the subject line "ILOVEYOU," with an attachment that, when opened, damages files and then sends itself again to email addresses on the computer (Middleton, 2017). Creating and executing one of the most damaging viruses in world history, the perpetrators were located in the Philippines, but no specific law against this action existed in the Philippines at the time so the accused were not brought to trial.

In 2001, cyberattackers also created a new infection technique, making it easier to increase their victim counts. Attackers no longer had to convince unsuspecting users to download files for an attack to occur, now attackers only had to convince users to click on a link to go to an infected website.

There were also larger scale attacks during this time. For example, between 2005 and 2007, one cybercriminal stole data from at least 45.7 million credit cards used by stores such as T.J. Maxx, which cost the parent company approximately $256 million (Middleton, 2017). The most famous hacking group in the world, Anonymous, was also created during this period in 2003, further complicating cybersecurity efforts (Buchan, 2016). The group is known for concerted efforts of organized hacking into government and organization computer systems.

Although cyberattacks grew to unprecedented levels as more people owned and stored vital information on their personal computers, there were also some important advancements in cybersecurity. In 2000, OpenAntivirus, the first open-source antivirus engine was made available to the public. Avast also launched a free antivirus software, which grew the company to more than 20 million users in just 5 years. Anti-malware was added to VirusScan and cybersecurity was added to operating systems, creating another layer of protection. Furthermore, as smartphones became more widely used, antivirus applications were also developed for Android and Windows (Patil & Jadhav, 2014).

Strengthening Cybersecurity

Throughout the 2000s, as more and more people had personal computers at home and smartphones in their pockets, leading to exponential growth in the number of people on the World Wide Web, cyberattacks also continued to dramatically increase. In fact, in 2007 there were almost five million total strains of malware. By 2015, just 8 years later, that number had grown to one million new malware threats developed every single day across the world (Harrison & Pagliery, 2015).

Large-scale attacks and data breaches became public with increasing calls for stronger cybersecurity more in the forefront than ever (Dixon, 2018). For example, a Saudi hacker published the details of over 400,000 credit card accounts in 2012 (Middleton, 2017). Hackers broke into Yahoo! in 2013 stealing the accounts and personal information of over 3 billion users (Kan, 2016). Former CIA employee Edward Snowden copied and leaked classified information from the National Security Agency (NSA) revealing sensitive information to the public and foreign enemies (Verble, 2014). In 2020, multiple DDoS attacks on New Zealand's stock market caused it to close for days (Choudhury, 2020). Phishing emails have also become one of the favored methods used by attackers recently, bypassing cybersecurity technology and instead taking advantage of human errors and naivety (Parsons et al., 2019).

In one of the most devastating attacks in cybersecurity history, the WannaCry ransomware, known as the first "ransomworm," gained access to computers through their Windows operating system, then took control of the computer. In order to regain access, users were told they had to pay money in Bitcoin as a ransom. This unprecedented ransomworm gained access to hundreds of thousands of computers across 150 countries in 2017 (Chen & Bridges, 2017).

And just like the rest of cybersecurity history, with increasing attacks comes more advancements in cybersecurity. Simple passwords have now been largely replaced with more complicated passwords or even multi-factor authentication (MFA). Virus scanning software continues to strengthen and web-access firewalls have been developed creating safer online environments for users (Cornier, 2019).

With predictions that cybercrime damages will reach $6 trillion by 2021, the global cybersecurity market is estimated to be worth $173 billion in 2020 and is projected to grow to $270 billion by 2026 (Columbus, 2020). The race between attackers and cybersecurity to outsmart each other continues (Klymenko et al., 2020).

Summary

The history of cybersecurity is fascinating; however, it is likely that we are still in the early stages of its development. In the emerging era of cybersecurity, one of the major initiatives to come out of this new national cybersecurity focus was the creation of a U.S. executive department in

2002—the Department of Homeland Security (DHS). Although the National Security Agency takes the lead when it comes to protection of federal computers, DHS is responsible for the protection of civilian computers. With continued and rapid technological advances by both computer experts and cybercriminals, the field of cybersecurity will continue to change and evolve to meet those demands.

References

Avolio, F. (1999). Firewalls and internet security, the second hundred (internet) years. *The Internet Protocol Journal, 2*(2), 24–32.

Buchan, R. (2016). Cyber warfare and the status of Anonymous under international humanitarian law. *Chinese Journal of International Law, 15*(4), 741–772.

Cerf, V. G. (2009). The day the internet age began. *Nature, 461*(7268), 1202–1203.

Chen, Q., & Bridges, R. A. (2017). Automated behavioral analysis of malware: A case study of WannaCry ransomware. In *2017 16th IEEE International Conference on Machine Learning and Applications* (ICMLA) (pp. 454–460).

Choudhury, S. R. (2020). New Zealand stock exchange halts trading for third day in a row. *CNBC.* https://www.cnbc.com/2020/08/27/new-zealands-exchange-faced-ddos-attacks-this-week.html.

Clarke, R. A. (2016). The Risk of Cyber War and Cyber Terrorism. *Journal of International Affairs, 70*(1), 179–181.

Columbus, L. (2020). 2020 roundup of cybersecurity forecasts and market estimates. *Forbes Magazine.* https://www.forbes.com/sites/louiscolumbus/2020/04/05/2020-roundup-of-cybersecurity-forecasts-and-market-estimates/?sh=7dcdcfa381d7

Condron, S. M. (2007). Getting it right: Protecting American critical infrastructure in cyberspace. *Harvard Journal of Law & Technology, 20*(2), 403–422.

Cornier, K. (2019). Cybersecurity ramps up as electronic infrastructure surges. *EE: Evaluation Engineering, 12,* 32.

Dixon Jr., H. B. (2018). Is hacking the new normal? *Judges' Journal, 57*(1), 36–37.

Federal Computer Systems Protection Act, H.R. 8766, 95th Cong. (1977). https://www.congress.gov/bill/95th-congress/house-bill/8766

Fuller, C. J. (2019). The roots of the United States' cyber (in)security. *Diplomatic History, 43*(1), 157–185.

Garber, L. (1999). Melissa virus creates a new type of threat. *Computer, 32*(6), 16–19.

Gillies, J. M., Gillies, J., & Cailliau, R. (2000). *How the Web was born: The story of the World Wide Web.* Oxford University Press, USA.

Griffith, D. S. (1990). The Computer Fraud and Abuse Act of 1986: A measured response to a growing problem. *Vanderbilt Law Review, 43,* 453.

Guice, J. (1998). Looking backward and forward at the internet. *Information Society, 14*(3), 201–211.

Harrison, V., & Pagliery, J. (2015). Nearly 1 million new malware threats released every day. *CNN Business.* https://money.cnn.com/2015/04/14/technology/security/cyber-attack-hacks-security/index.html

Johnson, T. C., & Hunter, R. D. (2017). Changes in homeland security activities since 9/11: An examination of state and local law enforcement agencies' practices. *Police Practice & Research, 18*(2), 160–173.

Kan, M. (2016). Yahoo data breach affects at least 500 million users, company says. *PCWorld, 34*(10), 48–49.

Kemp, R. L. (2012). Homeland security in America: Past, present, and future. *World Future Review (World Future Society), 4*(1), 28–33.

Klymenko, O. A., Gutsalyuk, M. V., & Savchenko, A. V. (2020). Combating cybercrime as a prerequisite for the development of the digital society. *Janus.Net: E-Journal of International Relations, 11*(1), 18–29.

Landwehr, C. E. (2010). History of US government investments in cybersecurity research: A personal perspective. In *2010 IEEE Symposium on Security and Privacy* (pp. 14–20).

Levy, S. (1984). *Hackers: Heroes of the computer revolution.* Nerraw Manijaime/Doubleday.

Linden, T. (1976). Operating system structures to support security and reliable software. *ACM Computing Surveys, 8*(4), 409–445.

Macewan, N. F. (2008). The Computer Misuse Act 1990: Lessons from its past and predictions for its future. *Criminal Law Review, 12,* 955–967.

Middleton, B. (2017). *A history of cyber security attacks: 1980 to present.* CRC Press.

Parsons, K., Butavicius, M., Delfabbro, P., & Lillie, M. (2019). Predicting susceptibility to social influence in phishing emails. *International Journal of Human-Computer Studies, 128,* 17–26.

Patil, B. V., & Jadhav, R. J. (2014). Computer virus and antivirus software: A brief review. *International Journal of Advances in Management and Economics, 4*(2), 1–4.

Raymond, E. S. (1996). *The new hacker's dictionary* (3rd ed.). MIT Press.

Rivest, R. L., Shamir, A., & Adleman, L. M. (1983). *Cryptographic communications system and method (U.S. Patent No. 4,405,829).* U.S. Patent and Trademark Office.

Roesener, G., Bottolfson, C., & Fernandez, G. (2014). Policy for US cybersecurity. *Air & Space Power Journal, 28*(6), 38–54.

Rosenzweig, R. (1998). Wizards, bureaucrats, warriors, and hackers: Writing the history of the internet. *The American Historical Review, 103*(5), 1530–1552.

Szor, P. (2005). *The art of computer virus research and defense.* Pearson Education.

Verble, J. (2014). The NSA and Edward Snowden: Surveillance in the 21st century. *ACM SIGCAS Computers and Society, 44*(3), 14–20.

Young, R., Zhang, L., & Prybutok, V. R. (2007). Hacking into the minds of hackers. *Information Systems Management, 24*(4), 281–287.

Understanding Geo-Localization and Cyberspace Detection

Marlon Mike Toro-Alvarez

Introduction

The cyberspace environment changed the traditional perspective of locating human beings, their assets, and ways of interaction. Multiple actors are in the middle of a chain of steps to move data from one point to another. The difference between the contexts of computer crime and street crime presents a challenge to cybercrime investigators. Indeed, traditional crime does not face the same complexity (Catlett et al., 2018). For instance, a homicide scene has a corpse in a physical address. It is possible to have details of the interactions between offender and victim in some squared meters; thus, investigators close that area with yellow tape.

But in cybercrime, it is impossible to surround a specific area. However, with applicable knowledge, it is possible to detect where to tie some "digital yellow tape." The following chapter introduces important concepts about internet architecture, networking, time, and techniques to deal with the spatial features of cyberspace within an investigation process.

Anti-Spatial and Temporal Features of Cyberspace

The internet has multiple communication protocols that allow connected devices to send data amongst them. Those protocols can also identify other devices in a simple yet complex way according to network size or offered services.

Considering that the internet is a global network, this huge amount of coverage will motivate more to ponder new features related to connectivity and user location (Wong et al., 2007). In the case of a simple victimization, one offender can be in the same city as the victim. Thus, space is easy to determine. However, when a cybercriminal sends malicious code from another country, located on a different continent, this code could be utilizing a low-cost computer infrastructure available in a third world country to attack thousands of potential victims from various countries all at the same time (see Figure 11.1).

The registered time of attack is completely different compared to the recovered records in the other parts of this cybercrime scenario. Not only is the time difference between Asian countries different from eastern European countries, but it is also an even bigger gap when compared to big cities in America. In addition, the same American continent has different time zones. This complex time puzzle is what cybercrime investigators consider an "anti-spatial" characteristic of cyberspace.

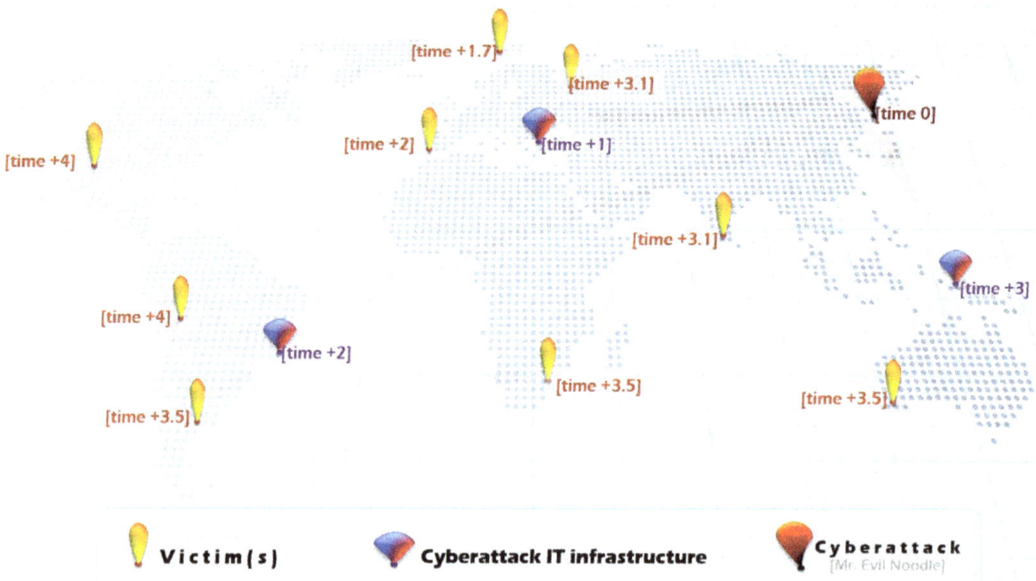

FIGURE 11.1 Time and location of cyberattack

Anti-spatiality is related to time features and vice versa. Those two considerations of cyberspace are key factors when an investigator has to testify in court to defend a cybercrime investigation. Some cases are lost when investigators assume that everyone knows about the special conditions of the internet and the other networks where cybercriminals interact. Hence, it is highly advisable to introduce the cyberspace context to any division within the criminal justice system (Choi & Toro-Alvarez, 2017). Nevertheless, explaining cyberspace conditions requires some technical knowledge and the capability to put technical language into laymen's terms. For that reason, it is important to learn about how the internet's architecture works. This includes identifying how the websites are indexed in servers and how machines as well as users interpret domain names. However, before talking about website domain services, it is vital to know about network addressing.

IP Addressing

The internet uses multiple computer protocols to allow data communications. Some of those protocols are the Transmission Control Protocol (TCP) and the Internet Protocol (IP). The IP is the main network communication protocol, which provides an identification and location system for connected devices and traffic routes (Kim & Ryu, 2010). That connectivity is possible because of a logical identification. One of those identifiers is the IP address which is associated with one network adapter (e.g., a computer card to transfer data by signals and connect the device to a network).

IP address structure depends on the version of the Internet Protocol. For instance, an IP address in the fourth version (IPv4) is a legible numerical code, while an IP address in the sixth version (IPv6) is identified by its hexadecimal chain of characters (see Figure 11.2). An IPv4 uses 32 bits while IPv6 reaches 128 bits. Additionally, IPv6 prepares itself for the current scarce amount of IP addresses in the fourth version (IPv4) (Comer & Soto, 1996). However, the main purpose behind talking about IP addresses is that identifying an address allows investigators to locate a connected device and identify exactly who was using that device.

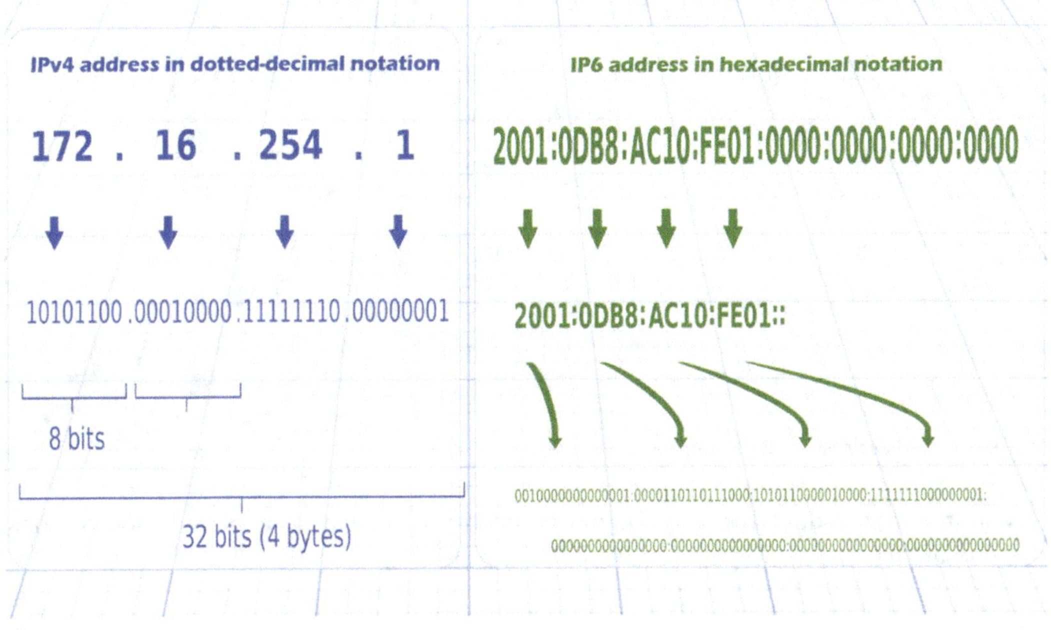

FIGURE 11.2 IP addresses in both versions

It is important to recognize that without an IP address, there is no connectivity to cyberspace. From a desktop computer to a phone or a wireless transfer via Bluetooth, all those connections require IP addresses. This kind of logical identification is required to be able to send and receive data and to be able to give and receive services on the internet. Additionally, the amount of IP addresses is limited by public or private usages. Hence, there are different control points and architectural services to establish their usages and configurations. Some of those controls validate that a computer is identifiable to receive communication, provide layers of connections, and establish internet routing. One of those services is the domain name system (DNS), which the current chapter introduces below.

Finally, in digital forensic investigations, the IP address is one of the main sources of information. In simple terms, the IP address provides the identification of the user, place of connection, internet service provider, connection data, and other information. To access this public information, researchers can access domain registration websites such as www.lacnic.net through any browser. Once on the website, the suspicious IP address can be validated and from there the data of the internet service provider (ISP). For example, if we have the IP 172.217.2.206, the next step is to validate the registered physical address of the person or organization related to that IP address.

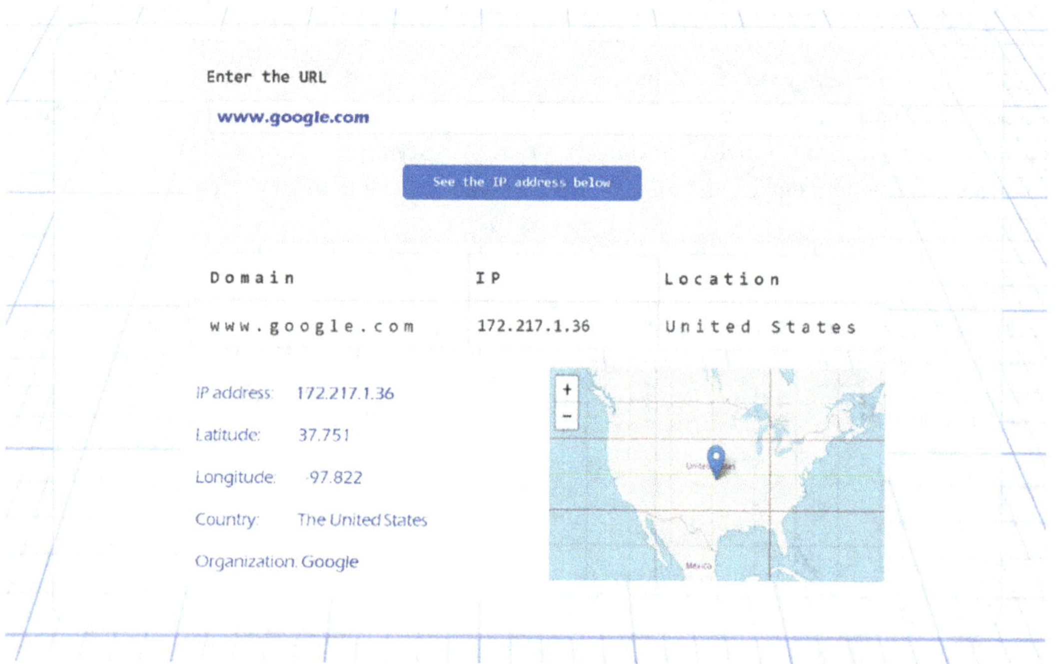

FIGURE 11.3 IP address verification

Within this procedure, the investigator can identify a possible place to visit, some persons to interview, and the request of possible search warrants.

Domain Name System (DNS)

The domain name system is the best friend of every user in cyberspace. This friendship is based on its interpretation functionality. While IP addresses are complicated to read (binary or hexadecimal versions), DNS translates those strings of numbers to legible website addresses (Gómez, 2010). They are known as Uniform Resource Locator (URL) or a web address. For instance, the IP address of a Google server is 172.217.2.206 but the user does not know it. The user knows to type www.google.com (i.e., Google's URL). Then, DNS translates the typed URL, looks for the IP address of the Google server, and sends the communication requirement to 172.217.2.206. This process is completely invisible to an internet user, but it is necessary to understand the functionalities in order to not identify a false positive regarding the IP address of a potential suspect.

Once a website address is typed on an internet browser, the communication is established with multiple devices (e.g., network active devices, servers, firewall) and by the structure of several communication services. Then multiple IP addresses arrive in a single communication procedure (see Figure 11.4).

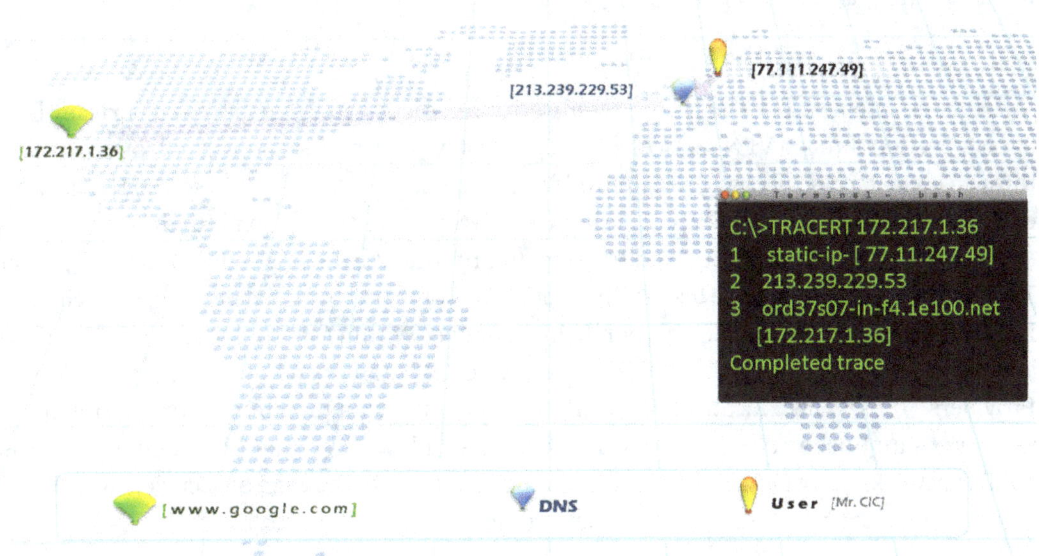

FIGURE 11.4 IP tracking and DNS identification

It should be noted that DNS groups are two categories of servers: name servers and solution servers. Name servers store information and respond to information requests. They resolve process codes and manage algorithms to find a name server that has the information the user is looking for. The functions of these servers are petitions combinations or communication requirements which can be prioritized by the communication structure (Mockapetris & Dunlap, 1988). Unfortunately, this great decision-making capacity can configure relationships of trust that can be exploited by cybercriminals. However, the objective of this chapter is to talk about geolocation instead of cyberattacks. In this context, a connected computer location by identifying a single IP address is confusing considering that there are many IP addresses that allow the connection to be established on a network. This architecture of the internet and cyberspace configures new forms of geography to analyze (Weaver et al., 2014). However, for the time being, this chapter has only addressed visible user identification, but many spoofing or IP hiding services are available for many uses. Some uses focus on digital security and other services that allow anonymity, privacy, and can affect the investigation of a cybercrime (Choi & Toro-Alvarez, 2017).

Talking about domains on the internet also allows us to infer a set of locations that can be logically related. The final parts of a URL (e.g., .org from www.something.org) and an email account (e.g., @email.org from someone@email.org) are identified as the domain and can provide valuable information in an investigation. There are generic domains and country domains and others can combine both descriptions. For example, among the generic domains, we find those ending in .com, .edu, .gov, .gob, .ext, .mil, .net, and .org. And among the country domains are country code top-level domains (ccTLD) or geographic top-level domains—for example, the characters .at are reserved for Austria reserves, .au for Australia, and Andorra is identified with .ad (Pouzzner, 2004).

In this way, the domain @usdoj.gov implies a U.S. government institution, while an address www.somename.edu.ad would give us information about an educational institution located within a European country.

VPN Services

A virtual private network (VPN) is a safer way of interacting in cyberspace. The privacy level of a connection through VPN services refers specifically to keeping secret what is communicated through connected devices and keeping hidden the messages from other viewers in the network environment (Kim et al., 1999). Technically, the privacy offered corresponds to the management of information isolation through addressing and routing during communication (Ferguson, 1998). VPN services allow the user to connect to the company's internal network with customizable access to applications and corporate information. Within this kind of service, VPN users are able to change the visibility of their own device's IP address and provide Internet Service Providers (ISPs) a fake location. VPNs are the extension of a private network over a public network or an uncontrolled network. Indeed, VPNs allow extending private networks over public networks, they provide information integrity, increase the level of security, and allow a higher level of confidentiality in end-to-end communication (Jiménez, 2014).

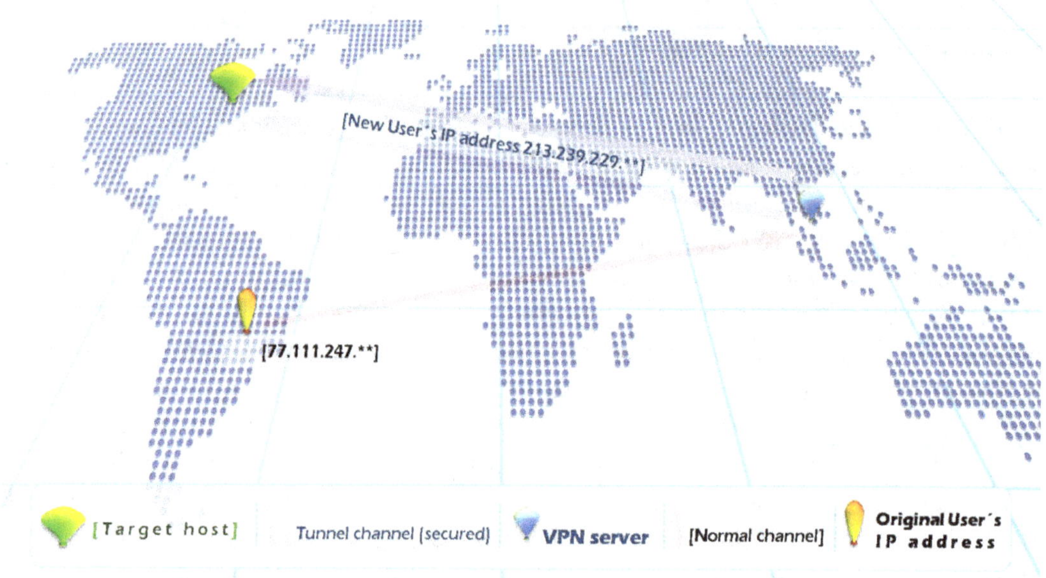

FIGURE 11.5 Sample of virtual private network (VPN) service

Source: Created by Mike Toro using Pencil2D and Microsoft PowerPoint.

This level of privacy protects against internet tracking related to geographical features but also blocks initial investigation procedures (Bellovin, 1995). Nevertheless, cybercrime investigators can require a VPN service provider to provide the communication route in order to document possible digital evidence under a cybercrime investigation. Some IP addresses from some VPN providers and host countries are getting commonly known. Thus, identifying an IP address from a VPN server allows the investigator to then identify the VPN provider and so on.

Timestamps in Cyberspace

Another mechanism to identify the temporal feature of a cyberspace interaction is a timestamp. Computer devices and networks leave a mark related to time which can confirm the relation of a suspect with an offense and the specific device used to commit it. Likewise, a timestamp is recognized as a mark made in a physical environment (bit pattern) and a temporal interpretation (interpretation of each bit pattern), to finally allow a computer to record a performed interaction. This unit of time is recorded in operating systems, databases, and websites, among others (Dyreson & Snodgrass, 1994).

In terms of internet browsing, it is important to consider that once a user visits a website, it activates the site cookies that correspond to the visitor records. Here, a trace of the interactions generated by the visitor through their browser and the website server is produced. Within the server log files the date and time, location of the web browser, reference links, frequencies of visits, and click selections are stored. These records are automatically created and are used to collect statistics from the websites. The timestamps stored within these records allow investigators to have the exact time information in which the visitor accesses a website, as well as to help identify a website visitor and the moment in which the visitor session expired or was supplanted. The latter applies when the suspect claims to have been the victim of identity theft or loses access to credentials (Glommen & Barrelet, 2004).

Some systems handle a level of detail of days in the timestamp format (e.g., DB2), while other systems use a format in seconds to cover a time representation of 120 years. However, the DB2 timestamp format reduces this detail to microseconds. This type of data requires 10 bytes and increases the time range to manage up to 10,000 years. Another well-known format is SQL2 which requires 27 bytes to handle a level of detail of microseconds and maintains time management like the DB2 timestamp format. The administration of these time units is accomplished through mechanisms that allow a computer to keep the current time accurate, calibrated to fractions of a second.

The use of timestamps allows an investigator to correlate the events within an investigation and to define whether the events correspond to isolated instants in time, periods determined by a simultaneous duration in time but without correlation, or intervals that are defined as correlated between two events. Likewise, it is important to know the logic of certain operations (Dyreson & Snodgrass, 1994) that can occur within timestamps:

- **Primitive constructors.** Operations that generate a timestamp without taking into account any previous timestamp.
- **Constructors.** Operations that are generated by taking a timestamp as input. Those operations choose between the first or last event between two events to generate the timestamp.
- **Mutators.** Operations that modify the timestamp and cause an event to move for a specified duration.
- **Observers.** Operations that take a timestamp as input and return another type of data. Here, the system decides if one event is before the other and if the event converts a timestamp value to a specific calendar value.

Obviously, a discussion of timestamps requires knowing about different date formats. According to Ripley and Hornik (2001) there are two common ways of presenting date formats: the ISO 8601 standard and the Portable Operating System Interface (POSIX). For its part, ISO 8601 defines the protocol to represent the Gregorian calendar by using the following syntax description:

- full-date = date-fullyear "-" date-month "-" date-mday
- date-fullyear = 4DIGIT
- date-month = 2DIGIT; 01-12
- date-mday = 2DIGIT; 01-28, 01-29, 01-30, 01-31 based on;
- month/year time-hour = 2DIGIT; 00-23
- time-minute = 2DIGIT; 00-59
- time-second = 2DIGIT; 00-58, 00-59, 00-60
- based on leap second; rules time-secfrac = "." 1*DIGIT
- time-numoffset = ("+" / "-")
- time-hour ":" time-minute
- time-offset = "Z" / time-numoffset
- date-time = full-date "T" full-time
- full-time = partial-time time-offset
- partial-time = time-hour ":" time-minute ":" time-second [time-secfrac]

POSIX defines calendar time based on the number of seconds since the early 1970s in the UTC time zone. This timestamp style is no longer accurate as it inserts 22 leap seconds since 1970, and these must be discarded. Computers store the number as a 32-bit signed integer, allowing times from the early years of the 20th century to 2037 (Ripley & Hornik, 2001).

The following table shows the applications of these timestamp formats as well as the adaptation from some commercial software.

Knowing about these different formats allows the investigator to decrease the likelihood of making interpretation mistakes based on just one timestamp format. It is important to know about the technical conversion among the different formats when it comes to explaining the chronological sequences of steps during a custody chain or digital evidence collection (Choi & Toro-Alvarez, 2017). Additionally, this technical approach will make investigators capable of cooperating with other agencies considering the multiple formats and frameworks among cybercrime investigation forces.

The Spatial Dimension of a Digital Connection

The spatial dimension is related to the jurisdiction of the authorities within the criminal justice system. This jurisdiction refers to the general authority of a specific court to accept and resolve a dispute. Traditionally, the authority of a court is based on geographic factors, such as the location of a defendant's residence or place of business, where a crime or misdemeanor occurred, or the address of a disputed property. A court can only consider accepting a lawsuit if the court has the authority to decide this type of matter. The concept of jurisdiction includes two specific legal reflections: (1) Does the judicial authority have the capacity to listen to the type of problem that has been presented to it? AND (2) Can the court enforce its decision on the parties to the lawsuit? (Choi, 2015).

TABLE 11.1 Application of Timestamps Formats

APPLICATION	TIMESTAMP FORMAT
SO Windows	d o D, M o m, YY o yyyy. Standard ISO 8601 or Gregorian calendar
SO Linux	POSIX 1234567890
SO Mac	POSIX i.e. 1234567890 o d o D, M o m, YY o yyyy
SO DOS	d o D, M o m, YY o yyyy. Standard ISO 8601 or Gregorian calendar
BD Oracle	TIMESTAMP [(fractions_of_seconds)] date(DD/MM/YYYY)
BD SQL Server	TIMESTAMP [(fractions_of_seconds)] date(DD/MM/YYYY) It includes Oracle/Ingress
BD PostgreSQL	TIMESTAMP [(fractions_of_seconds)] stores date and hour date(DD/MM/YYYY), stored from 4713 BC to 32767 AD Time (stores time of day) Additionally, it allows European UTC+1, UTC+2, and not European time zones. POSIX
BD MySQL	TIMESTAMP [(fractions_of_seconds)] date(DD/MM/YYYY) It includes Oracle/Ingress

Source: Adapted from Ripley & Hornik (2001), Date-time classes, R News, 1(2), 8–11.

As mentioned above, cyberspace is different from the physical world due to its temporal and spatial characteristics. Difficulties in prosecuting computer-related criminal activities arise because many of the assets involved are intangible and do not fit well with the specific statutes of traditional crime. While traditional crimes known as street crimes have been based on geographic boundaries, cybercrime is not limited to a specific site. For example, a cybercriminal may be in country X, using a server located in country Y, and executing malicious code on several victim computers in one or more Z countries.

The internet created a global jurisdiction. Due to its "anti-spatiality," the law must determine if an action carried out on the internet is controlled by the law of the country where the website used to victimize is located, where the internet provider is located, where the user resides, or if a completely new jurisdictional area must be generated exclusively for cyberspace (Choi, 2008).

Therefore, the re-examination and re-evaluation of traditional legal principles have become important tasks for legal professionals because they constantly encounter jurisdictional problems associated with cyberspace. This is where academics and professionals need to be clear about criminal conduct and the means used by perpetrators in order to find common ground amidst the various ways to tackle cybercrime and computer-based crime. This includes going beyond the legislation of each nation or even state regulation in countries with federated systems of government, such as Argentina, the United States of America, Germany, and Mexico (Choi & Toro-Alvarez, 2017).

Taking Advantage of Metadata

According to Senso and Piñero (2003), metadata is not a new term. It was used by Jack Myers in the 1960s to describe data sets. The first adequate meaning for the academic world was the "data on the data" because the metadata provides minimal information necessary to identify an information resource. In Myers' work, he describes the inclusion of descriptive information about the context, quality, and condition or characteristics of the data. However, metadata, as a concept, provides more information than the term indexing. When talking about "data about data," the existence of clues about the context and content, author, and technical specifications of place and time are also affirmed (Pasquinelli, 2003).

For their part, Senso and Piñero (2003), defined metadata as information about a publication as opposed to its content. In other words, the metadata not only includes the visible description but also contains relevant information such as materials, price, manufacturing conditions, and use. Similarly, Ercegovac (2005) states that metadata describes the attributes of a resource, taking into account that this resource can be built on geospatial information, visual resources, or software implementations. Although they may have different levels of specificity or structure, the main objectives are the same: describe a resource, report on conditions of use, and facilitate interoperability.

Cathro (1997) was one of the first to implement this consideration and emphasize location functionality and data processing data. This implementation allows an investigator to locate an offender in a simple but appropriate way while maintaining compliance with the criteria of the criminal justice system. This chapter will later show a real case where the use of metadata was applied to capture a cyber predator.

Metadata Applications

Returning to the contributions of Cathro (2009), some examples of metadata are the catalog of a database, the summary of a document, the heading of a multimedia file (image, video, or audio), IP or DNS addresses, and the heading of email messages. As illustrated, the study and application of metadata is varied (see Table 11.2), which is why the present work focuses on the application of metadata which allows two functionalities: (1) the description of the content of a file and (2) the location information and accessibility of a file collected as possible digital evidence.

Simply, an investigator can access metadata of a file and find relevant information about an offender, time, and place where the file was made. The simplest way to access metadata information is to open the properties of analyzed files (see Figure 11.6) and explore the different categories of information available according to the type of file. For example, image files can provide us with geographical information and the camera model used to produce the image (Naaman et al., 2004).

Digital Evidence Analysis

Once the digital evidence is recovered, researchers need to know what type of data the file contains to advance the investigation. Relying on the final part of the file name, that is, relying on the file extensions, can provide information about the possible content of the file. For example, .mp3 extensions indicate that the files contain music or voice messages and .jpg extensions

TABLE 11.2 Different Uses of Metadata

TYPE	DESCRIPTION	CYBERCRIME INVESTIGATION EXAMPLE
Administrative	Management of information resources	• Digital evidence acquisition • Copyright information • Information storage • Version control
Descriptive	Representing data and information resources	• Support data searching • Specialized index • Collecting user notes
Preservation	Saving and protecting information	• Usage conditions • Record of data storage
Technical	Related to technical usage of data sources	• Information about used hardware • Elaboration details including geo-localization • Login details (including passwords, encryption, and usernames) • System timestamp

Source: Adapted from Senso & Piñero (2003), El concepto de metadato: Algo más que descripción de recursos electrónicos, Ciência da Informação, 32(2), 95–106.

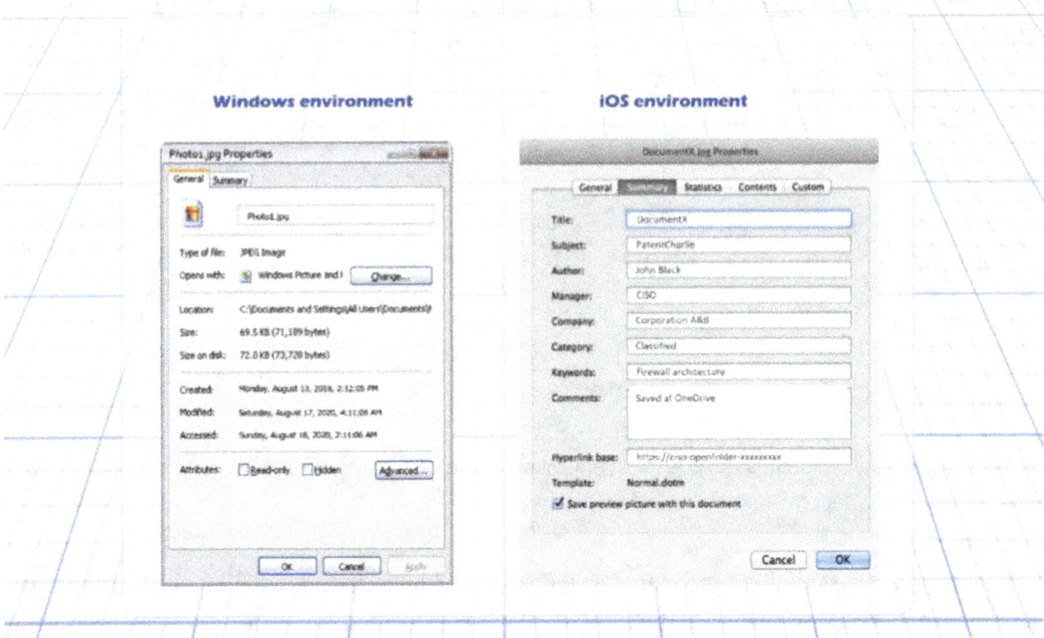

FIGURE 11.6 Properties of a file

indicate graphic files. But cybercriminals can exclude a file extension or assign any extension to any file type (e.g., naming a text file with an .mp3 extension). File extensions may even be hidden or unsupported by other operating systems, so analysts should not assume the accuracy of file extensions.

To avoid research disorientation, it is important to analyze the file header with a hex viewer. There, the type of stored data can be identified as well as other very useful metadata for research. File headers contain information about a file and possibly file positioning, compilation, and owner data. Figure 11.7 demonstrates how the file header has a file signature that shows the type of data it contains. The example shows a file header, FFD8, which suggests that it is a .jpeg file. Note that a file header may be separated from the actual file data. Another technique to identify the type of data in a file is to display a simple histogram with the distribution of ASCII values as a percentage of all the characters in a file. For example, a point on the line 'space', 'a', and 'e' usually suggests a text file, but the consistency in the histogram must be thoroughly analyzed. In some cases, it can be confusing when finding the file in a compressed format. However, the researcher will look for similar patterns, and if no such patterns are found, that would suggest the files were modified to hide information or were encrypted.

FIGURE 11.7 Visualization in a hexadecimal editor program of a graphic file with .jpeg extension and valid value FFD8

Encryption causes challenges for analysts. Folders, volumes, individual files, or partitions may have been encrypted so that only the individual who performed the encryption can access the content. Fortunately, in a forensic laboratory, you can count on operating systems or computer programs to reverse the encryption. It can be easy to identify an encrypted file, but it is not so easy to decrypt it. By discovering the encryption programs installed on the system, examining the file header, or finding encryption keys (which are normally stored on other media), the encryption method can be determined. Often, it is impossible to decrypt because the authentication (i.e., password) to decrypt is not available or the encryption method is strong, but the file header can quickly provide the investigator with a lot of information.

File headers, rather than file extensions, should be used to identify file content types. Analysts should not assume that file extensions are accurate because users can assign any extension to a file (Brown, 2001). By looking at the headings of a file, you can identify the type of data stored in many files and better guide research, including detecting the exact location where a photo was taken, as shown in the following case study.

Applied Case: Online Child Pornography, Hunting the Photographer

As suggested within a 2019 report from the United Nations, use of the internet as a means to sexually exploit children as well as to sexually traffic children continues to increase both nationally and internationally at an exponential rate (Bouché & Bailey, 2020). Through the rapid development of technological innovations, information sharing continues to advance, which ultimately results in further enabling communication between online offenders. In other words, with the assistance of these technological advancements, online offenders are further encouraged to continue their criminal activities knowing that the risks of being caught remain low (Choi et al., 2019).

Furthermore, with the increase of cybercriminals utilizing darknet sites in efforts to offend while remaining anonymous, law enforcement officials continue to face many challenges when attempting to investigate these complex crimes (Chertoff, 2017). Therefore, in order to combat as well as prevent technology-facilitated crimes against children from occurring in the future, a collective approach should be implemented and enforced. This includes collaborative assistance from public and private sectors, international agencies, communities, friends, and families (Choi & Toro-Alvarez, 2017).

The following case was investigated and prosecuted in real life, but some names and affiliations are modified for academic purposes and on the operational reserve of the involved institutions.

On January 21, 2012, an agent from the Homeland Security Department at a U.S. embassy for a South American country delivered investigative information to the head of the investigative group against child pornography from that South American country.

The delivered communication informed local investigators about the identification of a foreign citizen who was willing to buy material for child sexual exploitation (child pornography) on the South American Caribbean coast. Based on this information, investigators asked the judicial authorities for court orders to obtain information from different financial service providers on the transactions carried out by the suspect. A U.S. company that makes online payments gave information related to two people who had received payments from the suspect. The investigation revealed that the residence address of one of those two vendors was located on the Caribbean coast in Colombia.

The special operation group coordinated the operation of entering the residence of the geo-located vendor according to the judicial authorization. Two weeks later the search was carried out and a surveillance order was issued on the second suspect. In the operation, the suspect was found at home, but no evidence of production or commercialization of pornography was found.

Although the operation was unsuccessful, it was known that the foreign suspect made transactions twice a month and the week of the search coincided with the week of one of the weekly transactions. In this way, the head of the investigation made use of the surveillance order and established that the suspicious seller left home after the raid and went to the outskirts of the city in an area where the suspected buyer had previously been seen.

A new operation was organized to seize the two suspects when they met. On the night of Saturday, February 12, 2012, the suspects were intercepted and a microSD storage device with a 4GB storage capacity was seized. The suspects were taken to the forensic laboratory to be identified according to criminal law, while the computer forensic laboratory would deliver a report of what was stored in the seized device. While 516 files were found, none of them matched to child pornographic material.

A second file analysis was carried out, this time the size of the 516 found files was contrasted with the amount of space used in the seized device. Investigators found an inconsistency in those amounts. A deep file search was performed and this time 12,304 files with a .dat extension were found. That file extension would suggest to investigators that those files were system files from some recently used software.

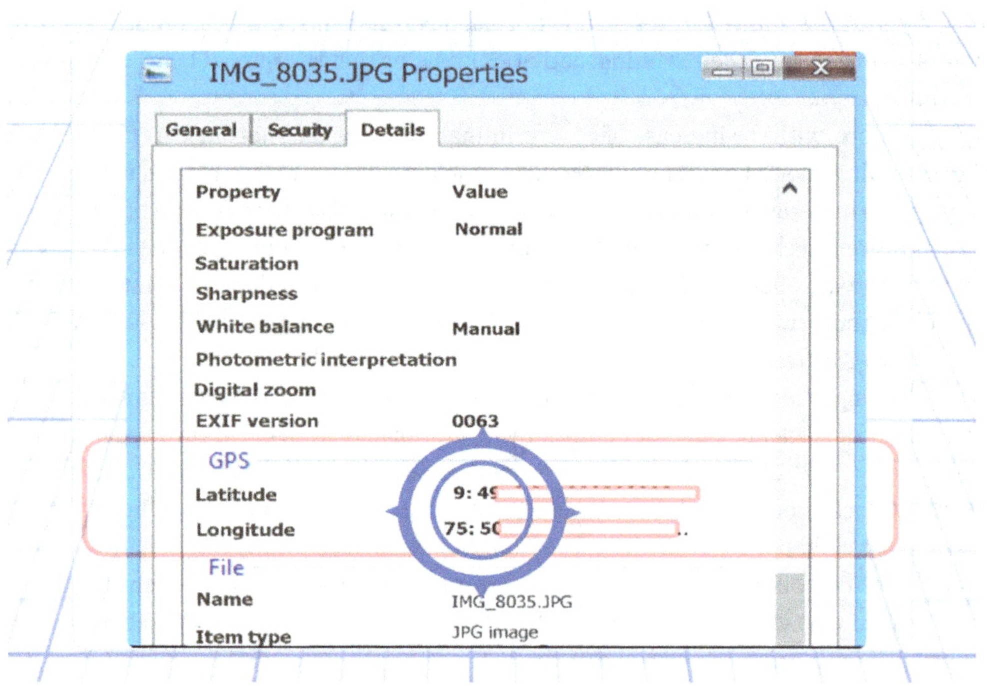

FIGURE 11.8 Photography geolocation properties box

Mr. CIC, one of the experts in the investigation team, recorded the techniques implemented and noted in his investigation notebook that he would change the file extension from .dat to .jpeg on all files. After this procedure, 299 photographs with incriminating material were made visible. Additionally, the investigator reviewed the metadata of some of the images and found coordinates from a camera geolocation system (GPS) which was active at the time the photographs were taken (see Figure 11.8).

The collected GPS information was converted to local geolocation and identified an address 21 minutes away from where the suspects were intercepted (see Figure 11.9). Another search operation was organized and authorized and at 04:00 on Sunday, February 13 of that year, alias

Mr. Evil was captured. The local police had captured the second person who had been identified in the financial transactions carried out with the suspicious buyer. Mr. Evil was found in the new residence with child pornography material, cameras, and elements used as scenery in the photographs analyzed hours earlier in the forensic computer lab.

In addition, buyer addresses were established in the southern tip of South America, and IP addresses that originated from the same country were identified. With this information, the location of other pornography vendors was confirmed and an investigation with the other country was opened. In this way, the ISP was asked to confirm the physical addresses for the IP addresses connected with the case and to provide biographical data for the persons who contracted the internet service with that ISP.

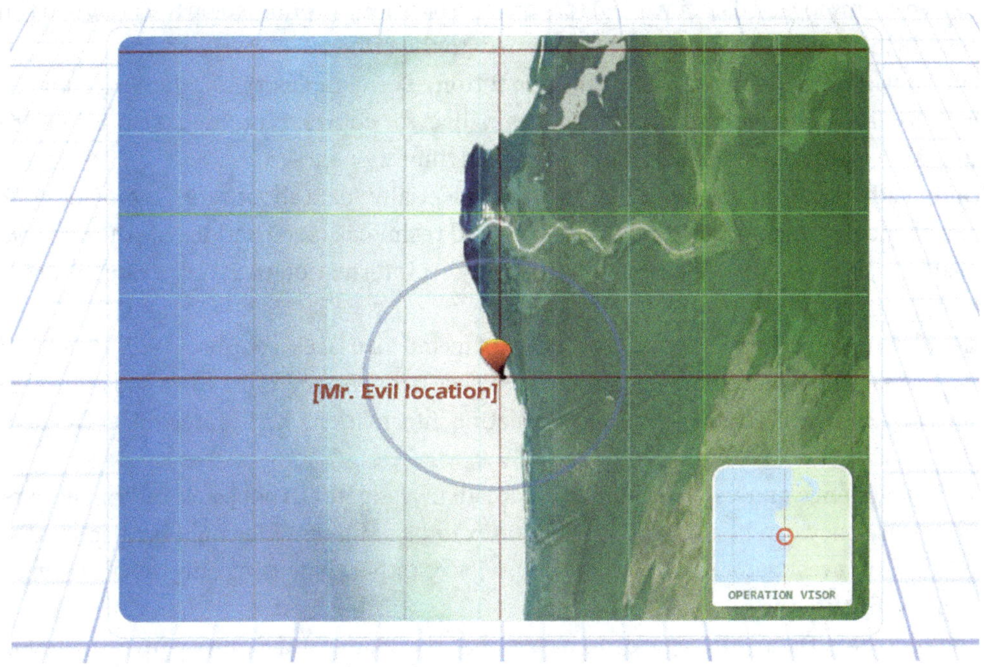

FIGURE 11.9 Second operation location

Finally, a total of 17 raids were carried out, 451 items were seized such as compact discs (CDs), hard drives (HD), tablets, PCs, cell phones, and USB flash drives. Four cyber-predators were captured and extradited to countries where the penalties for child pornography are strongest.

Practice in detecting and geolocating cybercriminals takes an extra effort in technical understandings of cyberspace. The synergy between different data sources has been shown to help reconstruct a strong picture of a cybercrime scenario (Negash et al., 2017).

Summary

Anti-spatial and temporal features of cyberspace: When a cyberattack occurs, the registered time is often found to be different compared to the records recovered in a cybercrime scenario. Due to global time zones, time differences between countries creates a complex time puzzle which cybercrime investigators refer to as the anti-spatial characteristic.

IP addressing: IP addresses are a vital source of information in a digital forensic investigation. They are essential because they allow the identification of the user, place of connection, ISP, and other connection data to be identified and utilized by an investigator.

Domain Name System: DNSs are extremely useful to everyone in cyberspace because they translate IP addresses into URLs which are also known as web addresses.

VPN services: VPNs offer a safer way to interact within cyberspace because they allow a user to change their own device's IP address which will provide their ISP with a fake location.

Timestamps in cyberspace: Within cyberspace, a user's device will document a timestamp after an interaction was performed. When a user accesses a website, the site's cookies start to log the time, date, and location from which the user interacted with the website. With the assistance of this logged data, investigators are able to trace a user based on the recorded timestamp, which can help to link a potential suspect to the crime.

The spatial dimension of a digital connection: The complexity of cyberspace makes it difficult for law enforcement to investigate as well as for courts to prosecute the crimes that occur within the cyber-world due to jurisdictional limitations.

Taking advantage of metadata: Metadata is essentially "data about data." It provides information about the context and content, author, and technical specifications of place and time in regards to a specific file. Investigators can use this information to gather evidence against a potential suspect.

Metadata applications: Four different types of metadata usages were discussed: (1) administrative (management of information resources); (2) descriptive (represents data and information resources); (3) preservation (saving and protecting information), and (4) technical (related to technical usage of data sources).

Digital evidence analysis: It is important for an investigator to not solely rely on file extensions because they can be tampered with. For example, an .mp3 file could be displayed as a .jpeg. Thus, an investigator should utilize a hex viewer rather than using only the file extension to properly identify the file type.

Applied case: Details in regards to a case that the Homeland Security investigated in a South American country. They utilized IP addresses and performed a deep file search on a confiscated SD card. After changing the file extensions from .dat to .jpeg, incriminating evidence was discovered. The metadata from these files helped to reveal coordinates of a camera geolocation system.

The following learning activity helps to consolidate the knowledge about geolocation and digital analysis during a cybercrime investigation. By using the information of the applied case, the reader will get involved in the challenges of another investigation stage.

Lab 11: Mr. Evil's Location—Geolocation Lab

After the successful police operation described earlier in the chapter, all efforts were concentrated on trial, even from the cybercriminal side. Hence, Mr. Evil asked for help from some organized criminal groups. He asked for false testimonies in order to contradict digital evidence about his location. However, Mr. CIC (our expert investigator) is not working alone. He needs your help to ensure that junior investigators are doing their best to collect evidence demonstrating that Mr. Evil was located in every place where victims and incriminatory material were found.

Please choose the correct option for each question:

1. Once the digital forensics laboratory receives the seized microSD storage device, which procedures should be deployed in order to identify the possible geolocation of Mr. Evil?
   a. Check for the smallest files
   b. Check the metadata from media files
   c. Analyze the physical features of the seized microSD
   d. None of the above

2. It seems like there is no media file related to the investigation. Which technique can the investigators deploy?
   a. Check the metadata from media files
   b. Analyze the physical features of the seized microSD
   c. Analyze the file header with a hex viewer looking for file extension modifications
   d. All of the above

3. Please describe another technique to check (without using specialized software) whether there are more media files on the seized microSD:

____

____

____

____

____

____

References

Bellovin, S. M. (1995, June). Using the domain name system for system break-ins. In *Proceedings of the Fifth USENIX UNIX Security Symposium*.

Bouché, V., & Bailey, M. (2020). The UNODC Global Report on Trafficking in Persons: An Aspirational Tool with Great Potential. *The Palgrave international handbook of human trafficking*, 163–176.

Brown, M. F. (2001). *Criminal investigation: Law and practice*. Butterworth-Heinemann.

Cathro, W. (1997). Metadata: An overview. A paper given at the standards Australia Seminar.

Cathro, W. (2009). Metadata: An overview. National Library of Australia Staff Papers.

Catlett, C., Cesario, E., Talia, D., & Vinci, A. (2018, June). A data-driven approach for spatio-temporal crime predictions in smart cities. *In 2018 IEEE International Conference on Smart Computing (SMARTCOMP)* (pp. 17–24). IEEE.

Chertoff, M. (2017). A public policy perspective of the Dark Web. *Journal of Cyber Policy, 2*(1), 26–38.

Choi, K. S. (2008). Computer crime victimization and integrated theory: An empirical assessment. *International Journal of Cyber Criminology, 2*(1).

Choi, K. S. (2015). *Cybercriminology and digital investigation*. LAB Scholarly Publishing LLC.

Choi, K., & Toro-Alvarez, M. M. (2017). *Cibercriminología: Guía para la investigación del cibercrimen y mejores prácticas en seguridad digital*. UAN Fondo Editorial.

Choi, K. S., Earl, K., Lee, J. R., & Cho, S. (2019). Diagnosis of cyber and non-physical bullying victimization: A lifestyles and routine activities theory approach to constructing effective preventative measures. Computers in Human Behavior, 92, 11–19.

Comer, D. E., & Soto, H. A. A. (1996). *Redes globales de información con Internet y TCP/IP* (Vol. 1). Prentice Hall.

Dyreson, C. E., & Snodgrass, R. T. (1994). Efficient timestamp input and output. *Software: Practice and experience, 24*(1), 89–109.

Ercegovac, Z. (2005). Towards user-centered displays of resources in global digital libraries. In *La dimensió humana de l'organització del coneixement* (pp. 148–162). Facultat de Biblioteconomia i Documentació.

Ferguson, P., & Huston, G. (1998). What is a VPN? White Paper available at https://www.potaroo.net/papers/1998-3-vpn/vpn.pdf

Glommen, C., & Barrelet, B. (2004). *Internet website traffic flow analysis using timestamp data* (U.S. Patent No. 6,766,370). U.S. Patent and Trademark Office.

Gómez, J. A. (2010). *Servicios en red*. Editex.

Jiménez Cely, R. (2014). Seguridad en redes VPN [Bachelor's thesis, Universidad Piloto de Colombia].

Kim, E. C., Hong, C. S., & Song, J. G. (1999, May). The multi-layer VPN management architecture. In Integrated Network Management. In *Proceedings of the Sixth IFIP/IEEE International Symposium on Integrated Network Management* (Cat. No. 99EX302) (pp. 187–200). IEEE.

Kim, Y. H., & Ryu, K. S. (2010). *IP addressing to support IPv4 and IPv6* (U.S. Patent No. 7,801,078). U.S. Patent and Trademark Office.

Mockapetris, P., & Dunlap, K. J. (1988, August). Development of the domain name system. In *Symposium Proceedings on Communications Architectures and Protocols* (pp. 123–133).

Naaman M., Song, Y. J., Paepcke, A., & Garcia-Molina, H. (2004). Automatic organization for digital photographs with geographic coordinates. In *Proceedings of the 2004 Joint ACM/IEEE Conference on Digital Libraries* (pp. 53–62).

Negash, L., Kim, S. H., & Choi, H. L. (2017). *Distributed unknown-input-observers for cyberattack detection and isolation in formation flying UAVs*. arXiv. https://arxiv.org/abs/1701.06325

Pasquinelli, A. (2003). *Information technology directions in libraries* [White paper]. Sun Microsystems.

Pouzzner, D. (2004). *Internationalized domain name system with iterative conversion* (U.S. Patent Application No. 10/296,105). U.S. Patent and Trademark Office.

Ripley, B. D., & Hornik, K. (2001). Date-time classes. *R News, 1*(2), 8–11.

Senso, J. A., & Piñero, R. (2003). El concepto de metadato: Algo más que descripción de recursos electrónicos. *Ciência da Informação, 32*(2), 95–106.

Weaver, N., Kreibich, C., Dam, M., & Paxson, V. (2014, March). Here be web proxies [Conference paper]. In *15th International Conference on Passive and Active Measurement* (pp. 183–192).

Wong, B., Stoyanov, I., & Sirer, E. G. (2007, April). Octant: A comprehensive framework for the geolocalization of internet hosts. Paper presented at the 4th USENIX Symposium on Networked Systems Design & Implementation, Cambridge, MA.

Credits

Cybersecurity Countermeasures

Sinchul Back and Jennifer LaPrade

Introduction

Cybersecurity measures can have a significant effect in countering cybercriminal and cyberterrorist activities if they are accurately executed and maintained along with properly installed, managed, and regularly updated firewalls; packet-sniffer software; antivirus software; access control lists; and user validation systems (MacKinnon et al., 2013).

This chapter begins with a stark overview of the cybersecurity landscape and continues with an in-depth review of cybersecurity concepts, which includes important cybersecurity terminology and a look at the different network security approaches that are generally utilized. This chapter then reviews the most-used cybersecurity countermeasures, such as firewall applications, intrusion-detection systems, encryption, virtual private networks, operating system hardening, and lastly, cyber-situational crime prevention. To better illustrate cyber-situational crime prevention, this chapter also explores the details of an empirical study which evaluates the effectiveness of the most common cybersecurity counter-measures on cybercrime prevention using the situational crime prevention approach. Overall, this chapter will provide readers with an understanding of cybersecurity countermeasures and guide readers through assignments to set up a honeypot and utilize steganography techniques.

Overview of Cybersecurity

Unfortunately, everyone and everything is now a target of cybercriminals. From smartphones, computers, and tablets, to baby monitors, hospital beds, and city infrastructures, our new reality is that all of our electronic devices and systems can be attacked and used as weapons against us. And worse, cybercriminals can gain hold of our devices without us even realizing it. Our smartphones and computers can be hacked to gain access to vulnerable and personal information that can then be used to harm and exploit us. Our compromised computers can become part of a bot army. Our CPU power can be taken over to steal Bitcoin to finance international terrorists. Thankfully, there are ways we can fight back and protect our systems and our personal information. These are cybersecurity countermeasures. This section will explore cybersecurity countermeasures that allow people and organizations to proactively defend against such attacks. This section will review basic cybersecurity concepts and cybersecurity terminology most used in cybersecurity systems.

The Concept of Cybersecurity

Cybersecurity System and Cybercrime Prevention

To effectively respond to cybercrime, cybersecurity systems strive to protect privacy, preserve data integrity, verify network users, and enable approved users to securely connect to internal networks (Holden, 2003). Figure 12.1 illustrates the typical structure of the cybersecurity system.

Even though multiple cybercrime prevention measures have been implemented throughout the world to protect against cybercrime attacks, empirical research on the effectiveness of such measures is limited. The research that has been conducted has focused mostly on applications of routine activities theory to cybercrime activities (e.g., Bossler & Holt, 2009; Choi, 2008; Cohen & Felson, 1979; Leukfeldt & Yar, 2016; Wilsem, 2011, 2013), but with mixed results. For example, Choi (2008) discovered that technical capable guardianship, such as antivirus software, was associated with cybercrime victimization, while Bossler and Holt (2009), Leukfeldt and Yar (2016), and Marcum and associates (2010) found that such guardianship did not relate to victimization of cybercrimes. In another study, Testa and associates (2017) examined the relationship between situational deterring cues and cyber-trespassers' behavior and found the cues did limit the criminal activities of those with basic hacking skills; however, this strategy did not deter cybercriminals with higher levels of hacking abilities.

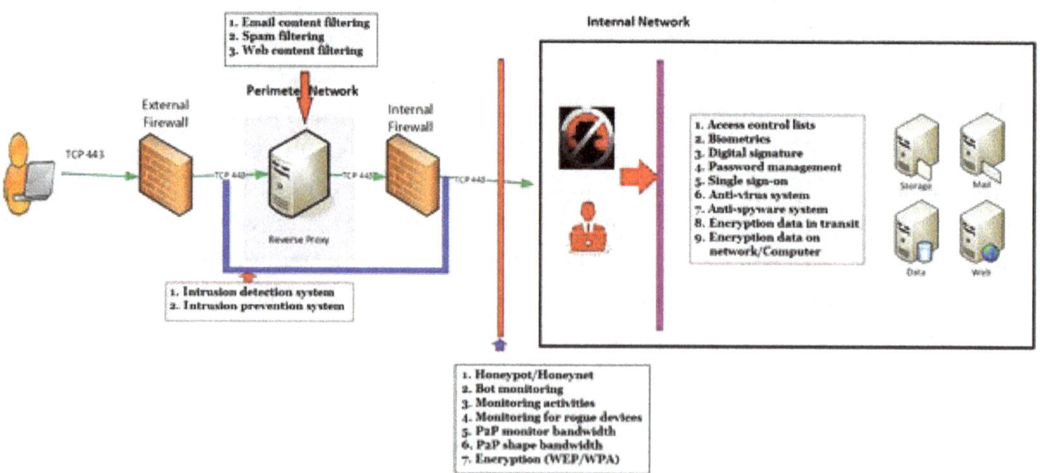

FIGURE 12.1 Diagram of cybersecurity system structure

Cybersecurity Terminology

Before moving further into the chapter, it is important to provide an understanding of the basic cybersecurity terminology used in the industry. The most commonly used terms are defined below (Kissel & Moon, 2014; Easttom, 2019).

Firewall. A firewall is a barrier between the internal network of a computer and the external network. Sometimes, the firewall can be a stand-alone server located outside the computer, such as a router, and sometimes the firewall comes in the form of software. The firewall's purpose is to filter traffic coming into and going out of the network to keep the internal network secure, often used in conjunction with a proxy server.

Proxy server. The proxy server is used to reduce exposure to attacks by hiding the computer's internal network IP address and, instead, present a single IP address to the outside world. Firewalls and proxy servers provide basic perimeter security. Sometimes the effectiveness of these devices can be strengthened with an intrusion-detection system (IDS).

Intrusion-detection system (IDS). An IDS monitors traffic searching for any suspicious activity that might indicate an unauthorized access.

Access control. Access control is another important cybersecurity measure that limits access to reduce unauthorized activity. Access control specific mechanisms can include specialized logon procedures, encryption, and any other method used to prevent unauthorized access.

Authentication. Serving as a subset of access controls, authentication is the process used to determine if the credentials given by a user or another system are sufficient to permit access to the network. For example, when a user logs in with a username and password, the security system verifies that username and password. If the information is authenticated, the user will be allowed access.

Auditing. Auditing is the process of reviewing logs, records, and procedures to determine whether they meet applicable standards.

Advanced persistent threat. Advanced persistent threat refers to an adversary that possesses sophisticated levels of expertise and significant resources which are used to create more opportunities for intrusion by using multiple attack vectors (such as cyber, physical, and deception).

Encryption. Encryption is the process of transforming plaintext into code to prevent unauthorized access.

Keylogger. Keylogger is software or hardware that monitors keystrokes and keyboard movements, usually covertly, to track actions by the user of an information system.

Private key. A private key is a cryptographic key that must be kept confidential and is used to enable the operation of an asymmetric (public key) cryptographic algorithm.

Public key. A public key is a cryptographic key that may be widely published and is used to enable the operation of an asymmetric (public key) cryptographic algorithm.

Secret key. A secret key is a cryptographic key used for both encryption and decryption, enabling the use of a symmetric key cryptography scheme.

Signature. A signature is a recognizable, distinguishing pattern used to prevent unauthorized access.

Symmetric cryptography. Symmetric cryptography is a branch of cryptography in which a cryptographic system or algorithms use the same secret key (a shared secret key).

Symmetric key. A symmetric key is a cryptographic key used to perform both the cryptographic operation and its inverse, for example to encrypt plaintext and decrypt ciphertext, or to create a message authentication code and verify the code.

Network Security Approach

There are several approaches to network security that an organization can choose to reduce unauthorized access to its systems. Once a particular approach, or paradigm, is chosen, this impacts all other cybersecurity decisions and mechanisms used to protect the network and sets the tone for the entire security infrastructure. There are two main ways that network security paradigms can be classified: (1) by the scope of the security measure taken (perimeter and layered) and (2) by the proactivity level of the system.

Perimeter Security Approach

In the perimeter security approach to network security, the magnitude of security efforts occurs in the network perimeter, such as firewalls, proxy servers, password rules, and other mechanisms, to reduce unauthorized network access. However, with this approach, very little is done to secure the system inside the network. Therefore, the perimeter is more secure, but once inside the perimeter, the network is more vulnerable. This approach is generally used by organizations with smaller budgets or more inexperienced network personnel (Hennin et al., 2007; Kissel & Moon, 2014). It is possible this paradigm is sufficient for smaller organizations without sensitive data; however, a more robust security approach is generally needed for larger corporate institutions.

Layered Security Approach

The layered security approach is generally stronger than the perimeter security paradigm and therefore serves as the preferred method whenever possible. Using this approach, not only is the perimeter secured, but the systems within the network are also secured, such as all servers, workstations, routers, and hubs (Yildiz et al., 2009). For example, using this paradigm, system network administrators can divide the network into segments and secure each segment separately, so that in the case of unauthorized access into the perimeter, only one section may be affected instead of the entire network.

Hybrid Security Approach

Some organizations also use the hybrid security approach, which combines elements of the perimeter and layered security paradigms (Bhatele et al., 2012). In fact, most networks generally fall along a continuum with elements of more than one security paradigm. For example, an institution can have a network that is predominantly passive on the perimeter but layered once inside the network, or one that is primarily perimeter focused but also has some level of protections once inside. It can be helpful to consider approaches to network security by using a Cartesian coordinate system, with the x-axis representing the level of passive-active approaches and the y-axis representing the range from perimeter to layered defense. The most desirable hybrid approach is a dynamic layered paradigm.

Cybersecurity Countermeasures

As stated, cybersecurity countermeasures are the capstone of computer security. This section will introduce you to network security threats and specific countermeasures most commonly used in computer and network system defense. It begins with an introduction to the fundamentals of firewalls followed by firewall practical applications. This section continues with explanations of intrusion-detection systems, encryption fundamentals, operating system hardening, as well as defending against malicious attacks.

Firewall Practical Applications

Types of Firewalls

As stated previously, firewalls serve as a barrier between the external world and the internal network of a computer (Bowen et al., 2012). This section will explore four types of firewalls: packet filtering, application gateway, circuit-level gateway, and stateful packet inspection. First, the packet-filtering firewall is the most basic type of firewall. When a packet-filtering firewall, also known as a screening firewall, is utilized, each incoming packet is examined. These firewall types can filter packets based on packet size protocol, source IP address, and many other characteristics. This is a useful type of firewall; however, there are some disadvantages, including the inability to examine the packet or compare it to previous packets, which can result in a ping flood or SYN flood. Furthermore, packet-filtering firewalls do not offer user authentication because they look at the packet header alone for information. Therefore, thousands of packets could possibly come from the same IP address in a short period of time, and a screened host would not identify that this pattern is unusual and could indicate a DoS attack on the network.

The second type of firewall is the stateful packet inspection (SPI) firewall, which is an improvement on basic packet filtering because this type of firewall not only examines each entering packet but also examines data from previous packets, providing more context which can better identify potential attacks. This makes these firewalls far less susceptible to ping floods, SYN floods, and spoofing. The third type of firewall most commonly used is the application gateway, which is a program that runs on a firewall. The application gateway assesses the client application and server-side application, then determines if it should be permitted. It is named application gateway because it operates by negotiating with various types of applications to allow their traffic to enter past the firewall. Application gateway enables the administrator to allow access only to specified types of applications, such as web browsers or FTP clients.

The final type of firewall most commonly used is the circuit-level gateway firewall, which is similar to application gateways but is considered more secure and generally implemented on more sophisticated equipment. User authentication is also used, but it happens earlier in the process than application gateways. To illustrate, an application gateway checks the client application first to see if access should be granted then as a second step it checks the user authentication. Using circuit-level gateways, checking the user's logon ID and password is the first step, so every individual user must be verified before moving forward to establish a connection.

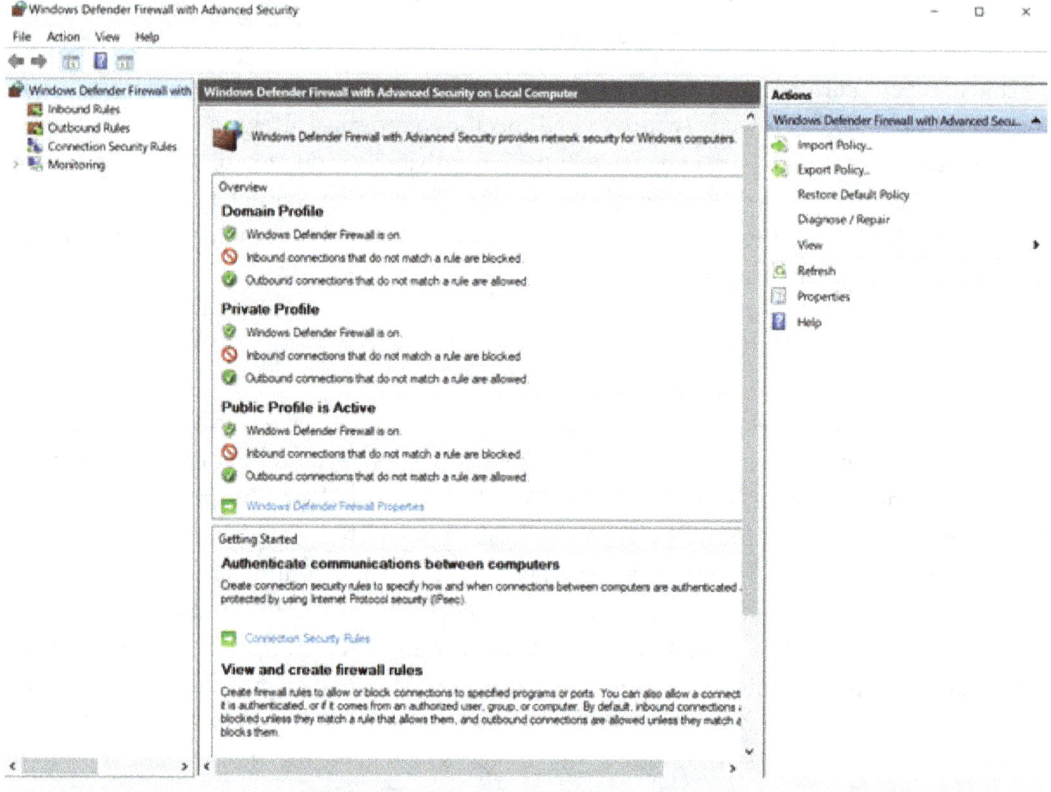

FIGURE 12.2 Windows 10 firewall

Implementing Firewalls

Not only are there various types of firewalls, but there are also different ways to implement firewall systems, which specifically refers to how the firewall is set up in relation to the network it serves to protect. The most common types of firewall implementation are (1) network-host based, (2) dual-homed host, (3) router-based firewall, and (4) screened host.

First, using a host-based implementation system means that the firewall is installed as software directly on the system. However, this means that if the underlying operating system is weak, the firewall software may not operate properly. A second type of firewall implementation is the dual-homed host, which means a firewall runs on a server and the network administrator indicates which packets to route through the firewall and how to route them. Systems inside and outside the firewall can communicate with the dual-homed host but cannot communicate directly with each other. Third, using the router-based firewall implementation system, a vendor customizes a router with a firewall that fits the customer's need. The customer then installs the router between the network and external internet connection, which can substantially increase the firewall's effectiveness. Finally, the screened host method of firewall installation is a combination of firewalls, such as a bastion host and a screening router, creating a dual firewall that can be very effective when filtering traffic and preventing unauthorized access.

Intrusion-Detection System (IDS)

An intrusion-detection system (IDS) is created to expose breaches and alert the network administrator when suspicious activity is detected. An IDS seeks patterns that may reveal an attempted attack in all incoming and outgoing activity on the machine or firewall. Unusually large flows of packets from the same IP address in a short amount of time, possibly indicating a DoS attack, can also be detected by the IDS. Two major approaches can decrease the likelihood of a successful attack such as (1) preemptive blocking and (2) anomaly detection (Bace & Mell, 2001; Jabez & Muthukumar, 2015). Preemptive blocking is used to prevent intrusions before they occur by detecting any warning signs and immediately blocking access to that user or IP address. For example, if the IDS identifies early footprinting stages of an upcoming attack, the IDS will deny access to the source of that activity.

When a particular IP address is regularly problematic, then the IP address can be blocked at the firewall. Using software, anomaly detection searches for unusual behavior, makes note of that behavior, and creates a log which is sent to the network administrator. The software identifies unusual behavior by collecting normal usage patterns, so it can flag any behavior outside of those patterns. Unusual activity can be detected by using threshold monitoring, resource profiling, user/group work profiling, and executable profiling.

Implementing IDSs

This section introduces the most common IDSs.

Snort. Snort, which is software installed on a server to monitor incoming traffic, is one of the most popular open-source IDSs available. Snort operates in one of three modes: sniffer, packet logger, and network intrusion-detection. Sniffer mode displays a continuous stream of all packet contents coming across the system. Packet logger mode is similar to sniffer mode; however, the packet contents are written to a text file log rather than displayed in the console. In network intrusion-detection mode, Snort uses a heuristic approach to detecting anomalous traffic, meaning it is rules based and learns from experience.

Cisco Intrusion-Detection and Prevention. In addition to their firewalls and routers, Cisco has several models of intrusion detection, each serving a different purpose, with the Firepower 9000 series being one of the most well known. All models include malware protection as well as sandboxing.

Honeypots

Another type of cybersecurity countermeasure is the honeypot, which is generally a single machine with the purpose of simulating a server or an entire network. If a hacker is able to gain unauthorized access, the hacker is attracted to the honeypot rather than the actual system or network. Since the honeypot is a fake machine, any user attempting to connect is suspicious and considered a possible hacker; therefore, software on the honeypot will track and identify the intruder (Peter & Schiller, 2011). The honeypot system can entice the hacker to stay connected long enough to trace the origins of the attack. Figure 12.3 illustrates the honeypot concept.

Implementing Honeypots

Several methodologies are utilized for honeypot solutions such as Specter, Symantec Decoy Server, Intrusion Deflection, and Intrusion Deterrence. Specter is a software honeypot solution,

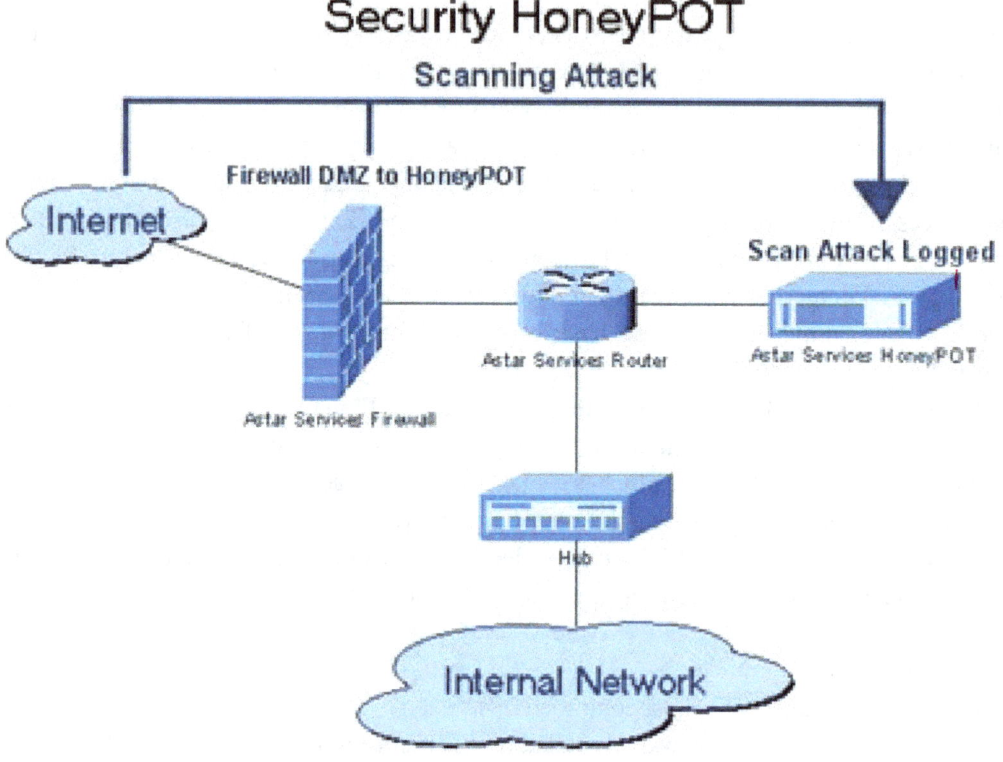

FIGURE 12.3 Honeypot diagram

which is installed on a dedicated computer to emulate the major internet protocols such as HTTP, FTP, POP3, SMTP, to appear as a fully functioning server to potential hackers. Secondly, Symantec Decoy Server is the first Symantec honeypot product, which executes an IDS which then monitors the network for signs of intrusion. Intrusion deflection then tricks the hacker into believing the intrusion was successful, when in reality, the hacker has really entered a specially designed environment to observe, track, and trace these malicious suspects.

Intrusion deterrence involves simply trying to make the system seem like a less attractive target. Therefore, it is important to make any potential reward from a successful attack appear more difficult than it is worth. This approach includes methods such as camouflage, which attempts to reduce the perceived value of the current system. Another possible way to deter intruders is by increasing the perceived risk of being caught, using methods such as displaying warnings of active monitoring. The perception of the security of a system can be drastically improved, even when the actual security remains the same.

Encryption Fundamentals

As discussed in previous chapters, cybercrime specialists must be able to explain encryption concepts, describe the history of encryption and modern encryption methods, and use some simple decryption techniques to effectively implement defensive operations. Encryption is a vital part of any network security strategy (Stallings et al., 2012). No matter how secure the network, if the data is not encrypted, then that data is vulnerable. Encryption is now so widely used that even most basic home wireless routers offer encryption. Symmetric encryption refers to those methods where the same key is used to encrypt and decrypt the plaintext.

It is sometimes necessary to lengthen a key to make it stronger, which is referred to as key stretching. Using this method, the key is put through an algorithm that will stretch it or make it longer. There are two widely used key stretching algorithms:

- **PBKDF2** (Password-Based Key Derivation Function 2) is part of PKCS #5 v. 2.01. It applies some function to the password or passphrase along with salt (adding random data for extra security) to produce a derived key.
- **bcrypt** is used with passwords, and it essentially uses a derivation of the Blowfish algorithm, converted to a hashing algorithm, to hash a password and add salt to it.

Public-key encryption is essentially the opposite of single-key encryption. With any public-key encryption algorithm, one key is used to encrypt a message and another is used to decrypt the message. The public key can be freely distributed so that anyone can encrypt a message to send to you, but only you have the private key to decrypt the message. Furthermore, a digital signature refers to the utilization of asymmetric cryptography, in reverse order. This can be useful to verify the sender of a message. For example, if an employee receives an unexpected email from their boss offering a week off with pay, the employee may want to verify that the message did indeed come from their employer and is not a prank. This can be verified using the digital signature method.

Another method of encryption is steganography, which refers to the art and science of writing hidden messages in such a manner that only the sender and intended recipient even know the message is there. If the message is hidden in another file such as a digital photo or audio file, then it cannot be detected, and no one will attempt to capture and decipher the data. Also, steganalysis can analyze an image to detect hidden messages, one of which is the Raw Quick Pair (RQP) method. Steganalysis of audio files examines noise distortion in the carrier file, which could indicate the presence of a hidden signal.

Virtual Private Networks

Virtual private networks (VPNs) are a common way to connect remotely and securely to a network. A VPN creates a private network connection over the internet to connect remote sites or users together. Instead of using a dedicated connection, a VPN uses virtual connections routed through the internet from the remote site or user to the private network. All transmissions are encrypted, resulting in a secure connection. A VPN allows a remote user to have network access just as if the user were local to the private network. Because most organizations have increased numbers of employees traveling and working from home, remote network access has become an important security concern. Since users want access, and administrators want security, the VPN is the current standard for providing both. To accomplish its purpose, the VPN must emulate a direct network connection, which means it must provide both the same level of access and the same level of security as a direct connection (Bhat et al., 2016). To accomplish this task, data is encapsulated, or wrapped, with a header that provides routing information allowing it to transmit across the internet to reach its destination. This creates a virtual network connection between the two points. The sent data is also encrypted, thus making that virtual network private.

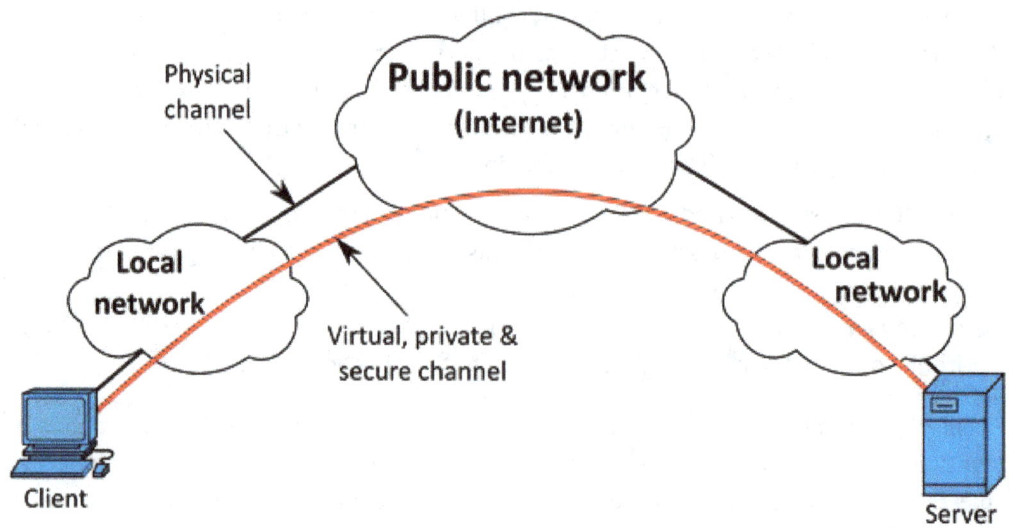

FIGURE 12.4 VPN technology

Using VPN Protocols for VPN Encryption

Encryption of VPNs can be achieved in multiple ways. The two most commonly used methods for this purpose are Point-to-Point Tunneling Protocol (PPTP) and Layer 2 Tunneling Protocol (L2TP).

Point-to-Point Tunneling Protocol (PPTP)

PPTP is a tunneling protocol that enables an older connection protocol, PPP (Point-to-Point Protocol), to have its packets encapsulated within Internet Protocol (IP) packets and forwarded over any IP network, including the internet itself. When connecting users to a remote system, encryption is vital to security, but you must also authenticate the user. PPTP supports two separate technologies for user authentication: Extensible Authentication Protocol (EAP) and Challenge Handshake Authentication Protocol (CHAP). EAP provides a framework for several different authentication methods such as passwords, challenge-response tokens, and public-key infrastructure certificates. Alternatively, CHAP is a three-part handshaking procedure. After the link is established, the server sends a challenge message to the origination client machine. The originator responds by sending back a value calculated using a one-way hash function. The server checks the response against its own calculation of the expected hash value. If the values match, the authentication is acknowledged; otherwise, the connection is usually terminated.

Layer 2 Tunneling Protocol (L2TP)

L2TP is an extension or enhancement of the PPTP that is often used to operate virtual private networks over the internet. L2TP offers support for other authentication methods such as

- **EAP**
- **CHAP**

- **MS-CHAP:** A Microsoft-specific extension to CHAP to provide functionality available on the LAN to remote users while integrating the encryption and hashing algorithms used on Windows networks.
- **PAP:** Password Authentication Protocol (PAP) is the most basic form of authentication. With PAP, a user's name and password are transmitted over a network and compared to a table of name-password pairs. Typically, the passwords stored in the table are encrypted.
- **SPAP:** Shiva Password Authentication Protocol (SPAP) is a proprietary version of PAP. With SPAP, the username and password are both encrypted when they are sent.
- **Kerberos:** Kerberos is one of the most well-known network authentication protocols. It works by sending messages back and forth between the client and the server. Because the actual password is never sent, it is impossible for someone to intercept it.

Internet Protocol Security (IPSec)

IPSec is a technology used to create virtual private networks. IPSec is used in addition to the IP protocol that adds security and privacy to TECP/IP communication. IPSec is incorporated with Microsoft operating systems as well as many other operating systems. For example, the security settings in the Internet Connection Firewall that ships with Windows XP and later versions enables users to turn on IPSec for transmissions. IPSec is a set of protocols developed to support secure exchange of packets. For IPSec to work, the sending and receiving devices must share a key, an indication that IPSec is a single-key encryption technology.

Operating System Hardening

Operating system hardening is vital to achieve a more secure network (Choi et al., 2018). Most network administrators know to properly configure each machine and server for optimal security; however, doing the same with operating systems, software, and applications is frequently ignored.

For example, most Windows systems come with default user accounts and groups, which can be a starting point for intruders who want to crack passwords for those accounts and thereby gain access to a server or network. Simply renaming or disabling some of these default accounts can improve network security. The Windows 10 Accounts screen is displayed in Figure 12.5. From this dialog box, you can add accounts, delete accounts, or change accounts. The default

FIGURE 12.5 Accounts in Windows 10

administrator account has administrative privileges, and hackers frequently seek to obtain the logon information for an administrator account. Guessing a logon is a two-fold process of first identifying the username and then the password. Using default accounts allows the hacker to bypass the first half of this process.

Furthermore, a number of registry settings affect how the TCP/IP stack handles incoming packets. Setting these properly can help reduce your vulnerability to DoS attacks. This process—stack tweaking—prevents the redirection of packets, changes the timeout on connections, and alters how Windows handles TCP/IP connections. Applying such security settings to a host machine can be a tedious task for even the most experienced information security officer; however, the best way to simplify this type of operating system hardening is to use security templates because they contain hundreds of possible settings that can control a single computer or multiple computers. Security templates can control areas such as user rights, permissions, and password policies, and they enable administrators to deploy these settings centrally by means of Group Policy Objects (GPOs).

Thus, proper configuration of the operating system can make many hacking techniques more difficult and make a system more resistant to DoS attacks. Setting up appropriate policies for users and accounts can make hacking into those accounts much more difficult. Policies should cover issues such as appropriate password length, password type, and password age/history.

The following section will discuss the concept of situational crime prevention in alignment with cybersecurity countermeasures with an in-depth look into a relevant empirical study.

Cyber-Situational Crime Prevention

The situational crime prevention approach suggests that crime can be prevented by environmental settings that directly and indirectly impact criminals' perceptions of efforts, risks, rewards, provocations, and excuses (Cornish & Clarke, 2003; Welsh & Farrington, 2004). Alongside these three theories, Clarke developed situational crime prevention (SCP) theory with 16 original opportunity-reducing techniques in 1980 (Clarke, 1997). Clarke defines "situational crime prevention as opportunity-reducing measures that (1) are directed at highly specific forms of crime; (2) involve the management, design, or manipulation of the immediate environment in as systematic and permanent a way as possible; (3) make crime more difficult and risky, or less rewarding and excusable as judged by a wide range of offenders" (Schneider, 2014, p. 45).

Critics including Wortley (2001), however, argued that the existing classification needs to be revised with four types of precipitator—prompts, pressures, permissions, and provocations—in order to reflect the relative neglect of other situational forces for criminal decision making and situational prevention (Cornish & Clarke, 2003). In line with that criticism, Cornish and Clarke (2003) revised and proposed a classification of 25 SCP techniques (see Table 12.1). The SCP theory consists of five major categories: (1) increasing the effort of the offender by target hardening; (2) increasing risks to the offender; (3) reducing rewards to the offender; (4) removing people's excuses to commit crimes; and (5) reducing the provocations of the offender (Cornish & Clarke, 2003; Schneider, 2014).

Cyber-Situational Crime Prevention Techniques

As discussed above, SCP techniques can be extended to cyber settings to enhance the cyber-crime preventative framework. For example, automobiles with target hardening techniques (e.g., steering column locks) are less likely to be burglarized than those without those techniques (Schneider, 2014). In the virtual world, online users with cyber target hardening techniques (e.g., firewall systems) might be less likely to experience cyber-intrusions and unauthorized access. Firewall systems are hardware or software that (1) allows traffic for an established connection and (2) denies traffic for malicious packets containing false information to protect a network from unauthorized access and malicious attacks (Holden, 2003). These examples provide support that situational crime prevention techniques cannot only be applied to reduce crime in the physical world but can also be applied to reduce cybercrime. Specifically, higher education institutions can especially utilize situational crime prevention techniques to protect valuable and sensitive data with such tools such as firewalls; encryption; cybersecurity training of faculty, staff, and students; and strong password management systems.

It must be noted that scholars now utilize the 25 SCP techniques to explain criminal offending and crime prevention in the physical world, whereas the original 16 SCP techniques are still mainly applied and perhaps most relevant to conducting research on cybercrime and crime prevention in cyberspace. According to Hinduja and Kooi (2013) who provided an application of 16 SCP to information security, although the 25 SCP techniques contribute more to identifying methods that explain situational precipitators and opportunity contributors, many of these elements were irrelevant for preventing cybercrime incidents. Also, because of the available variables, the data set analyzed in this study is more apt to the 16 SCP techniques than the 25 SCP techniques. As a result, the 16 cyber-SCP techniques (shaded in gray in Table 12.1) are utilized in this study to examine the relationships between cybercrime and cyber-SCP activities. In total, 46 cybercrime prevention measures were grouped according to 16 cyber-SCP techniques, with these 16 techniques divided according to one of the four categories of general means—increase the efforts, increase the risks, reduce rewards, and remove excuses.

Cyber-Situational Crime Prevention and the Breadth of Cybercrimes

In accordance with existing literature, there should be a link between cybersecurity measures, crime prevention activities, and cybercrimes. Thus, the purpose of this empirical study is to evaluate the effectiveness of common cybersecurity countermeasures on cybercrime prevention. Additionally, this study extends the SCP framework from existing research by providing cyber-SCP techniques. In this regard, this study particularly focuses on suggesting concrete concepts of cyber-SCP techniques based on SCP and other scholars' application of SCP to information security. Furthermore, this study demonstrates how these cyber-SCP techniques can be specifically applied to building effective cybercrime prevention strategies. With this reasoning, this study comprehensively measures four elements of opportunity-reducing techniques: increase effort, increase risks, reduce reward, and remove excuses. Thus, the following research questions are used to guide the analysis: (1) Is there a relationship between the use of cyber-SCP techniques and cybercrime incidents? (2) Is there a relationship between the use of cyber-SCP techniques and the breadth of cybercrime types?

To address these research questions, first, this study inductively hypothesizes there will be a positive or negative relationship between cyber-SCP techniques and cybercrime incidents. Additionally, this study hypothesizes that higher education institutions with higher level applications

TABLE 12.1 Cyber-Situational Crime Prevention Techniques

OPPORTUNITY-REDUCING STRATEGIES	CYBER-SCP TECHNIQUES	CYBERCRIME PREVENTION MEASURES
Increase efforts	1. Target hardening	• Firewall: perimeter • Firewall: interior • internal firewall • patch computers
	2. Access control	• Digital signatures • Password management • Single sign-on • Access control list
	3. Deflecting offenders	• Honeypot (i.e., identifying malicious hackers) • Honeynet (i.e., identifying bots/zombies)
	4. Controlling facilitators	• Reference check • Criminal background check • Identity management • Role-based access control
Increase risks	5. Entry/exit screening	• Intrusion-detection system • Intrusion-prevention system • Anti-virus • Anti-spyware • Use content filtering • Email content filtering • Spam filtering • Web content filtering
	6. Formal surveillance	• Bot monitoring • Monitor activity • Monitor for rogue devices
	7. Surveillance by employees	• Employees mandatory training • Full-time IT officer
	8. Natural surveillance	• Peer-to-peer technology: monitor bandwidth • Peer-to-peer technology: shape bandwidth
Reduce rewards	9. Target removal	• Encryption data on hard drive • Encryption backup data for off-site storage • Monitor use of backup media (e.g., USB drives)
	10. Identifying property	• Information asset classification
	11. Reducing temptation	• Level of sensitive information sharing • Physical separation
	12. Denying benefits	• Encryption (e.g., wep, wp) • Encryption data in transit (pki, ssl, https) • Encryption data on network or computers

(continued)

TABLE 12.1 Cyber-Situational Crime Prevention Techniques (*Continued*)

OPPORTUNITY-REDUCING STRATEGIES	CYBER-SCP TECHNIQUES	CYBERCRIME PREVENTION MEASURES
Remove excuses	13. Rule setting	• User agreement • Acceptable use policy/laws
	14. Stimulating conscience	• Warning banners on website • Codes of ethics
	15. Controlling disinhibitions	• Warning violators • Suspension • Dismissal • Restricted access to network
	16. Facilitating compliance	• Cybersecurity education for staff, faculty, and students

Source: Created by Sinchul Back, along with sources from Clarke (1992, 1995, 1997), Cornish and Clarke (2003), Beebe and Rao (2005), and Hinduja and Kooi (2013).

of cyber-SCP will have experienced a lower breadth of cybercrime types. In correspondence with these empirical investigations, the application of SCP to cybersecurity prevention efforts offers several potential benefits. First, this empirical exploration of cyber-SCP techniques can demonstrate how crime prevention frameworks can be extended to digital/cyber settings in addition to physical settings. Second, exploring the applicability of the cyber-SCP framework to cybercrimes can give us more insight into the nature of cybersecurity vulnerabilities and suitable tactics to protect academic institutions' assets against motivated cyber adversaries. Lastly, identifying an effective cyber-SCP framework can help stakeholders—academia, law enforcement, policymakers, private sector—create a practical roadmap for improving cybersecurity strategies.

Data

This study uses data derived from the "Impact of Information Security in Academic Institutions on Public Safety and Security in the United States, 2005–2006 (ICPSR 21188)." In this original study, four data sets (quantitative field survey data, qualitative one-on-one interview data, subject 1 network analysis data, and subject 2 network analysis data) were collected to develop practical policies or cost-effective controls for critical information security in academic institutions. Six hundred higher education institutions in the United States were randomly selected from the Department of Education's National Center. These academic institutions were asked to participate in the quantitative survey by postcard, telephone, and email in 2005–2006. The quantitative survey data was collected from 72 universities (12 percent response rate). While the small sample size and age of the data are both limitations of this study, this data set may be the only existing data set that researchers can currently utilize for an empirical cybersecurity study on American higher eduction institutions (Holt et al., 2015); therefore, the data is still valuable as a starting point to examine this topic.

Measures

Dependent Variables

The experience of cybercrimes was created by asking the question: "Which of the following types of cybercrimes has your institution experienced within the past 12 months?" To answer the question, respondents chose from the following lists: (1) denial of service; (2) website defacement; (3) unauthorized access to information, systems, or networks; (4) exposure of private information; (5) theft of private information; (6) theft of intellectual property; (7) sabotage; (8) fraud; (9) bot hosting; and (10) copyright infringement. The original responses were coded as a 1 if they experienced the type of cybercrime and a 0 if the cybercrime type was not experienced. In addition, the 10 items were put into an additive scale (Cronbach's alpha = .72) that ranged from 0 to 8, with 8 indicating the university experienced all eight cybercrime incidents within the past 12 months.

Independent Variables

To measure independent variables, participants were asked to respond to the following questions: "Please describe the cyber-SCP techniques that your academic institution has implemented or not implemented (Please select one response for each technique below—implemented [1], not implemented [0])."

1. **Target hardening:** firewall perimeter, firewall interior, internal firewall, and patching computer.
2. **Access control:** digital signatures, password management, single sign-on, and access control list.
3. **Deflecting offenders:** honeynet and honeypot.
4. **Controlling facilitators:** reference/criminal background check, identity management, and role-based access control.
5. **Entry/exit screening:** intrusion-detection/prevention system, anti-virus/spyware, and email/spam/web content filtering.
6. **Formal surveillance:** bot monitoring, monitoring activity, and monitoring for rogue devices.
7. **Surveillance by employees:** employees mandatory cybersecurity training, and full-time IT officer.
8. **Natural surveillance:** P2P monitor bandwidth, and P2P shape bandwidth.
9. **Target removal:** encryption data on hard drive, encryption backup data off-site storage, and using backup media.
10. **Identifying property:** information asset classification.
11. **Reducing temptation:** sharing sensitive information with federal agencies, and physical separation.
12. **Denying benefits:** encryption (e.g., WEP, WPA), encryption data in transit (PKI, SSL, HTTPS), and encryption data on network or computers.
13. **Rule setting:** requesting user agreement for cybersecurity policy/laws.
14. **Stimulating conscience:** warning banners on website, and login banner when users access.
15. **Controlling disinhibitions:** warning violators, suspension, dismissal, and restricted access to network.
16. **Facilitating compliance:** cybersecurity education for staff, faculty, and students.

Control Variables

Three control variables were included in the multivariate analysis: higher education institution's type of control, degree of urbanization, and highest degree, which could all possibly affect SCP framework capabilities and cyber-victimization. Respondents were asked: "How would you characterize your institution's type of control?" The item for type of control is coded 0 = private, 1 = public. Respondents also were asked: "How would you characterize your institution's degree of urbanization?" The item for degree of urbanization is coded 1 = large town/small town/rural, 2 = mid-size city/urban fringe of mid-size city, and 3 = large city/urban fringe of large city. Lastly, respondents were asked: "How would you characterize your institution's highest degree?" The item for highest degree is coded 1 = associates, 2 = bachelors, 3 = masters, 4 = doctoral, and 5 = doctoral and first-professional.

Results

Based on the original SCP and other scholars' application of SCP to information security, 46 cyber-SCP measures were initially established to transplant the concept SCP to cybercrime and cybersecurity. Next, these 46 cyber-SCP measures were utilized to explore the relationship between cyber-SCP techniques and the occurrence of cybercrime type. Accordingly, only 29 cyber-SCP measures, which were statistically significant in the Phi correlation analysis for cyber-SCP techniques and cybercrime types, were displayed in the descriptive statistics (Table 12.2). Lastly, 16 cyber-SCP measures, which were the most significant measures from each of the 16 cyber-SCP techniques, were employed in the Poisson regression models in order to investigate the association between cyber-SCP techniques and the 10 major cybercrime types. Descriptive analysis was performed to describe the sample characteristics and responses to the candidate variables. Table 12.2 provides the descriptive statistics (i.e., means, standard deviations, and number of sample) for each of the dependent variables and all other variables described below.

Multivariate Models

More importantly, as a next step, the relationships between cyber-SCP techniques and the breadth of cybercrime types were assessed. This was considered with five Poisson regression equations. As a goodness of fit measure, Kolmogorov-Smirnov test is not significant (Asymp. Sig: .208); thus, Poisson regression is deemed fit to analyze the dependent variable. Also, Pearson chi-square value/df is close to 1, which indicates the model is a good fit for the data analysis. Finally, the omnibus test is statistically significant, which shows that the full model with all the independent variables is a major improvement over the intercept/baseline model.

The results for these equations are shown in Table 12.3. The variables for target hardening (internal firewall), entry/exit screening (spam filtering), and reducing temptation (physical separation) were statistically related to the breadth of cybercrime types. Specifically, target hardening ($\exp(B) = .51$, $p < .01$), entry/exit screening ($\exp(B) = .59$, $p < .05$), and reducing temptation ($\exp(B) = .46$, $p < .001$) were negatively associated with the breadth of cybercrime types. In contrast, stimulating conscience ($\exp(B) = 1.60$, $p < .05$) was positively associated with the breadth of cybercrime types. In sum, the evidence found throughout these analyses indicated that three cyber-SCP techniques—target hardening, entry/exit screening, and reducing temptation—may be the key components to preventing certain types of cybercrime activities or are at least the most commonly used.

TABLE 12.2 Descriptive Statistics

VARIABLES	MEAN	SD	N
Dependent			
DDoS attack	.50	.51	72
Website defacement	.22	.41	72
Unauthorized access	.38	.48	72
Exposure of private information	.21	.40	72
Theft of private information	.15	.36	72
Theft of intellectual property	.04	.20	72
Cyber-sabotage	.08	.27	72
Internet fraud	.07	.25	72
Bot hosting	.50	.51	72
Copyright infringement	.54	.50	72
Cybercrimes by count	2.69	2.16	72
Independent/Control			
Internal firewall	.70	.45	68
Digital signature	.09	.29	65
Honeynet	.10	.30	68
Reference check-IT	.86	.33	69
Criminal check-IT	.60	.49	65
Identity management	.40	.49	66
Role-based access control	.55	.50	67
Intrusion prevention	.16	.37	65
Anti-virus	.97	.16	70
Spam filtering	.84	.36	71
Web content filtering	.10	.30	69
Bot monitoring	.26	.44	69
Monitoring rogue devices	.25	.44	70
Full-time IT staff	.43	.49	71
P2P monitor bandwidth	.75	.43	70
P2P shape bandwidth	.60	.49	70
Encryption data on hard drive	.09	.29	65

(continued)

TABLE 12.2 Descriptive Statistics (*Continued*)

VARIABLES	MEAN	SD	N
Encryption data for off-site	.23	.43	69
Information identity	.25	.43	72
US-CERT	.21	.40	72
US Department of Education	.65	.47	72
US Internal Revenue Service (IRS)	.57	.49	72
Physical separation	.46	.50	69
Encryption data in transit	.75	.43	69
Encryption data on network/computer	.27	.44	69
Students agree to cybersecurity policy	.46	.50	71
Affiliates agree to cybersecurity policy	.25	.43	71
Warning banners	.65	.47	70
Restricted access to network	.66	.47	72
Cybersecurity education	.36	.48	72
Type of control	.50	.50	72
Urbanization	2.25	.83	72
Highest degree	3.2	1.4	72

Discussion

The malfunction or total loss of an information system from academic institutions can cause a tremendous amount of economic damage and is a massive security threat to the United States. Although scholars have begun to examine the applicability of the SCP theoretical framework in preventing cybercrime, no previous study has empirically assessed the relationships between SCP techniques and cybercrimes. In an effort to fill this gap in the literature, the present study addressed common cybersecurity measures and portrayed the concept of cyber-SCP techniques. In addition, this study explored the relationships between cyber-SCP activities and cybercrime types through bivariate analyses. Lastly, this study empirically investigated the applicability of 16 forms of cyber-SCP techniques in preventing cybercrime in the virtual world under the following categories: (1) increase effort, (2) increase risks, (3) reduce reward, and (4) remove excuses.

The findings of this study indicate that certain associations exist between cyber-SCP techniques and each of the cybercrime types. In fact, the results are mostly consistent with the existing literature (e.g., Clarke, 1997; Welsh & Farrington, 2004) pertaining to SCP theory—increasing a criminal's effort and risk, and removing the rewards of crimes are substantially associated with crime prevention. Specifically, the results of this study lend support for the

TABLE 12.3 Multivariate Models of Sixteen Cyber-SCP Techniques to Breadth of Cybercrime Types from Poisson Regressions

<table>
<thead>
<tr>
<th rowspan="2">OPPORTUNITY-REDUCING STRATEGIES</th>
<th rowspan="2">CYBER-SCP TECHNIQUES</th>
<th rowspan="2">CYBERCRIME PREVENTION MEASURES</th>
<th>MODEL 1 (N = 60)</th>
<th>MODEL 2 (N = 69)</th>
<th>MODEL 3 (N = 66)</th>
<th>MODEL 4 (N = 69)</th>
<th>MODEL 5 (N = 57)</th>
<th></th>
<th></th>
<th></th>
<th></th>
<th></th>
</tr>
<tr>
<th>EXP(B)</th>
<th>SE</th>
<th>EXP(B)</th>
<th>SE</th>
<th>EXP(B)</th>
<th>SE</th>
<th>EXP(B)</th>
<th>SE</th>
<th>EXP(B)</th>
<th>SE</th>
</tr>
</thead>
<tbody>
<tr>
<td>Increase Efforts</td>
<td>*Target hardening*</td>
<td>Internal firewall</td>
<td>.44***</td>
<td>.16</td>
<td></td>
<td></td>
<td></td>
<td></td>
<td></td>
<td></td>
<td>.51**</td>
<td>.21</td>
</tr>
<tr>
<td></td>
<td>*Access control*</td>
<td>Digital signature</td>
<td>1.18</td>
<td>.23</td>
<td></td>
<td></td>
<td></td>
<td></td>
<td></td>
<td></td>
<td>1.91*</td>
<td>.34</td>
</tr>
<tr>
<td></td>
<td>*Deflecting offender*</td>
<td>Honeynet</td>
<td>.75</td>
<td>.26</td>
<td></td>
<td></td>
<td></td>
<td></td>
<td></td>
<td></td>
<td>.93</td>
<td>.32</td>
</tr>
<tr>
<td></td>
<td>*Controlling facilitators*</td>
<td>Identity management</td>
<td>.72</td>
<td>.17</td>
<td></td>
<td></td>
<td></td>
<td></td>
<td></td>
<td></td>
<td>.91</td>
<td>.21</td>
</tr>
<tr>
<td>Increase Risks</td>
<td>*Entry/exit screen*</td>
<td>Spam filtering</td>
<td></td>
<td></td>
<td>.86</td>
<td>.20</td>
<td></td>
<td></td>
<td></td>
<td></td>
<td>.59*</td>
<td>.24</td>
</tr>
<tr>
<td></td>
<td>*Formal surveillance*</td>
<td>Bot monitoring</td>
<td></td>
<td></td>
<td>1.33</td>
<td>.16</td>
<td></td>
<td></td>
<td></td>
<td></td>
<td>1.09</td>
<td>.23</td>
</tr>
<tr>
<td></td>
<td>*Surveillance by employees*</td>
<td>Full-time IT staff</td>
<td></td>
<td></td>
<td>1.38*</td>
<td>.15</td>
<td></td>
<td></td>
<td></td>
<td></td>
<td>.91</td>
<td>.24</td>
</tr>
<tr>
<td></td>
<td>*Natural surveillance*</td>
<td>P2P monitor bandwidth</td>
<td></td>
<td></td>
<td>1.03</td>
<td>.20</td>
<td></td>
<td></td>
<td></td>
<td></td>
<td>1.04</td>
<td>.31</td>
</tr>
<tr>
<td>Reduce Rewards</td>
<td>*Target removal*</td>
<td>Encryption backup data for off-site</td>
<td></td>
<td></td>
<td></td>
<td></td>
<td>1.23</td>
<td>.18</td>
<td></td>
<td></td>
<td>1.37</td>
<td>.22</td>
</tr>
<tr>
<td></td>
<td>*Identifying property*</td>
<td>Information identity</td>
<td></td>
<td></td>
<td></td>
<td></td>
<td>1.34</td>
<td>.16</td>
<td></td>
<td></td>
<td>1.33</td>
<td>.24</td>
</tr>
<tr>
<td></td>
<td>*Reducing temptation*</td>
<td>Physical separation</td>
<td></td>
<td></td>
<td></td>
<td></td>
<td>.46***</td>
<td>.17</td>
<td></td>
<td></td>
<td>.74</td>
<td>.25</td>
</tr>
</tbody>
</table>

(continued)

TABLE 12.3 Multivariate Models of Sixteen Cyber-SCP Techniques to Breadth of Cybercrime Types from Poisson Regressions (*Continued*)

OPPORTUNITY-REDUCING STRATEGIES	CYBER-SCP TECHNIQUES	CYBERCRIME PREVENTION MEASURES	MODEL 1 (N = 60)		MODEL 2 (N = 69)		MODEL 3 (N = 66)		MODEL 4 (N = 69)		MODEL 5 (N = 57)	
			EXP(B)	SE	*EXP*(B)	SE	*EXP*(B)	SE	*EXP*(B)	SE	*EXP*(B)	SE
	Denying benefits	Encryption data on network or computers					1.39*	.16			.92	.23
Remove Excuses	*Rule setting*	Students agreed policy							.75	.16	.91	.20
	Stimulating conscience	Warning banners							1.41*	.16	1.60*	.23
	Controlling disinhibition	Restricted access to network							1.70**	.18	.82	.24
	Facilitating compliance	Cybersecurity education							1.13	.16	.90	.23
	Control variables	Control by Pub/Pri/Mil									.75	.20
		Urbanization									.83	.13
		Highest degree									1.40***	.10
		Pearson x^2 value/df	1.32		1.68		1.36		1.64		1.06	
		Omnibus test sig.	.000		.024		.000		.005		.000	

continued use of target hardening, entry/exit screening, and reducing temptation in directly or indirectly preventing crime in the online setting.

The results suggest that denying benefits (e.g., encryption data in transit and encryption data on network or computer) measures were positively related to some types of cybercrime. Although existing literature argued that reducing the rewards (encryption data in transit and encryption data on network or computer) can reduce cybercrime incidents, the findings do not confirm this prediction. The most likely explanation for this result is that encryption is easily hacked by interception tools for eavesdropping or impersonation of decryption methods (Holden, 2003). According to Holden (2003), transport method encryption cannot offer a high level of security against cyber-trespassing and eavesdropping. Consequently, one might expect a higher likelihood of cybercrime victimization experiences, instead of reducing the occurrence of the cybercrime. In short, these findings suggest that this prevention technique needs to be re-examined and improved to appropriately prevent cybercrime incidents in the future.

In line with the findings from Testa et al.'s (2017) study, these results demonstrate that a warning banner on websites did not deter cyber-perpetrators from committing crimes. In fact, the academic institutions who implemented warning banners on websites were more likely to experience cybercrime incidents (e.g., bot hosting and copyright infringement) than the academic institutions without it. The current study was not able to reveal exactly why cyber-perpetrators committed cybercrimes against the academic institutions despite the display of warning banners. However, based on the studies of Testa et al. (2017) and Pogarsky (2002), this study can provide a possible explanation. In this regard, cybercriminals who have a high criminal efficacy and strong level of confidence can easily evade detection; therefore, they increasingly commit cybercriminal activities to achieve their goals, despite seeing sanction signs (Testa et al., 2017). In a broad sense, future research should consider uncovering the exact mechanisms that drive cybercriminals to continue their criminal behaviors in the presence of sanctioned threats in order to appropriately apply conscience-stimulating strategies to the digital realm as an effective opportunity-reducing technique.

Policy Implications

With these thoughts in mind, it is important to discuss the policy implications of this study. Consistent with the application of SCP measures in physical environments, the continued use of (1) increasing the efforts (target hardening), (2) increasing the risks (entry/exit screening) of committing cybercrime, and (3) reducing the rewards (reducing temptation) can be effective ways to prevent cybercrime in online settings in higher education institutions. First, to increase the efforts of crime, target hardening techniques are represented as a feasible crime prevention strategy in most ordinary street crimes as well as cybercrimes (Clarke & Eck, 2018). For example, target hardening (e.g., steering column locks) reduces burglary by making properties physically harder to break into. Likewise, firewall systems as a target hardening technique block access to the target or victim; thus, they can actively prevent cyber-intrusion and cyber-theft by making information systems and facilities among higher education institutions harder to penetrate. In other words, target hardening with firewall measures can lead cybercriminals to perceive crime opportunities (temptation) as less attractive because the offender needs more effort to successfully break the law in the digital realm. Consequently, firewall systems provide strong digital guardianship; therefore, we must keep improving and executing this capstone of cybersecurity techniques for maximum efficacy in higher education institutions.

Second, offenders tend to worry more about the risks of being arrested than about the results if they are caught (Clarke & Eck, 2018); therefore, increasing the risks of being apprehended, especially with entry/exit screening techniques, may be effective to deal with crime in both the physical and online settings. Metal detectors and screeners are regarded as effective entry/exit screening techniques in that they allow only certain individuals admittance to the physical property of an organization (Hinduja & Kooi, 2013). In online settings, entry/exit screening systems such as spam filtering and intrusion-detection systems can help to enhance a cybercriminal's risk of being apprehended. To better enhance the existing entry/exit screening systems in higher education institutions, artificial intelligence technology can be considered with the existing cyber-entry/exit screening systems. This is because artificial intelligence can enforce real-time detection or filtering, and then it is able to send quick alerts to cybersecurity staff and US-CERT team members using an artificial neural network and knowledge-based intrusion detection and filtering. Along with self-learning capabilities, artificial neural networks and knowledge-based intrusion techniques can be utilized to quickly identify suspicious and malicious behavioral patterns in cyberspace (Vieira et al., 2010). In short, as explained above, increasing the risk of arrest can discourage cyber-perpetrators from committing cybercrime against higher education institutions since the cybercriminals will feel afraid of being caught via cutting-edge entry/exit screening systems. Thus, we can prudently pursue the application of artificial intelligence to improve the existing cybersecurity systems in order to effectively combat cyberattacks.

Third, offenders are always seeking the rewards of crimes; hence, if they do not see any benefit, they are less likely to commit crimes. To prevent street crimes, law enforcement officials advise people to hide their valuable assets so that criminals cannot see their property as targets. Similar to crime prevention strategies for street crimes, techniques for removing the temptation (e.g., physical separation) of critical information and facilities in higher education institutions might decrease the opportunities of perpetrators accessing these assets simply because they are not aware of whether or not valuable properties exist. Therefore, cybercrime can be discouraged when a potential offender does not perceive a situation as a criminal opportunity (the rewards of crimes) because they cannot see any attractive targets (values) through physical separation of critical information and facilities in higher education institutions.

Summary

In sum, this chapter explored cybersecurity countermeasures such as firewalls, honeypots, intrusion-detection systems, encryption, and operating system hardening to reduce cyberthreats and create more secure technological environments. Additionally, this chapter presented an empirical study on the effectiveness of some cybersecurity measures in the environment of higher-education institutions. More specifically, the study explored the applicability of ideas—cyber-SCP techniques—drawn from situational crime prevention theory to cybercrime. It is important to note that these policy implications will have both theoretical and practical benefits to creating future cybercrime prevention strategies. Theoretically, research on criminal opportunities can lead to a better understanding of how and why cybercrimes occur in particular cyber-environments. On the practical side, it can lead to a new approach: applying situational crime prevention to cybercrime control. Particularly, this chapter provides further direction for cybercrime prevention strategies along with increasing the efforts, increasing the risks, and reducing the rewards of crimes. Therefore, we need to employ a defensive posture to optimally configure our systems; to employ patch management; and to share, receive, and act on threat indicators through the cutting-edge cybersecurity countermeasures.

Lab 12.1: Setting Up a Honeypot

In this project, as a cyber-intelligence analysist and/or forensic investigator, you will learn how to set up a honeypot. Mr. CIC wanted to set up a honeypot in order to trap Mr. Evil Noodle and other cyber-offenders who try to gain unauthorized access to information systems of the University of CIC. To set up a honeypot, here's what you need to do:

1. Open Kali Linux.
2. From the left sidebar, click the Terminal icon.
3. In the Terminal window, execute the following command to get the IP address:

   - ifconfig

4. Note the inet IP address for lo which will be required later.
5. In the Terminal window, execute the following command to change the directory to the pentbox folder:

   - cd pentbox-1.8/

6. In the Terminal window, execute the following command to run pentbox:

   - ./pentox.rb

7. In the Terminal window, type **2** and press Enter to select the Network tools option.
8. In the Terminal window, type **3** and press Enter to select the Honeypot option.
9. In the Terminal window, type **2** and press Enter to select Manual configuration.
10. In the Terminal window, type 443 and press Enter to open port.
11. In the Terminal window, type **Caught You!!** and press Enter to enter the false message.
12. In the Terminal window, type **y** and press Enter twice to save log.
13. In the Terminal window, type **y** and press Enter for not activating beep sound.
14. You will observe that a honeypot is activated on port 443.
15. From the left sidebar, click the Firefox ESR icon.
16. In the address bar, type **https://<ip address>** and press Enter. (You should take the IP address that you obtained at step 3).
17. You will observe the message as Secure Connection Failed.
18. Switch back to the Terminal window and you will observe INTRUSION ATTEMPT DETECTED. Press Ctrl + C to stop.

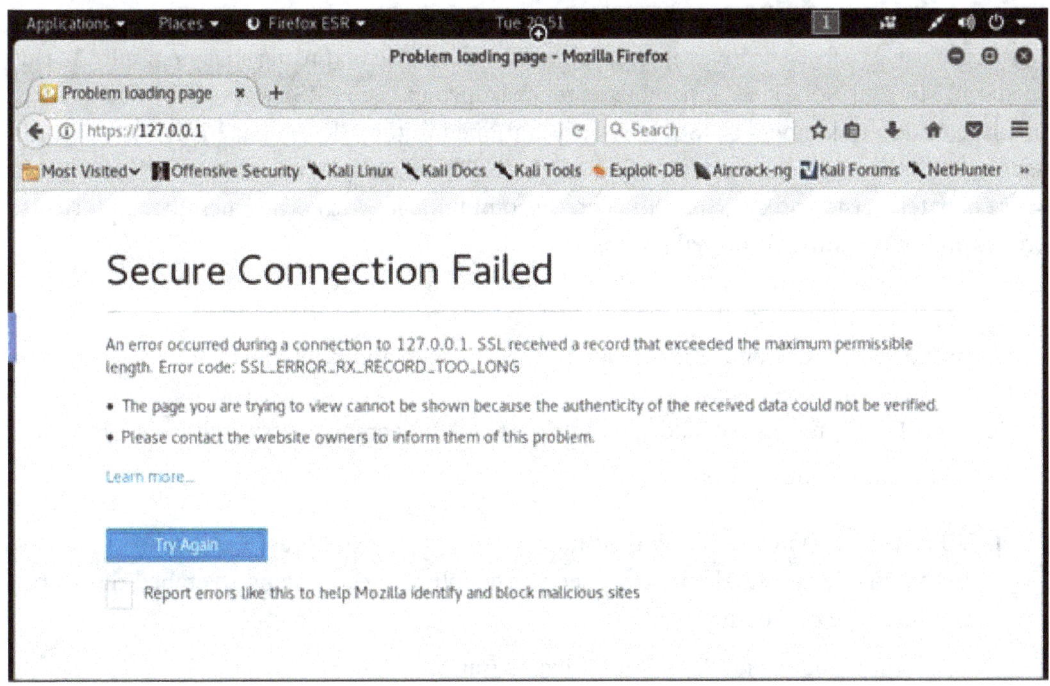

FIGURE 12.6 Honeypot activated

FIGURE 12.7 Secure connection failed

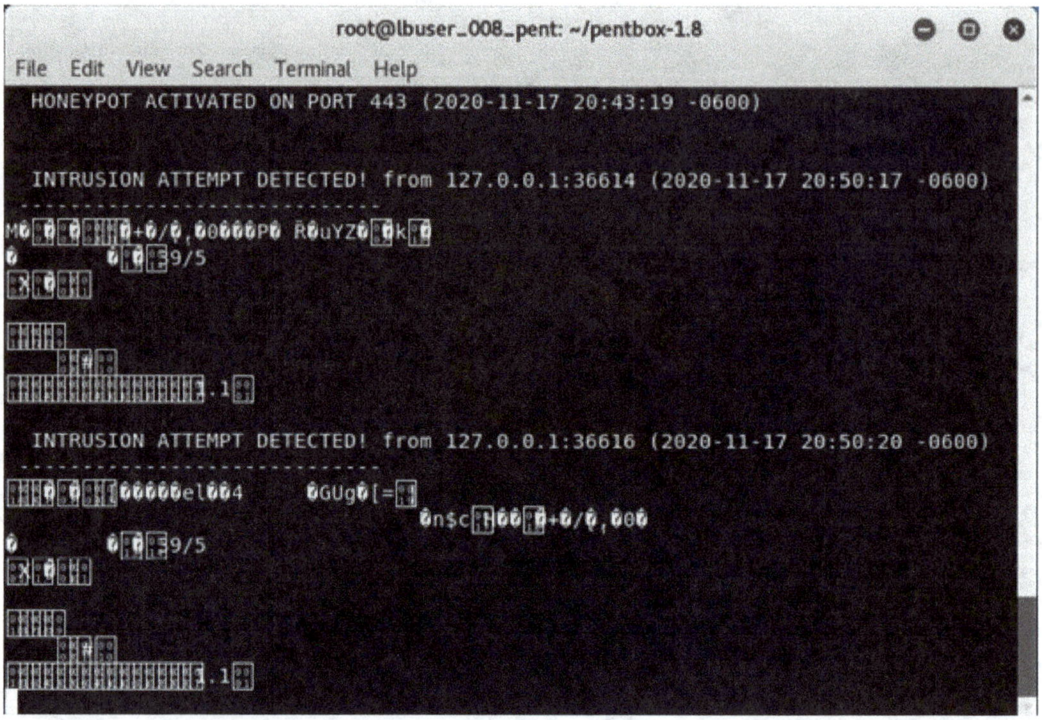

FIGURE 12.8 Intrusion attempt detected

Lab 12.2: Use of Steganography Technique

In this project, as a forensic investigator, you will learn to hide text using steganography. In this scenario, Mr. Evil Noodle as a member of terrorist group (i.e., ISIS) wanted to send a secret message to other members of ISIS currently stationed in the UK. He has hidden the text file secret.txt in the image picture.jpg using steghide. This hands-on experience will give you a sense how terrorists and/or criminals secretly communicate with other members of terrorist groups and/or organized criminal groups.

1. From the desktop, open PuTTY.
2. In the PuTTY Configuration dialog box, click Open and type the details to log into the terminal window.
3. In the PuTTY terminal window, execute the following command to install the steghide tool:

   - apt-get install steghide

4. When asked "Do you want to continue? [Y/N]," type **y** and press Enter.
5. In the PuTTY terminal window, execute the following command to embed the secret. txt file cyber the picture.jpg file:

   - cd /root/Documents/Lab_Files/Steganography

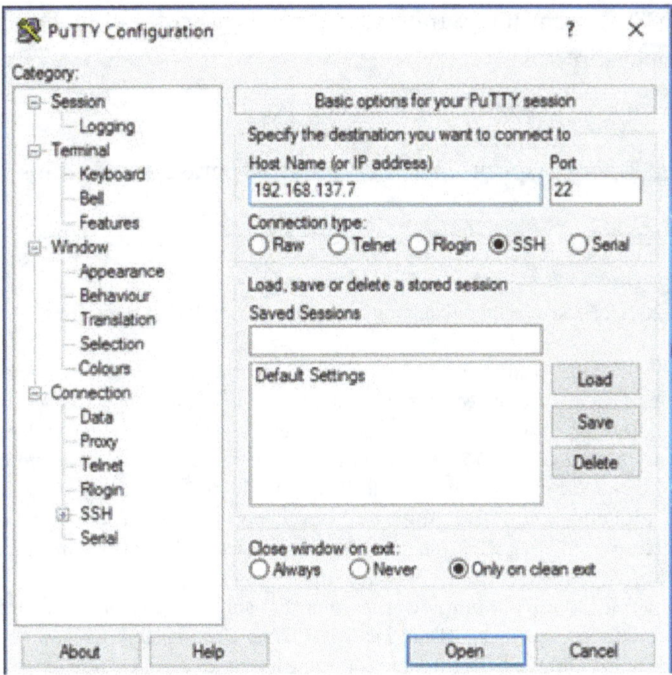

FIGURE 12.9 PuTTY configuration

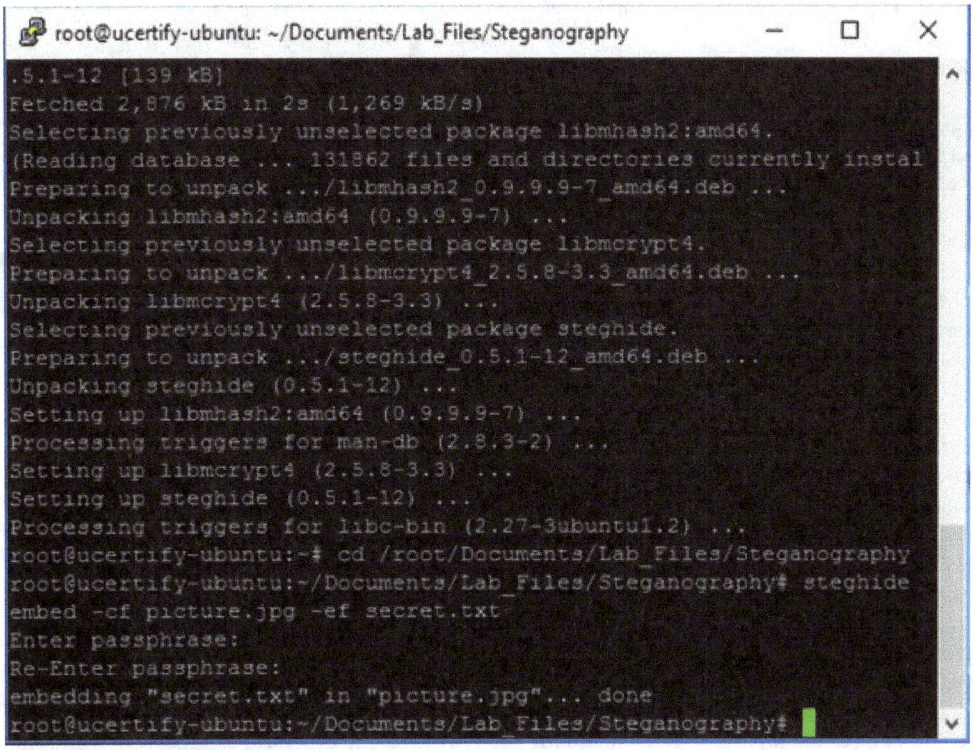

FIGURE 12.10 PuTTY terminal window for steganography

6. In the PuTTY terminal window, execute the following command to go to Steganography directory:

- steghide embed -cf picture.jpg -ef secret.txt

7. When asked Enter passphrase and Re-Enter passphrase, type details to log into.

References

Bace, R., & Mell, P. (2001). *NIST special publication on intrusion detection systems*. Booz-Allen & Hamilton Inc.

Beebe, N. L., & Rao, V. S. (2005). Using situational crime prevention theory to explain the effectiveness of information systems security. In *Proceedings of the 2005 SoftWars Conference, Las Vegas, NV* (pp. 1–18).

Bhat, A. Z., Al Shuaibi, D. K., & Singh, A. V. (2016, September). Virtual private network as a service—A need for discrete cloud architecture. In *2016 5th International Conference on Reliability, Infocom Technologies and Optimization (Trends and Future Directions) (ICRITO)* (pp. 526–532). IEEE.

Bhatele, K., Sinhal, A., & Pathak, M. (2012, August). A novel approach to the design of a new hybrid security protocol architecture. In *2012 IEEE International Conference on Advanced Communication Control and Computing Technologies (ICACCCT)* (pp. 429–433). IEEE.

Bossler, A. M., & Holt, T. J. (2009). On-line activities, guardianship, and malware infection: An examination of routine activities theory. *International Journal of Cyber Criminology, 3*(1), 400–420.

Bowen, P., Hash, J., & Wilson, M. (2012). Information security handbook: A guide for managers. *NIST National Institute of Standards and Technology* (Special Publication 800-100).

Choi, K. S. (2008). Computer crime victimization and integrated theory: An empirical assessment. *International Journal of Cyber Criminology, 2*(1).

Choi, K. S., Yang, C. H., & Kwak, J. (2018). System hardening and security monitoring for IoT devices to mitigate IoT security vulnerabilities and threats. *KSII Transactions on Internet & Information Systems, 12*(2).

Clarke, R.V., (1992). *Situational crime prevention: Successful case studies*. Albany, N.Y.: Harrow & Heston.

Clarke, R. V. (1995). Situational crime prevention. *Crime and justice, 19*, 91–150.

Clarke, R. V. (1997). A revised classification of situational crime prevention techniques. *Crime Prevention at a Crossroads*. Anderson.

Clarke, R. V., & Eck, J. E. (2018). Crime analysis for problem solvers in 60 small steps. In *Center for Problem-Oriented Policing*. https://cops.usdoj.gov/RIC/Publications/cops-w0047-pub.pdf

Cohen, L. E., & Felson, M. (1979). Social change and crime rate trends: A routine activity approach. *American Sociological Review, 44*(4), 588–608.

Cornish, D. B., & Clarke, R. V. (2003). Opportunities, precipitators and criminal decisions: A reply to Wortley's critique of situational crime prevention. *Crime Prevention Studies, 16*, 41–96.

Easttom, C. (2019). *Computer security fundamentals*. Pearson IT Certification.

Ferrari, E. & Thuraisingham, B. (Eds.). (2005). *Web and information security*. IGI Global.

Hennin, S., Germana, G., & Garcia, L. (2007, May). Integrated perimeter security system. In *2007 IEEE Conference on Technologies for Homeland Security* (pp. 70–75). IEEE.

Hinduja, S., & Kooi, B. (2013). Curtailing cyber and information security vulnerabilities through situational crime prevention. *Security Journal, 26*(4), 383–402.

Holden, G. (2003). *Guide to network defense and countermeasures*. Course Technology Press.

Holt, T. J., Burruss, G. W., & Bossler, A. (2015). *Policing cybercrime and cyberterror*. Carolina Academic Press.

Jabez, J., & Muthukumar, B. (2015). Intrusion detection system (IDS): Anomaly detection using outlier detection approach. *Procedia Computer Science, 48*, 338–346.

Kissel, R., & Moon, H. (2014). *Small business information security: The fundamentals*. NIST Internal or Interagency Report (NISTIR) 7621 Rev. 1 [Draft]. National Institute of Standards and Technology.

Leukfeldt, E. R., & Yar, M. (2016). Applying routine activity theory to cybercrime: A theoretical and empirical analysis. *Deviant Behavior, 37*(3), 263–280.

MacKinnon, L., Bacon, L., Gan, D., Loukas, G., Chadwick, D., & Frangiskatos, D. (2013). Cyber security countermeasures to combat cyber terrorism. In Babak Akhgar & Simeon Yates (Eds.), *Strategic intelligence management* (pp. 234–257). Butterworth-Heinemann.

Marcum, C. D., Higgins, G. E., & Ricketts, M. L. (2010). Potential factors of online victimization of youth: An examination of adolescent online behaviors utilizing routine activity theory. *Deviant Behavior, 31*(5), 381–410.

Peter, E., & Schiller, T. (2011). *A practical guide to honeypots.* Washington University.

Pogarsky, G. (2002). Identifying "deterrable" offenders: Implications for research on deterrence. *Justice Quarterly, 19*(3), 431–452.

Schneider, S. (2014). *Crime prevention: Theory and practice.* CRC Press.

Stallings, W., Brown, L., Bauer, M. D., & Bhattacharjee, A. K. (2012). *Computer security: Principles and practice.* Pearson Education.

Testa, A., Maimon, D., Sobesto, B., & Cukier, M. (2017). Illegal roaming and file manipulation on target computers. *Criminology & Public Policy, 16*(3), 689–726.

Vieira, K., Schulter, A., Westphall, C., & Westphall, C. (2010). Intrusion detection for grid and cloud computing. *IT Professional, 12*(4), 38–43.

Welsh, B. C., & Farrington, D. P. (2004). Surveillance for crime prevention in public space: Results and policy choices in Britain and America. *Criminology & Public Policy, 3*(3), 497–526.

Wilsem, J. V. (2011). 'Bought it, but never got it': Assessing risk factors for online consumer fraud victimization. *European Sociological Review, 29*(2), 168–178.

Wilsem, J. V. (2013). Hacking and harassment—Do they have something in common? Comparing risk factors for online victimization. *Journal of Contemporary Criminal Justice, 29*(4), 437–453.

Wortley, R. (2001). A classification of techniques for controlling situational precipitators of crime. *Security Journal, 14*(4), 63–82.

Yildiz, M., Abawajy, J., Ercan, T., & Bernoth, A. (2009, December). A layered security approach for cloud computing infrastructure. In *2009 10th International Symposium on Pervasive Systems, Algorithms, and Networks* (pp. 763–767). IEEE.

Credits

Dark Web Investigation

Marlon Mike Toro-Alvarez

Introduction

Cyberspace is not only made up of the internet but also of other non-indexed computer connections like the deep web and clandestine markets that require additional technical configurations to navigate (Choi & Toro-Alvarez, 2017). This collection of clandestine markets and services is known as the dark web.

The markets of this darknet are anonymous and allow users of these sites to buy weapons, false identities, child pornography, and illegal drugs, including synthetic drugs. The anonymity of the markets on the darknet makes it difficult for authorities to prosecute these cybercriminal behaviors. An example of this problem was the cybercriminal investigation against the darknet market Silk Road (Choi, 2015).

Silk Road marketed illegal drugs such as heroin and cocaine, and facilitated various criminal activities, for example, contract murder. The Federal Bureau of Investigations (FBI) rated Silk Road as the most sophisticated and extensive online criminal market. According to Christopher Tarbell, the lead FBI investigator, Silk Road was an online black market and the first modern darknet market, better known as a platform to sell illegal drugs. Silk Road operated as a hidden service from a network configuration known as the onion router (Tor). This chapter will show more details about Tor architecture (Chaabane et al., 2010).

Most of the products were delivered via postal mail. These shipments followed Silk Road's guide to specially pack shipments to avoid detection by authorities (Greenberg, 2013). In March 2013, the site had more than 10,000 products for sale, 70% of which were drugs. These were grouped under the labels stimulants, psychedelics, prescriptions, precursors, others, opioids, ecstasy, dissociatives, and steroids (Choi & Toro-Alvarez, 2017). In the first days of Silk Road's operation, its creator and administrators instituted terms of service that prohibited the sale of any item whose purpose was to "harm or defraud." But with business profitability and user demand, Silk Road also included child pornography, fake driving licenses, stolen credit cards, and weapons of any kind (Chen, 2011).

From February 6, 2011, to July 23, 2013, Silk Road involved more than 146,000 buyers and 3,800 sellers. There were approximately 1,229,465 completed transactions, and the total revenue generated from these sales was 9,519,664 Bitcoins with total commissions collected on sales of 614,305 Bitcoins, equivalent to approximately $1.2 billion in revenue and $79.8 million in commissions based on the 2015 Bitcoin value. (District Court Southern District of New York, 2015).

According to the location of the registered users, the United States was at the top of the list, with the United Kingdom next on the list, and then Australia, Germany, Canada, Sweden, France, Russia, Italy, and the Netherlands. However, the anonymity offered by the darknet allowed 27% of the users in this illegal market to avoid geolocation. It should be noted that during the period between May 24 and July 30, 2013, 1,217,218 messages from different continents were sent through the Silk Road private messaging system (Oremus, 2013).

However, this market was dismantled. Investigative work utilizing techniques such as undercover agent in cyberspace made the difference against anonymity within the darknet. In October 2013, the Silk Road site was turned off and its founder and owner was captured. The FBI arrested Ross William Ulbricht on October 2, 2013, at 3:15 p.m. at the Glen Park Library in San Francisco. The captured was accused of a drug trafficking conspiracy, hacking, money laundering, and the attempted murder of six people. Prosecutors alleged that Ulbricht paid some users $730,000 to commit these murders, although the killings could not be proven in court. However, according to the documentations of this case, in January and February 2013, the detainee received photographs of a person being tortured and of the body of a person (District Court Southern District of New York, 2015).

Ulbricht used the alias "Dread Pirate Roberts" or "DPR." He led an illegal marketplace that provided an anonymous online haven for multiple crimes and over a million users worldwide (Choi et al., 2014). The judicial process after Ulbricht's arrest included information the FBI obtained from an image of the Silk Road server system. Although it was a very difficult task in 2013, it reminds us that techniques and tactics can be applied to capture cybercriminals even on the darkest network.

Deep Web and Darknet

A significant portion of internet users legitimately seek to connect and configure services to protect their privacy (Castells, 2002). These users seek to prevent websites from tracking their data and to keep relevant sources of information secret, either during a journalistic interview with a witness or to protect the process of a technology patent. That protection is achieved by configuring computer systems that deviate from the traditional operating architecture of the internet. In this way, users enter the deep web with the specific purpose of protecting their communications, identities, and the integrity of the shared information (Le Blond et al., 2013).

The deep web is a great platform that can provide the answer to many academic research challenges, copyright protection, as well as shield criminal justice processes where witnesses need to be protected (Halevy et al., 2006). These applications motivate researchers to not mistake the deep web for dark web. While the positive uses of the deep web can help educate society, the opportunity for criminal services in the dark web demonstrates its complex threat against coexistence in cyberspace.

The existence of cybercriminal markets located within the dark web means that we must understand its functional architecture in order to approach investigation techniques beyond the traditional tracking on the internet (Choi & Toro-Alvarez, 2017).

Tor Architecture

The onion router (Tor) or onion routing program was developed at the U.S. Naval Research Laboratory. In the mid-90s, it easily became the most widely used computer technology in the world to preserve anonymity online (Dingledine et al., 2007). The user's web history, online postings, instant messages, and other communication can hardly be traced back to the user.

Tor works as a network made up of thousands of volunteer nodes, also called "relays." A relay is a computer within Tor, which is listed in the main directory. This computer receives internet signals from another relay and passes that signal to the next relay on the route. For each connection request (for example, a query through TorSearch, the darknet's oldest running search engine for Tor hidden services) the path is generated randomly. None of the relays keep records of these connections, so there is no way for any relays to report the traffic they have handled. The Tor network (or simply Tor) is made up of approximately 7,000 relays and 3,000 bridges.

When the user connects to the Tor network, for example, through the Tor browser, all the data that the user sends and receives passes through this network. User data travels through a random selection of nodes. Tor encrypts all that data multiple times before it leaves the user's device. This includes the IP address of the next node in the sequence. An encryption layer is removed each time the data reaches another node until the final exit node is reached. This process is named onion routing. This means that no one, not even the people running the nodes, can see the content of the data or where the data is going.

A bridge is a hidden relay, which means it is not listed in the main Tor relay directory. Bridges are provided to users who cannot access Tor with the normal configuration. In other words, when the user is using a connection with a configured proxy (a kind of supervisory intermediary between the user's computer and the internet gateway) to block Tor traffic, he or she can use a bridge to overcome that limitation.

Finally, the last relay on the route is the output node (see Figure 13.1). The egress node is the only part of the network that connects to the server that the user is trying to access, and therefore it is the only bit that the server sees and can only record the IP address of that relay. Anyone who intercepts the data will not be able to easily trace it back to the user. Usually, only the input or output node can be determined, but never both. This makes it almost impossible to track user activity and browsing history. All relays and bridges are run by volunteers who donate part of their bandwidth and computing capabilities to support Tor's functional architecture (Syverson et al., 2004).

Tor is configured in this way to allow an internet user to browse cyberspace anonymously by hiding their internet address (IP address) from the website and search engines accessed through Tor. Similarly, Tor hides network traffic from anyone at the end of the connection. An observer will only see that the user is connected to Tor, and will not see any other websites or communications that interact with the user's computer.

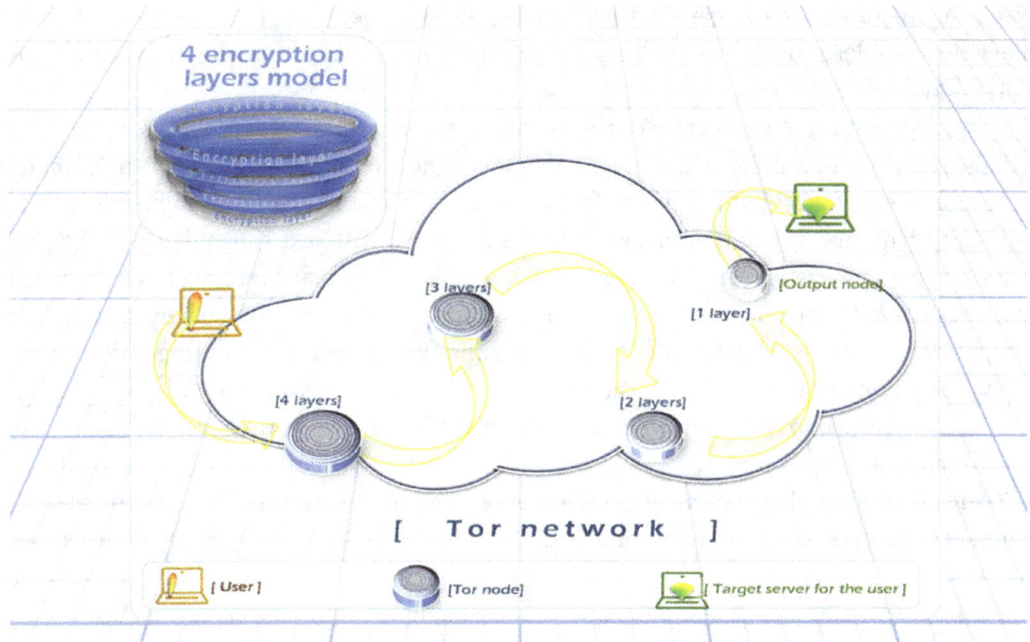

FIGURE 13.1 Model of a Tor network connection

Additionally, Tor offers other hidden services in the form of "onion" sites and an instant messaging server. "Onion" sites are websites hosted on Tor servers. These sites are hidden, generating random routes from "introductory points" on the network. In this way, users access the sites, but they do not determine the location of the servers that host them. In conclusion, web browsing within the Tor browser is completely anonymous, but other activities on the user's computer are not. Connecting other applications and services to the Tor network requires the configuration of other applications that will be mentioned later in this chapter.

Although Tor has many legal uses, the darknet it houses is also where markets for illicit goods and services gather, as well as blogs and forums for extremist groups. This is the reason the darknet is particularly suited to crime and has a reputation as the black spot of cyberspace. To access Tor, you need to use a different browser. The next section of this chapter will discuss details about the available access, browsing, and anonymity applications (Mulazzani et al., 2013).

Tor Setup

The Tor browser is the easiest and most popular way to use Tor from a personal or desktop computer. It is based on the traditional internet browser Firefox and works as a basic version of any other web browser. It is like a "plug and play" browser. No special or additional settings are required to start browsing anonymously after the initial setup. The browser is a custom version of Mozilla Firefox and is therefore seen and used like any other web browser. Personalization is designed to leave no trace of your web browsing on the computer (Eckersley, 2010).

Installation is simple, the user only downloads the compressed file for the operating system they use, be it Windows, macOS, or Linux. Then the user must extract into a local folder. After executing, the user can navigate to the content they are looking for in complete anonymity. When the user closes the browser, all traces of navigation are erased from memory. Only bookmarks and downloads are left.

If you do not want to install the Tor browser, it is possible to use Tails. This is a "live operating system" that runs from CD or USB memory. The user can connect to Tails from the removable media to a computer just before the computer restarts. If the computer's start-up software (firmware) is configured correctly, the computer will load Tails instead of the computer's operating system. In this way, Tails is a perfect application to access a computer that does not belong to them and browse the web anonymously without leaving a trace. The computer's internal hard drive is not touched while the computer is running Tails and the computer's memory is erased with each restart. Also, temporary internet files uploaded to Tails are not burned to the CD or USB memory drive while in use, so they are also lost as soon as the computer is restarted.

There are other applications available if a user requires access from a different computer to a commercial computer, that is, from a phone, tablet, robotic application, Raspberry Pi device, etc.

Orbot

Users can access the Tor network on Android devices using Orbot. Orbot creates a Tor proxy on the user's device so that all internet traffic from the user's device passes through the Tor network. That means that all applications on the user's phone or tablet will also have their traffic routed through Tor (Vallina-Rodriguez et al., 2015).

FIGURE 13.2 Orbot

Orfox

To accompany Orbot, there is also a browser for Android devices that allows the user to browse the net using Tor. However, this only applies to web browsing in a browser. All other applications on the user's Android device will communicate through normal traffic lines without the benefit of anonymity provided by the onion router.

FIGURE 13.3 Orfox icon

Arm

To use this application, the user needs prior knowledge. Arm is a command-line-based monitor for a Tor relay. This application displays real-time information for a relay or bridge on the Tor network. In this way, the investigator can monitor the retransmission of the connected user by providing statistics, metrics, and performance reports. The user can know how many Tor users have accessed Tor through the user relay or how much available bandwidth is being used in support of Tor.

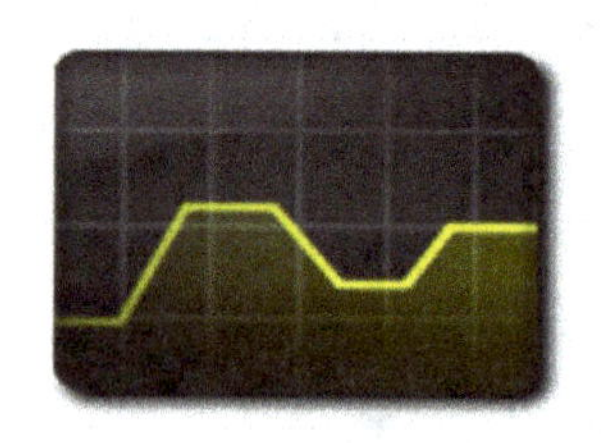

FIGURE 13.4 Arm icon

Other applications that can assist investigators are

Atlas

Atlas is a web application that provides information on the current status of relays on the Tor network. The researcher can type the name of a relay in the search box at the top of the site's graphical interface. You can also get a basic description of your current status. A researcher who wants a much more detailed report can select the alias of the relay and receive an explanation of all the information that applies to that particular node.

FIGURE 13.5 Atlas icon

Pluggable Transports

This tool changes the way the data stream of the user connected to Tor appears. This is another way to maintain the connection on this network. Some network architectures block the connection to Tor based on traffic, not the IP address of the relay or bridge being used for the connection. Pluggable Transports change the appearance of Tor traffic to look like normal traffic other than Tor to escape detection.

FIGURE 13.6 Pluggable Transports icon

Stem

This is where the library developers go to create programs to interact with Tor—for example, the interactions that Arm makes possible.

FIGURE 13.7 Stem icon

TorBirdy

This is an extension for running the cross-platform email client Mozilla Thunderbird on Tor. Some technicians consider it to be an additional button for Tor.

FIGURE 13.8 TorBirdy icon

Metrics portal

As the name implies, this is where the researcher gets metrics related to the Tor network, such as available bandwidth and the estimated size of the current user base. Any researcher who is interested in a specific topic can find very useful information from detailed statistics to on-demand metrics.

continued

Shadow

This application allows simulation of a network using the real Tor browser. An investigator can set up a lab with Shadow to assess how Tor can affect the user's network, without affecting the user's actual network. It is a highly recommended application to experiment with Tor and other programs before deploying controlled deliveries during an operation or investigation.

FIGURE 13.9 Shadow icon

Tor2web

This website gives access to users who do not want to use the Tor browser. In this way, internet users can access hidden Tor services but sacrifice anonymity. It should be noted that the anonymity of the websites accessed is not compromised. Tor2web is an alternative to make a preliminary approach to an underground market in darknet when you have limited bandwidth.

FIGURE 13.10 TOR2web icon

Tor Messenger

This instant messaging application is used by default on the Tor network. Tor Messenger is a chat with multiplatform functions that are very well accepted in the dark web community. In other words, it can be the gateway to many investigations.

FIGURE 13.11 Tor Messenger icon

txtorcon

This networking library for writing Python-based applications has all the tools to access Tor's circuits, streams, registry functions, and hidden services. The senior researcher with the ability to use it can change many paradigms regarding the limitations of an investigation in the dark web (Clarke et al., 2002).

Other "Deep-Nets"

JonDoFox

JonDoFox, another onion routing type anonymizer for web browsing, is a profile for Mozilla Firefox or Firefox ESR where the user's computer connects to a series of Mix operators that anonymize the user's web traffic and wrap it in various layers of encryption. An advantage of JonDoFox is the size of the network, which is considerably smaller than that of Tor due to its certification processes. In JonDoFox, a mixed operator is a user who passed a certification process in such a way that certain validation processes are skipped in the operation of the network (Westermann et al., 2011).

Freenet

According to Clarke et al. (2002), it is a network that bases its connectivity on the use of known computers or "friends" of the user—a different concept than using dedicated entry and exit points as in Tor. This friendly relationship creates a much higher sense of security and allows it to be considered as a file distribution service where encrypted files are stored on the hard drives of computers across the network. There is also the option to connect through strangers' computers; however, that is not the philosophy behind Freenet and is considered less secure than connecting to trusted friends' computers.

Dissent

The Dissent application offers anonymity, but its demands on bandwidth and equipment resources at the entry point are significant. This application works with symmetrical message sizes and combines them with a robust encryption algorithm. This way it does not optimize communication, but it provides anonymity which can cause a lot of problems in an investigation (Goldberg & Brewer, 2000).

FIGURE 13.12 JonDoFox icon

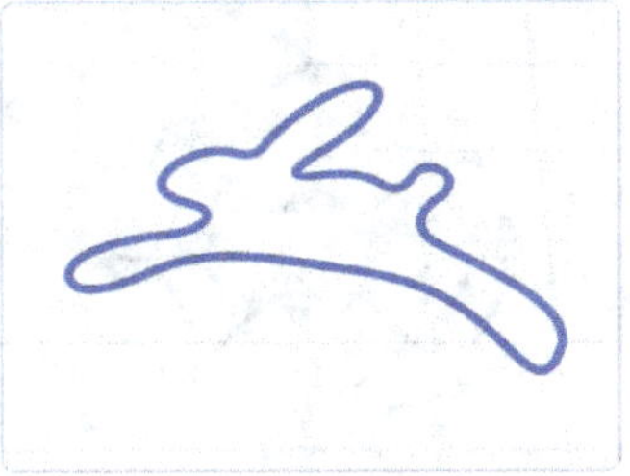

FIGURE 13.13 Freenet icon

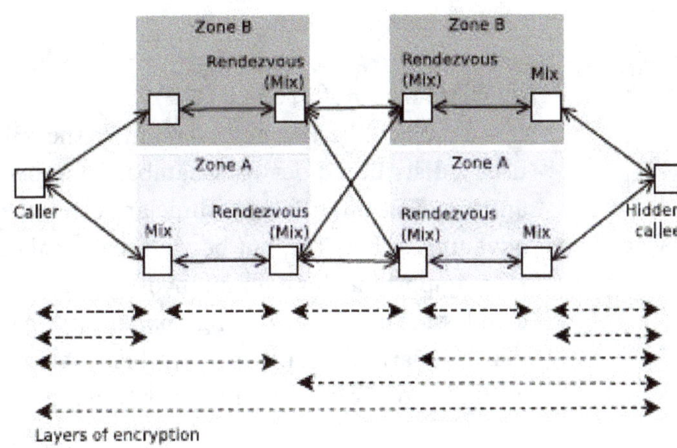

FIGURE 13.14 Aqua encryption circuit sample

Aqua

Anonymous Quanta, or Aqua, is a file-sharing network designed to be completely anonymous, but it allows metadata to be sent for network traffic. This functionality would allow tracking a user and the server with which the user is communicating. Aqua also has an IP telephony service known as Herd. This network can facilitate the investigative processes due to its inherent weaknesses (Le Blond et al., 2015).

FIGURE 13.15 Alpenhorn icon

FIGURE 13.16 GNUnet icon

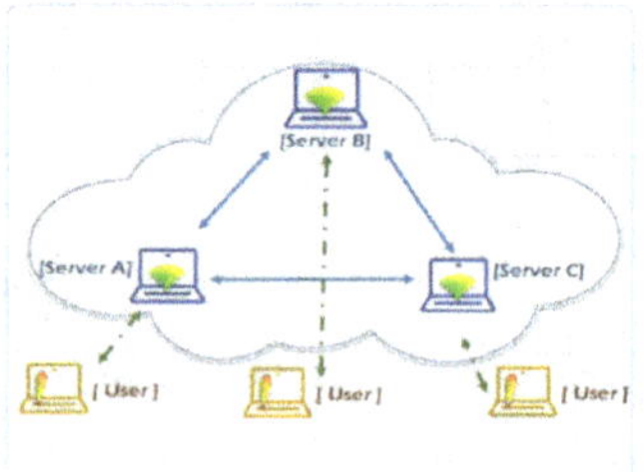

FIGURE 13.17 Riffle group sample

FIGURE 13.18 I2P icon

Alpenhorn (Vuvuzela)

According to Lazar & Zeldovich (2016) it is the second iteration of Vuvuzela, an anonymous communication app named for the horn that is normally used in soccer matches in Latin America and Africa. Alpenhorn is a metadata-free chat program that was designed to facilitate communication among millions of users. For now, this application has testing problems but there are already loyal users of this application.

GNUnet

GNUnet is a peer-to-peer file sharing tool that uses groups to hide the identities of the group members. A user can be indistinguishable from any other user unless that user has group initiator roles (Cutillo et al., 2009).

Riffle

Riffle is another application for sharing files anonymously but with better speed than any of the previously mentioned applications. Additionally, Riffle has no limitation on the file size and is focused on increasing the Tor project. This support consists of optimizing the Tor network and facilitating microblogging. Social networks Twitter and Pinterest are examples of this kind of blogging (Kwon et al., 2016).

Invisible Internet Project (I2P)

It is a project like Tor; however, unlike the onion project, I2P uses a distributed network database and peer selection for anonymous traffic. Some technicians claim that I2P has more advantages than Tor, but beyond technical discussions, the most important thing for an investigator is that I2P also has clandestine markets with great popularity among cybercriminals. Similarly, illicit I2P markets, being relatively new, are not as inhibited as drug dealers in Tor (Coelho et al., 2017).

Cyberagent Procedures

Beyond the advantages of the anonymity that is offered by the applications described in this chapter, all of them use the same transmission mechanics and need electronic equipment to retransmit. This argument leads us to reflect on the intrinsic vulnerability that each device can have. Those weaknesses can become gateways for a properly trained cyber-investigator (Casey, 2000).

One of those vulnerabilities is cybercriminals' use of Torrent. This computer application allows the sharing of files of considerable size but eliminates the protection of anonymity of the users and facilitates the tracking and location of the computers in communication. That is, an undercover agent in cyberspace can motivate a suspect to use Torrent during their interaction within a judicial investigation.

The above are some general recommendations to keep in mind when conducting investigations on the dark web. Next we discuss specific activities and a set of steps known as undercover agent blackjack in cyber-investigative practice.

Open-Source Intelligence

Open-source intelligence (OSINT) is a set of techniques and online resources that allow you to collect public information that will be useful to start an investigation, analyze business trends, or plan a police operation (Calof et al., 2008). When an investigator finds an open group on social media where they are trading in illicit property, the investigator may request to open a criminal case based on the information collected under OSINT. For their part, marketing companies perform OSINT tasks to analyze the relevance of launching a new service or product. And at the police level, the special operations team, before conducting a search of a cybercriminal's residence, can use OSINT to observe any evasive behavior on social networks associated with the targeted cybercriminal. That is, if the interaction on open sources on the internet shows calm, the police infer that the surprise of the operation has not been compromised and they can search the residence immediately.

The first documented case of the applicability of OSINT is the information collected in a book about German senior officers. This book, written by Berthold Jacob during the Second World War, caught the attention of the Nazi forces (Gestapo) because of its level of detail and accuracy. When the author was questioned by Gestapo, he stated that the source of the information had been the German press. Jacob had worked for years to record and organize information from the press to compile each page of the scandalous book. Finally, the Germans confirmed that there were no information leaks or espionage; on the contrary, they only found open-source material (OSINT) and the work of Jacob's analytical mind.

The information from social networks and mass media is the source for an OSINT task that begins with collecting information, then analyzing it, and finally processing it.

Another example of OSINT is performing an advanced search in a web search engine such as Google, DuckDuckGo, Internet Explorer, or Bing (Chauhan & Panda, 2015). Here you can use filters that deliver indexed information from someone or an organization. This development of OSINT tasks shows that it is low cost and does not require large information systems or advanced technological infrastructures. Here, it is important to remember that the darknet has countless open resources ranging from names to phone numbers and aliases. This can be extremely valuable information during any investigation.

The OSINT is a methodology applicable to investigations and is accessible by anyone with access to open sources of information or computer applications that automate the process of combining open information (Choi & Toro-Alvarez, 2017).

It should be noted that OSINT can also be developed on the darknet since there are open chat groups and social networking pages operating with services on the Tor network. There are also users who connect from Tor to these social networks such as Facebook and Reddit. Additionally, it is important to consider the kind of vocabulary used during the information-gathering process. If the investigator starts a conversation in Tor Messenger referring to their police profession, it is very unlikely that the native users of that application will interact with the investigator.

Therefore, the protocols for interviewing suspects are also applicable in these exercises in cyberspace and remind us that the techniques for collecting information are associated with the experience, human capacity, and ongoing training of investigators.

To conclude, the search for information in open sources can be defined as an investigation method that allows obtaining knowledge about possible criminal action from publicly accessible sources. These sources include websites, newspapers, blogs, magazines, social networks, or any repository of information the investigator can access without having to violate privacy protection mechanisms. Another important feature is that OSINT activities do not require a court order since they work on the analysis of public information.

At the end of this chapter, a practical case will be presented to reinforce the concepts discussed so far and the inclusion of the 21-step protocol known as investigator's blackjack.

Investigator's Blackjack

The 21 tasks that compose the practical protocol for collecting information in an environment associated with a cybercriminal investigation on the dark web are listed below. It should be noted that the baseline or start of the investigation is based on the results of the OSINT activities carried out by the investigators (Butler et al., 2016):

TASK-1: The investigator performs a keyword search associated with the criminal activities of concern of the agency where the investigator works.

TASK-2: The investigator records OSINT activity, includes search keywords, clues of detected criminal activity, and the list of preserved digital evidence.

TASK-3: The investigator reviews the metadata of the collected information and verifies the hash values of the preserved evidence.

TASK-4: The investigator verifies the privacy levels of each of the accounts and profiles associated with the detected criminal activity.

TASK-5: The investigator identifies the need to use an identity profile to access additional information and evidence.

TASK-6: The investigator requests authorization to deploy the undercover agent in cyberspace.

TASK-7: The investigator reports the willingness to be an undercover agent in cyberspace (UAC) and registers the principal investigator who will serve as the controlling agent.

TASK-8: Investigator requests court order for service providers to perform selective database searches.

TASK-9: The investigator verifies receiving the authorization to carry out activities as an undercover agent in cyberspace.

TASK-10: The investigator reports to the control agent and initiates activities as an undercover agent in cyberspace.

TASK-11: The investigator verifies the advances in the request for the selective search in databases.

TASK-12: The investigator records the information related to relays, IP addresses, connection logs, biographical data, and registration data.

TASK-13: The investigator requests a new selective search in databases with the new identified profiles.

TASK-14: Investigator reviews records for time zone conversion, ISP identification, and submits undercover agent activities to oversight by the controlling agent.

TASK-15: The investigator prepares a field investigator report disclosing advances in the investigation and new information collected.

TASK-16: The investigator records the information provided by the second selective search in the database.

TASK-17: The investigator records the interactions of the undercover agent with the investigated profiles.

TASK-18: The investigator records controlled delivery and legality control. Additionally, the investigator requests arrest warrants and search warrants as appropriate.

TASK-19: Investigator reports new information generated by controlled delivery/purchase and requests additional arrest warrants and search warrants.

TASK-20: The investigator coordinates the captures and raids with the special unit.

TASK-21: The investigator requests closing the controlled purchasing procedure, finishes the undercover agent figure, and reports the operational results achieved.

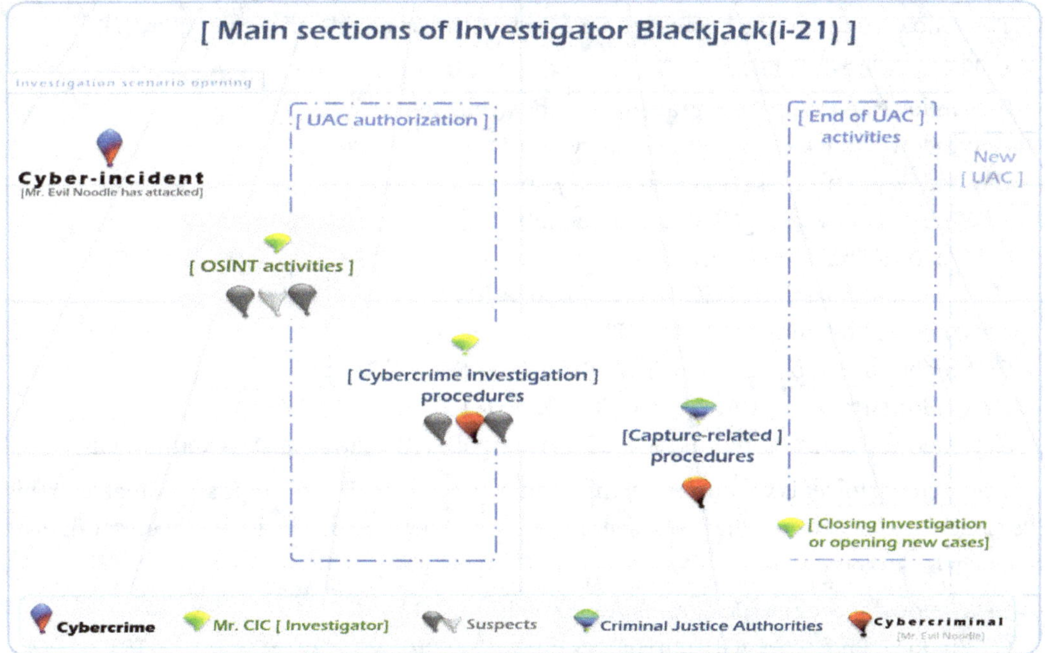

FIGURE 13.19 Main interactions of an undercover agent in cyberspace (UAC)

Investigator blackjack, also known as I-21, is a flexible protocol that serves as a guide, but it does not require the consecutive execution of the steps. I-21 permanently advises the recording of investigator activities to avoid deviation from the functions of an undercover agent in cyberspace (Choi & Toro-Alvarez, 2017). A vital aspect to consider here is the investigator's mental health since cyber-investigators are exposed to a large amount of undesirable content such as child pornography and "happy slapping" (see Chapter 9).

Traditional Digital Forensics and Tor

Silk Road was perhaps the most successful clandestine illicit drug market for cybercriminals. But this market was dismantled thanks to the activity of investigators and the coordinated work between law enforcement agencies in different parts of the world. In other words, it is possible to carry out investigations and operational work in the darknet using the combination of different techniques such as traditional digital evidence and OSINT techniques. Next, this chapter introduces two procedures based on forensic informatics to investigate whether a computer was used to operate in the darknet (Choi et al., 2014).

Tracking Tor Configuration in the Physical World

An investigator can demonstrate whether a computer has connected to the Tor network through the Tor browser. But first it is necessary to check whether that browser was installed; therefore, the investigator must be familiar with the file architecture of an operating system. In a Windows environment, the investigator must recognize the existence and location of the system files pagefile.sys, hiberfil.sys, swapfile.sys, and Ntuser.dat. It should be noted that the management of these files may require the use of specialized software for forensic computer labs or ethical hacking.

The following demo line begins with a file structure available on a computer that has not had the Tor browser installed.

```
C:\Windows\appcompat\Programs\Amcache.hve.LOG2
C:\Windows\Prefetch\AgGlFaultHistory.db
C:\Windows\Prefetch\AgGlFgAppHistory.db
C:\Windows\Prefetch\AgGlGlobalHistory.db
C:\Windows\Prefetch\AgRobust.db
C:\Windows\Prefetch\DLLHOST.EXE-5E46FA0D.pf
C:\Windows\Prefetch\MPCMDRUN.EXE-F401FBB4.pf
C:\Windows\Prefetch\SEARCHFILTERHOST.EXE-77482212.pf
C:\Windows\Prefetch\SEARCHPROTOCOLHOST.EXE-0CB8CADE.pf
C:\Windows\ServiceProfiles\NetworkService\AppData\Local\Temp\MpCmdRun.log
```

The Prefetch folder is very useful in this kind of procedure. Timestamps are available in this file structure that can give the investigator clues as to how long the device has been configured to navigate darknet.

```
C:\Windows\appcompat\Programs\Amcache.hve.LOG2
C:\Windows\Prefetch\AgGlFaultHistory.db
C:\Windows\Prefetch\AgGlFgAppHistory.db
```

C:\Windows\Prefetch\AgGlGlobalHistory.db
C:\Windows\Prefetch\AgRobust.db
C:\Windows\Prefetch\DLLHOST.EXE-5E46FA0D.pf
C:\Windows\Prefetch\MPCMDRUN.EXE-F401FBB4.pf
C:\Windows\Prefetch\SEARCHFILTERHOST.EXE-77482212.pf
C:\Windows\Prefetch\SEARCHPROTOCOLHOST.EXE-0CB8CADE.pf
C:\Windows\ServiceProfiles\NetworkService\AppData\Local\Temp\MpCmdRun.log

Additionally, inside this folder, it is suggested to analyze the strings "geoip", "everywhere", "http", "notificationuxbrowser", and "Tor". This analysis can be carried out using a hexadecimal file reader. The analysis allows finding matches from the file AgRobust.db.

C:\Windows\appcompat\Programs\Amcache.hve.LOG2
C:\Windows\Prefetch\AgGlFaultHistory.db
C:\Windows\Prefetch\AgGlFgAppHistory.db
C:\Windows\Prefetch\AgGlGlobalHistory.db
C:\Windows\Prefetch\AgRobust.db
C:\Windows\Prefetch\DLLHOST.EXE-5E46FA0D.pf
C:\Windows\Prefetch\MPCMDRUN.EXE-F401FBB4.pf
C:\Windows\Prefetch\SEARCHFILTERHOST.EXE-77482212.pf
C:\Windows\Prefetch\SEARCHPROTOCOLHOST.EXE-0CB8CADE.pf
C:\Windows\ServiceProfiles\NetworkService\AppData\Local\Temp\MpCmdRun.log

These matches display the name of the TorBrowser package installation file. Another very useful piece of evidence can be found within the Prefetch folder. Here, you can view the file TORBROWSER-INSTAL-5.0_ENUS.-37B68F50.pf (depending on the browser version). There is another .db file that is affected during the installation of the Tor browser. This file is found by analyzing the cache files modified during the installation. Specifically, this file will be in the directory folder where the Tor package has been installed.

An additional suggested procedure is to export the paging file entitled pagefile.sys and analyze the system memory dump which is stored in that file. "TorBrowser" is a suggested search term. The records that this action returns are immediate and show the directory that stored the executable file before installation. A forensic analysis in the memory dump allows for finding sufficient evidence on the TorBrowser installation files.

Other terms that are easily identifiable are "everywhere" and "notificationUxBroker." Finally, the investigator can access the NTUSER.DAT file and, by searching for the term "tor," will find hexadecimal sequences that show the existence of the Tor browser.

There are multiple traces left by the Tor browser in its installation even if the user has deleted the folder in which Tor was installed. This is due to the high sensitivity of the procedure of modifying system files such as ntuser.dat and pagefile.sys. If a cybercriminal wants to erase the data from these files, the cybercriminal would need anti-forensic knowledge to locate the files, modify the Prefetch folder, and delete the records that have been identified in this chapter as possible sources of evidence.

Traces of Tor Execution

Considering the installation evidence previously studied, it is worthwhile to search for traces associated with navigation on the darknet. A good practice is to dump the memory and copy various system-locked files such as swapfile.sys and pagefile.sys. It should be noted that the management of these files requires the use of specialized software.

The next step is to analyze the files with software that allows the hexadecimal display. There, the investigator can start searching for text strings that match the domain associated with Tor web pages. In other words, the investigator would type the term ".onion". During a cybercriminal investigation, this search returns not only the pages visited but also connection dates, connection language, version of the browser used, information on the operating system, files displayed (this includes photographs), and storage routes of directories on the visited .onion page.

As previously mentioned, interaction in cyberspace, either on the open internet or within the darknet through Tor, uses the same transmission mechanics and they need electronic equipment in order to retransmit. Investigators always hope to find evidence at different moments of interaction between installed software, the user's logging time records, and the geolocation metadata discussed in previous chapters. It is necessary to continue carrying out deep forensic computing activities and build good practices for cybercrime investigations. This kind of knowledge can help contain crime in all its manifestations, including in cyberspace.

Applied Case: Dismantling a Transnational Cybercriminal Organization

The following case was investigated and prosecuted in real life. Some names, dates, and affiliations are modified for academic purposes and on the operational reserve of the involved institutions.

An early cyber patrol in December 2018 within the darknet by an investigator from the International Observatory of Computer Crime (INTOCC) led to the detection of a forum on material related to nude boys and girls. Conversations located within this forum were open so an INTOCC investigator reported on the start of this OSINT work. During this process, two very active users were detected in the collection of accounts from the anonymous messaging application Tor Messenger. The aliases LOVEPEDOGIRL and DARKPONY also featured accounts from other social networks where the profile photos coincided and allowed the inference that they were the same two users of a child pornography forum on the Tor network. The investigators were invited by the two aliases and accessed other child pornography forums where buyers and sellers of illegal material from 25 different countries participated.

On Thursday, February 28, 2019, INTOCC informed the Police Cyber Center (C2P) of a South American country of the participation of 122 IP addresses in the distribution of multimedia material associated with child pornography. These IP addresses are associated with 14 internet service provider companies (ISPs) in that country. These 122 IP addresses would be part of the first delivery of the results obtained in the development of a transnational operation called Atarraya.

On Wednesday, April 3, 2019, the judicial authority of that South American country ordered the establishment of a methodological program to prioritize the operational objectives that led to the dismantling of the transnational organization dedicated to child pornography.

Under the investigative support of the judicial authorities, the C2P investigators used the information provided by the ISPs to determine the physical location of the two residences in two different cities. The head of the Special Unit Against Child Pornography and other Internet Violence (UESPVI) requested authorization to enable one undercover agent in cyberspace and two undercover agents for each geo-referenced city. By Monday, April 15, the three undercover agents identified a total of 19 residences where an unknown number of users would be accessing platforms hosted on the darknet.

On Thursday, April 18, the control agent of the undercover agent in cyberspace receives a surveillance and monitoring report and proceeds to make the request for 17 raids, while the USVI chief coordinates, with the special operations group (COPES), the raids and capture operation of eight individuals red-handed. The operation against the criminal organization known as "The 26" began at 3:00 p.m. on Sunday, April 21, 2019. Investigators rescued two children, ages 8 and 16, captured six cybercriminals, and seized 465 electronic devices including digital storage media, cameras, tablets, and laptops. These elements were analyzed in the C2P forensic computer lab. There, the investigators found more than 15,000 explicit images of child abuse. Additionally, they detected the use of applications such as ARES, KIK, Telegram, and Tor Browser.

Those captured were linked to previous investigations that included criminal activities of two communities of pedophiles identified in the instant messaging applications WhatsApp and Telegram. The findings also made it possible to link other connections through peer-to-peer networks that had buyers of illicit material in two European countries.

Operation Atarraya is still ongoing, and, thanks to trained and proactive investigators, it will never cease operations.

Summary

Deep web and darknet: The chapter introduces the most common applications for darknet and deep web browsing as well as presenting useful information for a formal investigation in darknet scenarios.

Tor architecture: The onion router (Tor) has a special structure for protecting anonymity, this architecture offers hidden services in a broad context. Specialized browser and computer applications are required to access these services.

Cyberagent procedures: Although darknet markets present challenges for cybercrime investigators, it is possible to detect interaction and connectivity weaknesses, which are gateways for well-trained investigators.

Open-source intelligence: OSINT is based on collecting information from social networks and mass media. It begins with collecting information, the investigator then analyzes and processes the collected intelligence. OSINT is also applicable in darknet investigations.

Investigator blackjack: The experiences of darkweb investigatotions are summarized in a set of 21 steps. This set is also known as the I-21.

Traditional digital forensics and Tor: Digital forensics includes multiple techniques that include detection of Tor browser traces by searching specific folders and files.

Tracking Tor configuration in the physical world: Detecting traces of Tor exploring is possible with techniques related to the system files such as pagefile.sys, hiberfil.sys, swapfile.sys, and Ntuser.dat.

Traces of Tor execution: The investigator can dump the memory and copy various system-locked files such as swapfile.sys and pagefile.sys. This technique requires the use of specialized software and different ways of visualizing and analyzing these locked files.

Applied case: Dismantling a transnational criminal organization: A real-life operation is presented as an example of how an investigator can deploy the techniques and procedures described in the chapter.

Mr. Evil is almost free—Tor digital evidence lab: This learning activity helps to consolidate the knowledge about darknet investigation. Three questions will be applied in a hypothetical case.

Lab 13: Mr. Evil Is Almost Free—Tor Digital Evidence Lab

Mr. CIC (our expert investigator) received an email from the cybercrime court. It seems that Mr. Evil's lawyer got testimony from an old friend, a local drug-dealer, who stated that he sold all of the incriminatory material—which is already in the possession of the court—related to Mr. Evil's case. Additionally, the fake witness said "Mr. Evil was not capable of setting up a computer program on his own laptop." Hence, Attorney Iron wrote Mr. CIC the following:

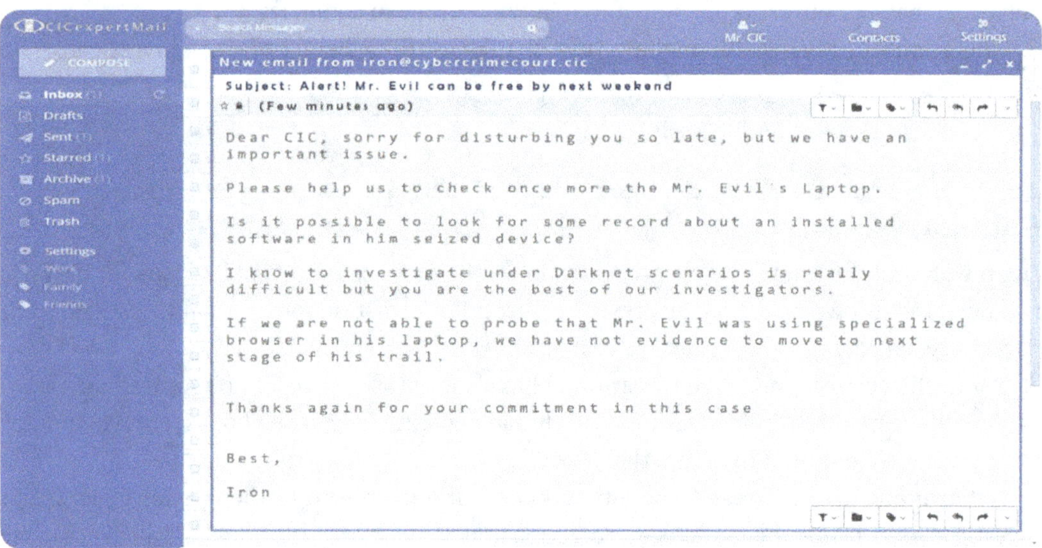

FIGURE 13.20 Email from Attorney Iron

Fortunately, Mr. CIC is not working alone. You are assigned to help him find the record in Mr. Evil's laptop. The seized evidence is coming from the forensic lab to your desk. Please remember that Mr. Evil's laptop has the Windows operating system.

1. Which of the following is a folder where you can find useful information about the Tor browser?
   a. C:\Windows\Settings\
   b. C:\Windows\Prefetch\
   c. C:\Windows\applocal\
   d. None of the above

2. An additional suggested procedure is to export the pagefile.sys paging file and analyze the system memory dump which is stored in that file. Is this statement correct?

    a. No, there is nothing in that file.

    b. No, pagefile.sys is not a Windows file.

    c. Yes of course. There is the evidence.

    d. No, the system memory dump is not a proper procedure in computer forensics.

3. A good practice is to dump the memory and copy various system locked files such as swapfile.sys and pagefile.sys. It should be noted that the management of these files requires the use of specialized software. Is it true that this requires specialized software?

    a. Yes, it is.

    b. No, and specialized software for this purpose is always licensed and expensive.

    c. No, pagefile.sys is easy to get just by browsing with Windows Explorer.

    d. No, there is no software for system memory dump.

References

Butler, B., Wardman, B., & Pratt, N. (2016, June). REAPER: An automated, scalable solution for mass credential harvesting and OSINT. In *2016 APWG Symposium on Electronic Crime Research (eCrime)* (pp. 1–10). IEEE.

Calof, J. L., Wright, S., & Fleisher, C. S. (2008). Using open source data in developing competitive and marketing intelligence. *European Journal of Marketing, 42*(7/8), 852–866.

Casey, E. (2000). *Digital evidence and computer crime.* Academic Press.

Castells, M. (2002). *The internet galaxy: Reflections on the internet, business, and society.* Oxford University Press.

Chaabane, A., Manils, P., & Kaafar, M. A. (2010, September). Digging into anonymous traffic: A deep analysis of the Tor anonymizing network. In *2010 Fourth International Conference on Network and System Security* (pp. 167–174). IEEE.

Chauhan, S., & Panda, N. K. (2015). *Hacking web intelligence: Open source intelligence and web reconnaissance concepts and techniques.* Syngress.

Chen, A. (2011). The underground website where you can buy any drug imaginable. *Gawker.* http://gawker.com/the-underground-website-where-you-can-buy-any-drug-imag-30818160

Choi, K. S. (2015). *Cybercriminology and digital investigation.* LAB Scholarly Publishing LLC.

Choi, K. S., Earl, K. J., Park. A., & Giustina, J. D. (2014) Use of synthetic cathinones: Legal issues and availability of darknet. *Criminal Justice Faculty Publications, 7,* 19–32.

Choi, K. S., & Toro-Alvarez, M. M. (2017). *Cibercriminología: Guía para la investigación del cibercrimen y mejores prácticas en seguridad digital.* UAN Fondo Editorial.

Clarke, I., Miller, S. G., Hong, T. W., Sandberg, O., & Wiley, B. (2002). Protecting free expression online with Freenet. *IEEE Internet Computing, 6*(1), 40–49.

Coelho, N., Fonseca, B., & Castro, A. (2017, June). Paranoid operative system methodology for anonymous & secure web browsing, doctoral project. *Atas da Conferência da Associação Portuguesa de Sistemas de Informação, 17*(17), 127–143.

Dingledine, R., Mathewson, N., & Syverson, P. (2007). Deploying low-latency anonymity: Design challenges and social factors. *IEEE Security & Privacy, 5*(5), 83–87.

District Court Southern District of New York. (2015). *District Court Southern District of New York v. Ross William Ulbricht,* 14 Cr. 68 (KBF). Doc 277. http://antilop.cc/sr/files/2015_06_30_ULBRICHT_sentencing.pdf

Eckersley, P. (2010, July). How unique is your web browser? In *International Symposium on Privacy Enhancing Technologies* (pp. 1–18). Springer, Berlin, Heidelberg.

Goldberg, I. A., & Brewer, E. (2000). *A pseudonymous communications infrastructure for the internet.* University of California, Berkeley.

Greenberg, A. (2013, September 5). Follow the Bitcoins: How we got busted buying drugs on Silk Road's black market. *Forbes*. https://www.forbes.com/sites/andygreenberg/2013/09/05/follow-the-bitcoins-how-we-got-busted-buying-drugs-on-silk-roads-black-market

Halevy, A., Rajaraman, A., & Ordille, J. (2006, September). Data integration: The teenage years. In *Proceedings of the 32nd International Conference on Very Large Data Bases* (pp. 9–16).

Kwon, A., Lazar, D., Devadas, S., & Ford, B. (2016). Riffle: An efficient communication system with strong anonymity. *Proceedings on Privacy Enhancing Technologies*, 2016(2), 115–134.

Lazar, D., & Zeldovich, N. (2016). Alpenhorn: Bootstrapping secure communication without leaking metadata. In *12th USENIX Symposium on Operating Systems Design and Implementation* (OSDI '16) (pp. 571–586).

Le Blond, S., Choffnes, D., Caldwell, W., Druschel, P., & Merritt, N. (2015, August). Herd: A scalable, traffic-analysis resistant anonymity network for VoIP systems. In *Proceedings of the 2015 ACM Conference on Special Interest Group on Data Communication* (pp. 639–652).

Le Blond, S., Choffnes, D., Zhou, W., Druschel, P., Ballani, H., & Francis, P. (2013). Towards efficient traffic-analysis resistant anonymity networks. *ACM SIGCOMM Computer Communication Review*, 43(4), 303–314.

Mulazzani, M., Reschl, P., Huber, M., Leithner, M., Schrittwieser, S., Weippl, E., & Wien, F. C. (2013, May). Fast and reliable browser identification with JavaScript engine fingerprinting. In *Web 2.0 Workshop on Security and Privacy* (W2SP) (Vol. 5).

Oremus, W. (2013). The bonehead mistake that brought down an online drug-dealing empire. *Slate*. http://www.slate.com/blogs/future_tense/2013/10/02/silk_road_s_dread_pirate_ross_ulbricht_asked_stack_overflow_question_under.html

Syverson, P., Dingledine, R., & Mathewson, N. (2004). Tor: The second-generation onion router. In *Usenix Security* (pp. 303–320).

Vallina-Rodriguez, N., Sundaresan, S., Kreibich, C., Weaver, N., & Paxson, V. (2015, May). Beyond the radio: Illuminating the higher layers of mobile networks. In *Proceedings of the 13th Annual International Conference on Mobile Systems, Applications, and Services* (pp. 375–387).

Westermann, B., & Kesdogan, D. (2011, February). Malice versus AN.ON: Possible risks of missing replay and integrity protection. In *International Conference on Financial Cryptography and Data Security* (pp. 62–76). Springer, Berlin, Heidelberg.

Credits

Cyber-Intelligence Operations and Incident Response

Sinchul Back

Introduction

Cyberthreat-intelligence analysts, cyber defenders, cybercrime investigators, and others who are tasked with understanding the security in which an organization operates need to know how to gather that information. This process is called reconnaissance or intelligence gathering. Information gathering is often a requirement of information security standards and laws. For example, the PCI-DSS standard calls for vulnerability scanning, requiring both internal and external network vulnerability scans at least quarterly and after any significant change. Gathering internal and external information about your own organization is typically considered a necessary part of understanding organizational risk, and implementing industry best practices to meet due diligence requirements is likely to result in this type of work. In this chapter you will explore how to gather information about an organization using passive open-source intelligence as well as active enumeration and scanning methods. You will also take a look at other important techniques, including packet crafting, capture, and inspection for information gathering, in addition to the role of code analysis for intelligence gathering and related techniques.

Furthermore, this chapter covers incident response methods and procedures that are taught in alignment with industry frameworks such as US-CERT's NCIRP (National Cyber Incident Response Plan) and Presidential Policy Directive (PPD) 41, *United States Cyber Incident Coordination*. It is ideal for candidates who have been tasked with managing compliance with state legislation and other regulatory requirements regarding incident response and for executing standardized responses to such incidents. Thus, this chapter introduces procedures and resources to comply with legislative requirements regarding incident response.

We begin with an overview of mobile devices and then explain mobile forensics procedures. You will learn how to identify and reconstruct file fragments, analyze file headers, and repair damaged file headers.

Overview of Cyber-Intelligence Operations

Cyber-intelligence operation is a rapidly growing field. This section briefly provides the concepts of intelligence, intelligence cycle, and cyberthreat intelligence. It also focuses on obtaining the intelligence along with planning, direction, and the generation of intelligence artifacts.

Understanding Intelligence

The government intelligence capacity involves a full range of activities and operations related to intelligence gathering, analysis, and sharing. Through systems and procedures, the intelligence agents and/or cybercrime investigators convert the information they acquire into clear, comprehensible intelligence and deliver it to end users as required. The intelligence cycle is a backbone of these operations. For example, the intelligence cycle begins with identifying key issues that interest policymakers and defining the answers they require in order to make educated decisions on action and policy. The steps of the intelligence cycle (Figure 14.1) include the following (Bullock et al., 2016, pp. 205–206):

- **Planning.** During the planning step, decisions are made regarding what types of information to collect and how to collect it.
- **Collection.** During the collection step, the intelligence agent gathers the raw data used to produce finished intelligence products.
- **Processing.** In the processing step, information that is collected is converted into a usable format, such as by language translation or decryption.
- **Analysis.** In the analysis step, intelligence officers analyze processed information to turn it into finished intelligence.
- **Dissemination.** In the dissemination step, intelligence products are provided to those who request or otherwise need them.

Intelligence gathering is performing reconnaissance against a target to gather as much information as possible to be utilized when penetrating the target during the vulnerability assessment and exploitation phases. The more information you are able to gather during this phase, the more vectors of attack you may be able to use in the future (Richelson, 2015). Open-source intelligence (OSINT) is a form of intelligence collection that involves finding, selecting, and acquiring information from publicly available sources and analyzing it to produce actionable intelligence (Richelson, 2015).

OSINT is employed to gather information to determine various entry points into an organization, which may be physical, electronic, and/or human. Actually, many organizations fail to take into account what information about themselves they place in public and how this information can be utilized by a malicious actor. Moreover, frequently many employees fail to take

FIGURE 14.1 The intelligence cycle

into account what information they release in public and how that information can be used to compromise them or their companies.

Understanding Cyber-Intelligence Operations

In most cyber-environments, we are dealing with incidents after the fact; we are detecting them after cyber adversaries achieve their objective. By the same token, many times the cyber adversaries stick around because they want to get more information over the course of months and years, but the cyber adversaries have typically already accomplished their initial goal by the time we detect them. Thus, the objective of the cyber-intelligence operation is to detect cyber intruders before they exploit and/or hit the networks and systems. In other words, you must "run faster" than the cyber adversary by observing attacks and integrating investigation tools.

There is no permanent definition of cyber-intelligence; however, Duvenage, von Solms, and Corregedor (2015) define cyber-intelligence as collecting and analyzing information regarding cyber adversaries and their methods; identifying, tracking, and predicting cyber actors' capabilities, intentions, and activities; and offering intelligence products that inform the decision making of policymakers. Moreover, the traditional intelligence cycle is adjusted and utilized in a cyber-intelligence operation. The cyber-intelligence cycle is a circular and repeated process to convert data into intelligence product to be disseminated to intelligence customers; therefore, it has the following steps (Vardangalos, 2016, p. 4).

Planning and direction. Determine the requirements and needs of the specific cyber-intelligence effort. Have defined goals and intentions in order to appropriately create any amount of intelligence out of available information.

Collection. Determine where and how to get the data and information to process. The environment to collect data and information can be honeypots, firewall logs, intrusion-detection system logs, etc.

Processing and exploitation. The conversion of collected information into something usable (how the data is stored or the actual parsing of data such as converting it to human readable information).

Analysis and production. Taking the data and turning it into an intelligence product. This is achieved through analysis and interpretation and thus is heavily dependent on the analyst. All produced reports should meet a defined intelligence need or goal from the planning and direction phase.

Dissemination and integration. Supplying the customer or user with the finished intelligence product. If the users cannot access this product or cannot use it, then it is useless and does not meet a goal.

Information Gathering and Vulnerability Identification

Cyber-intelligence analysts must be able to interrogate and fully understand their collection sources. Analysts do not have to be malware reverse engineers, as an example, but they must at least understand that work and know what data can be sought. This section discusses identifying key collection sources for analysis in alignment with the most commonly used techniques (e.g., OSINT).

Footprinting

The first step when gathering organizational intelligence is to identify an organization's footprint. Footprinting is used to create a map of an organization's networks, systems, and other infrastructure. This is typically accomplished by combining information-gathering tools with manual research to identify the networks and systems that an organization uses.

Active Reconnaissance

Information gathered during footprinting exercises is typically used to provide the targets for active reconnaissance. Active reconnaissance uses host-scanning tools to gather information about systems, services, and vulnerabilities (Scarfone et al., 2008). Specifically, network reconnaissance is employed to assess potential vulnerabilities in a computer network. Information gathered during the footprinting exercise is typically utilized to provide the targets for active reconnaissance. Figure 14.2 indicates performing reconnaissance on a network.

Scanning networks or systems can cause problems for the devices that are scanned. Some services may not tolerate scan traffic well, while others may fill their logs or set off security alarms when scanned. In doing so, as a cyberthreat-intelligence officer or cybercrime investigator, you should make sure you have legal authority from your organization before executing active reconnaissance. Scanning other organizations' systems may also be illicit without permission or may be prohibited by the terms of use of your internet service provider. Again, you must make sure before you conduct active reconnaissance.

```
root@lbuser_006_ckal: ~

File  Edit  View  Search  Terminal  Help
Nmap scan report for 10.0.0.1
Host is up (0.00040s latency).
Not shown: 997 filtered ports
PORT    STATE SERVICE
22/tcp open   ssh
53/tcp open   domain
80/tcp open   http
MAC Address: 00:50:56:90:67:14 (VMware)
Warning: OSScan results may be unreliable because we could not find at least 1 o
pen and 1 closed port
OS fingerprint not ideal because: Missing a closed TCP port so results incomplet
e
No OS matches for host
Uptime guess: 0.000 days (since Sun Nov 22 10:49:35 2020)
Network Distance: 1 hop
TCP Sequence Prediction: Difficulty=255 (Good luck!)
IP ID Sequence Generation: All zeros

Read data files from: /usr/bin/../share/nmap
OS detection performed. Please report any incorrect results at https://nmap.org/
submit/ .
Nmap done: 1 IP address (1 host up) scanned in 9.54 seconds
        Raw packets sent: 2074 (94.948KB) | Rcvd: 24 (1.716KB)
root@lbuser_006_ckal:~#
```

FIGURE 14.2 Performing reconnaissance on a network

Port Scanning and Service Discovery Techniques

Port scanning tools are designed to send traffic to remote systems and then gather responses that offer information pertaining to the systems and the services (Bartlett et al., 2007). It is essential to use these tools when gathering information about a network and the devices connected. Port scans, therefore, are the first task that most cyber-intelligence officers think of when preparing to assess a target. Port scanners include four major features—host discovery, port scanning and service identification, service version identification, and operating system identification.

An important part of port scanning is an understanding of common ports and services. Ports from 0 to 1023 are known as well-known ports or system ports, but there are quite a few higher ports that are commonly of interest when conducting port scanning. Ports ranging from 1024 to 49151 are registered ports and are assigned by the Internet Assigned Numbers Authority (IANA) when requested. Many are also used arbitrarily for services. Since ports can be manually assigned, simply assuming that a service running on a given port matches the common usage isn't always a good idea. In particular, many SSH and HTTP/HTTPS servers are running on alternate ports, either to allow multiple web services to have unique ports or to avoid port scanning that only targets their normal port. Analysis of scan data can provide information about what hosts are on a network, what services they are running, and clues about whether they are vulnerable to attacks. Next, the scan demonstrates the ports it found open, whether they are TCP or UDP, their state, and its guess about what service the port is. After the services listed are displayed, we get the MAC address. The final steps to note about this scan

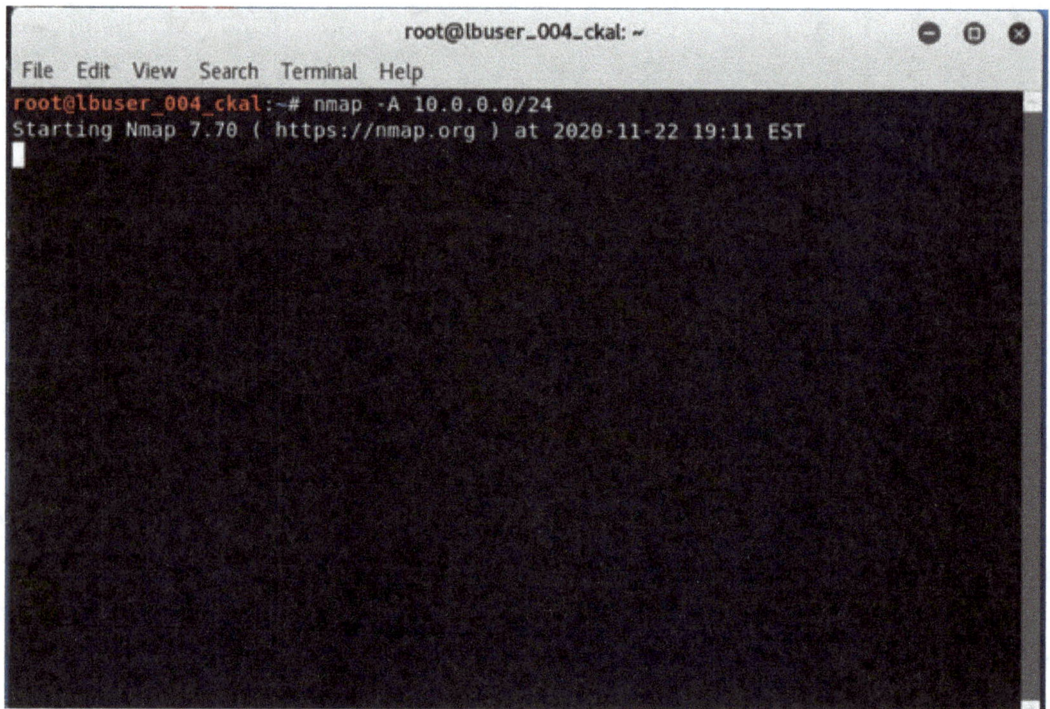

FIGURE 14.3 Performing the initial scan

are the time it took to run and how many hops there are to the host. This initial port scanning is completed in less than two seconds, which tells us that the host responded quickly and that the host was only one hop away. A more complex network path will show more hops. Scanning more hosts or additional security on the system or between the scanner and the remote target can also slow things down.

OS Fingerprinting

The ability to identify an operating system based on the network traffic that it sends is known as *operating systems fingerprinting*, and it can provide useful information when performing reconnaissance (Matsunaka et al., 2013). This is typically done using TCP/IP stack fingerprinting techniques that focus on comparing responses to TCP and UDP packets sent to remote hosts.

Service and Version Identification

The ability to identify a service can provide useful information about potential vulnerabilities, as well as verify that the service that is responding on a given port matches the service that typically uses that port. Service identification is usually done in one of two ways: either by connecting and grabbing the banner or connection information provided by the service or by comparing its responses to the signatures of known services.

Common Tools

Nmap is the most commonly used command-line port scanner, and it is an open-source tool. It provides a broad range of capabilities, including multiple scan modes intended to bypass firewalls and other network protection devices. In addition, it provides support for operating system fingerprinting, service identification, and many other capabilities. Using basic functionality of *nmap* we can provide the target system's hostname or IP address. Nmap also has an official graphical user interface, Zenmap, which offers additional visualization capabilities, including a topology view mode that provides information about how hosts fit into a network. *Angry IP Scanner* is a multiplatform (Windows, Linux, and macOS) port scanner with a graphical user interface. Unlike nmap, Angry IP Scanner does not provide detailed identification for services and operating systems, but you can turn on different modules called "fetchers," including ports, time to live, filtered ports, and others. Figure 14.4 illustrates performing Angry IP Scanner.

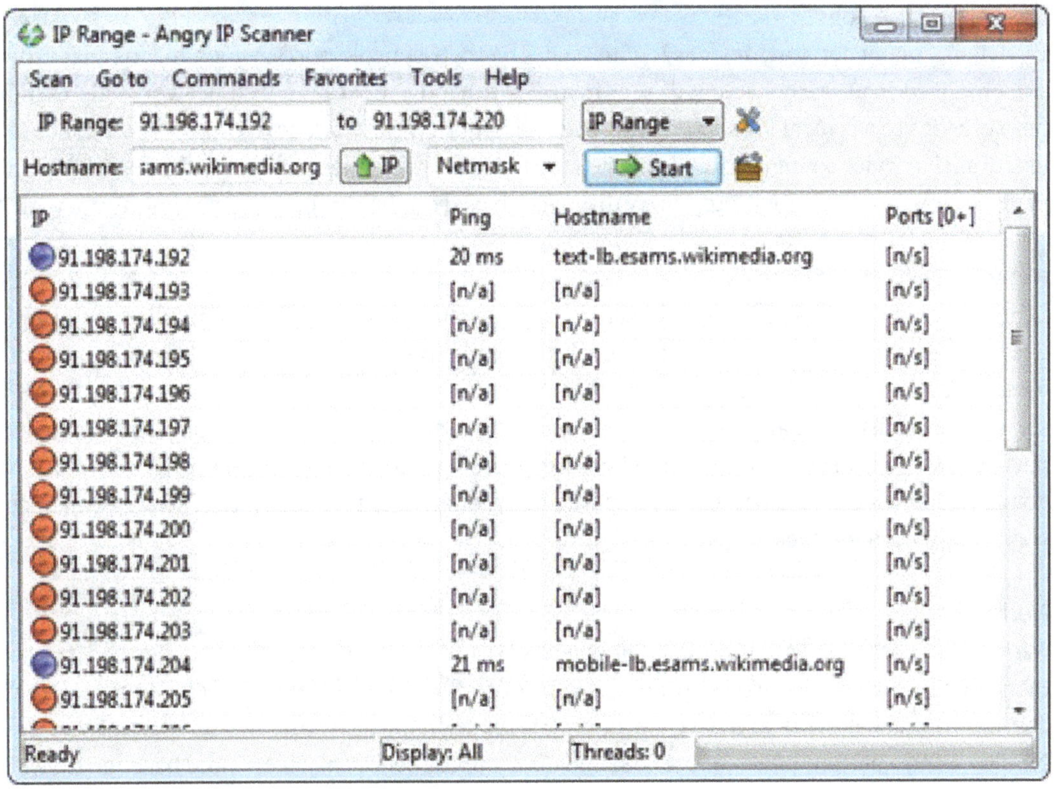

FIGURE 14.4 Angry IP Scanner

Passive Footprinting

Passive footprinting is far more challenging than active information gathering. Passive analysis relies on information that is available about the organization, systems, or network without performing your own probes. Passive footprinting typically relies on logs and other existing data, which may not provide all of the information needed to fully identify targets. Passive scanning tools capture wireless traffic being transmitted within the range of the tool's antenna (Scarfone et al., 2008). Most tools provide several key attributes regarding discovered wireless devices, including service set identifier (SSID), device type, channel, media access control (MAC) address, signal strength, and number of packets being transmitted. This information can be used to evaluate the security of the wireless environment, and to identify potential rogue devices and unauthorized ad hoc networks discovered within range of the scanning device. The wireless scanning tool should also be able to access the captured packets to determine if any operational anomalies or threats exist.

Log and Configuration Analysis

Log files can provide a treasure trove of information about systems and networks. If you have access to local system configuration data and logs, you can use the information they contain to build a thorough map of how systems work together, which users and systems exist, and how they are configured (Bul'ajoul et al., 2015; Lyytinen & Damsgaard, 2011).

Network Device Logs

By default, many network devices log messages to their console ports, which means that only a user logged in at the console will see them. Fortunately, most managed networks also send network logs to a central log server using the syslog utility. Network device logs are often not as useful as the device configuration data when the focus is intelligence gathering, although they can provide some assistance with topology discovery based on the devices they communicate with.

Network Device Configuration

Configuration files from network devices can be invaluable when mapping network typology. Configuration files often include details of the network, routes, systems that the devices interact with, and other network details. In addition, they can provide details about syslog and SNMP servers, administrative and user account information, and other configuration items useful as part of information gathering. As a cyberthreat-intelligence officer, this is useful information about an intruder, and it will be the starting point of an effective prevention strategy against social engineering attacks.

Netflows

Netflows is a Cisco network protocol that collects IP traffic information, allowing network traffic monitoring. Flow data is used to provide a view of traffic flow and volume. A conventional flow capture consists of the IP and port source and destination for the traffic and the class of service. Netflows and a netflow analyzer can assist in determining service problems and baseline typical network activity, and they can also be useful in detecting abnormal activities.

DHCP Logs and DHCP Server Configuration Files

DHCP, the Dynamic Host Configuration Protocol, is a client/server protocol that provides an IP address, the default gateway, and subnet mask for the network segment that the host will reside on. Conducting passive reconnaissance, DHCP logs from the DHCP server for a network can provide a quick way to scan many of the hosts on the network. If DHCP logs are combined with firewall logs, you can determine which hosts are provided with dynamic addresses and which hosts are using static addresses.

Firewall Logs and Configuration Files

Router and firewall configuration files and logs often contain information about both successful and blocked connections. This means that analyzing router and firewall ACLs (access control lists) and logs can provide useful information about what traffic is allowed and can help with topological mapping by identifying where systems are based on traffic allowed through or blocked by rules. Configuration files make this even easier, since they can be directly read to understand how systems interact with the firewall.

System Log Files

System logs are collected by most systems to provide troubleshooting and other system information. Log information can vary greatly depending on the operating system, how it is configured, and what services and applications the system is running. Log files can provide information about how systems are configured, what applications are running on them, which user accounts exist on the system, and other details, but they are not typically at the top of the list for reconnaissance. They are gathered if they are accessible, but most log files are kept in a secure location and are not accessible without system access.

Harvesting Data from DNS and Whois

The domain name system (DNS) is often one of the first stops when gathering information about an organization. Not only is DNS information publicly available, it is often easily connected to the organization by simply checking for Whois information about a website (Castelluccia et al., 2013). With that information, you can find other websites and hosts to add to your organization footprint.

DNS and Traceroute Information

DNS converts domain names like google.com to IP addresses or from IP addresses to human-understandable domain names. *Nslookup* is the command for this on Windows, Linux, and macOS systems. Once you know the IP address that a system is using, you can look up information about the IP range it resides in. That can provide information about the company or about the hosting services that they use. Nslookup provides a number of additional flags and capabilities, including choosing the DNS server.

Domains and IP Ranges

Domain names are managed by domain name *registrars*. Domain registrars are accredited by generic top-level domain (gTLD) registries and/or country top-level domain (ccTLD) registries. This means that registrars work with the domain name registries to provide registration services: the ability to acquire and use domain names. Registrars provide the interface between customers and the domain registries and handle purchase, billing, and day-to-day domain maintenance, including renewals for domain registrations. Registrars also handle transfers of domains, either due to a sale or when a domain is transferred to another registrar. This requires authorization by the current domain owner, as well as a release of the domain to the new registrar.

DNS Entries

In addition to the information provided using nslookup, DNS entries can provide useful information about systems simply through the hostname. A system named "AD4" is a more likely target for Active Directory-based exploits and Windows Server-specific scans, whereas hostnames that reflect a specific application or service can provide both target information and a clue for social engineering and human intelligence activities.

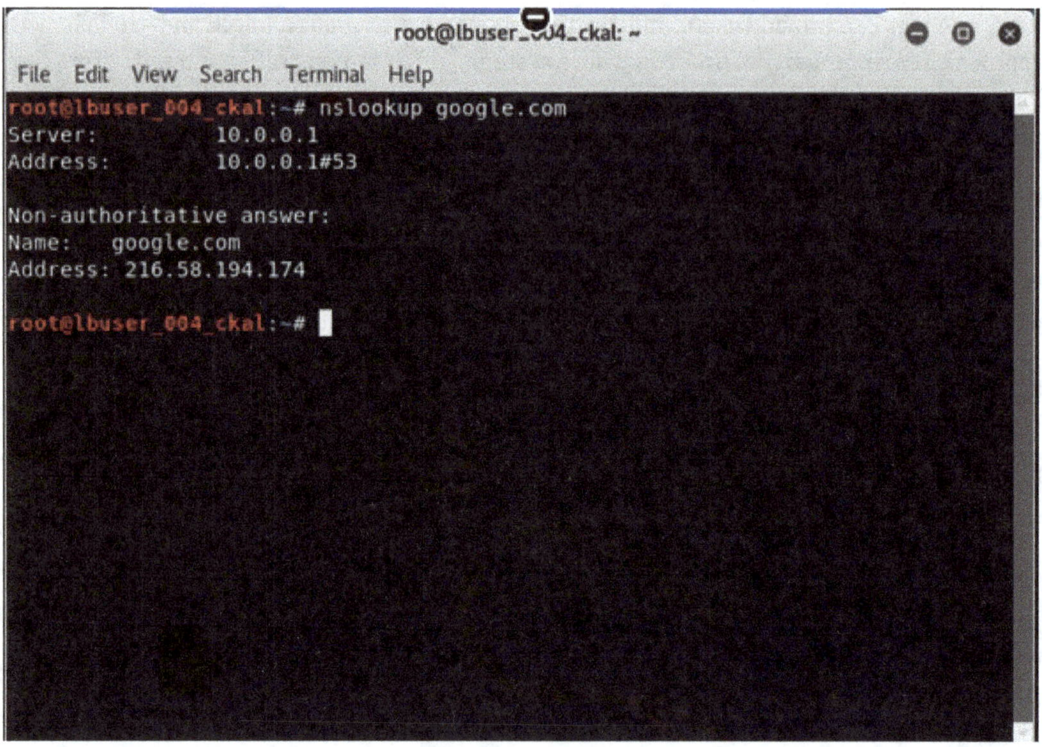

FIGURE 14.5 Nslookup for google.com

DNS Discovery

External DNS information for an organization is provided as part of its Whois information, providing a good starting place for DNS-based information gathering. Additional DNS servers may be identified either as part of active scanning or passive information gathering based on network traffic or logs, or even by reviewing an organization's documentation. Once you have found a DNS server, you can query it using dig or other DNS lookup commands, or you can test it to see if it supports zone transfer.

Zone Transfers

One way to gather information about an organization is to perform a zone transfer. Zone transfers are intended to be used to replicate DNS databases between DNS servers, which makes them a powerful information-gathering tool if a target's DNS servers allow a zone transfer. This means that most DNS servers are set to prohibit zone transfers to servers that aren't their trusted DNS peers, but cyber-intelligence officers and/or security analysts are likely to still check to see if a zone transfer is possible.

Whois

Whois allows you to search databases of registered users of domains and IP address blocks, and it can provide useful information about an organization or individual based on their registration information. Whois query allows you to search information such as the company's

WHOIS search results

```
Domain Name: GOOGLE.COM
Registry Domain ID: 2138514_DOMAIN_COM-VRSN
Registrar WHOIS Server: whois.markmonitor.com
Registrar URL: http://www.markmonitor.com
Updated Date: 2019-09-09T15:39:04Z
Creation Date: 1997-09-15T04:00:00Z
Registry Expiry Date: 2028-09-14T04:00:00Z
Registrar: MarkMonitor Inc.
Registrar IANA ID: 292
Registrar Abuse Contact Email: abusecomplaints@markmonitor.com
Registrar Abuse Contact Phone: +1.2083895740
Domain Status: clientDeleteProhibited https://icann.org/epp#clientDeleteProhibited
Domain Status: clientTransferProhibited https://icann.org/epp#clientTransferProhibited
Domain Status: clientUpdateProhibited https://icann.org/epp#clientUpdateProhibited
Domain Status: serverDeleteProhibited https://icann.org/epp#serverDeleteProhibited
Domain Status: serverTransferProhibited https://icann.org/epp#serverTransferProhibited
Domain Status: serverUpdateProhibited https://icann.org/epp#serverUpdateProhibited
Name Server: NS1.GOOGLE.COM
Name Server: NS2.GOOGLE.COM
Name Server: NS3.GOOGLE.COM
Name Server: NS4.GOOGLE.COM
DNSSEC: unsigned
URL of the ICANN Whois Inaccuracy Complaint Form: https://www.icann.org/wicf/
>>> Last update of whois database: 2020-11-23T05:33:10Z <<<
```

FIGURE 14.6 Whois query data for google.com

headquarters location, contact information, and its primary name servers. This information can provide you with additional hints about the organization by looking for other domains registered with similar information, email addresses to contact, and details you can use during the information-gathering process.

Information Gathering Using Packet Capture

Packet capture can provide huge amounts of useful information. Capture from a single host can tell you what systems are on a given network by capturing broadcast packets, and OS fingerprinting can give you a good guess about a remote host's operating system (Okada et al., 2013). If you are able to capture data from a strategic location in a network using a network tap or span port, you will have access to far more network traffic, and thus even more information about the network. Using packet capture can allow you to dig into specific responses or to verify that you did test a specific host at a specific time. Thus, packet capture can be used both as an analysis tool and as proof that a task was accomplished.

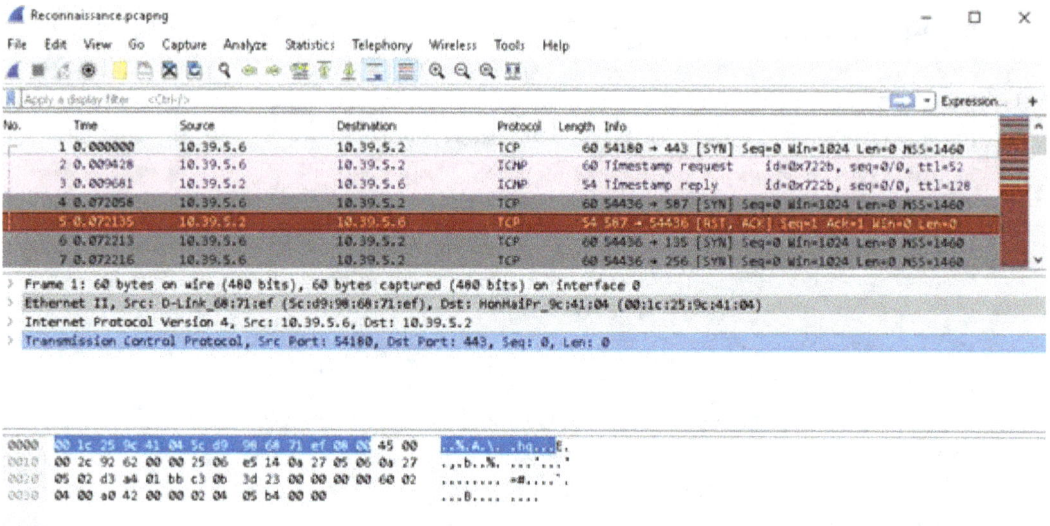

FIGURE 14.7 Packet capture data from an nmap scan

Incident Response

No matter how well an organization prepares its cybersecurity defenses, the time will come that it suffers a cybercrime incident that compromises the confidentiality, integrity, and availability of information or systems under its control. This section covers incident response procedures and analyzing symptoms for incident response that are explained in alignment with US-CERT's NCIRP (National Cyber Incident Response Plan) and Presidential Policy Directive (PPD) 41, United States Cyber Incident Coordination.

Handling a Cyber Incident

The incident response process has several phases (Cichonski et al., 2012; Department of Homeland Security, 2016). This section will explore incident response methods and procedures. The initial phase engages in building an incident response team and obtaining the necessary tools and resources. During preparation, in accordance with the results of risk assessments, we need to limit the number of incidents that will occur by selecting and conducting a set of controls. Detection of cybersecurity breaches is a pre-condition to alert the organization whenever a cyber-incident occurs. The organization can reduce the impact of the incident by containing it and recovering from it. During the containment eradication and recovery phase, activity may cycle back to detection and analysis. After the incident is properly handled, the organization issues an incident response report that includes the cause and cost of the incident and the methods the organization should take to proactively mitigate future incidents. Figure 14.8 illustrates the incident response life cycle: preparation, detection and analysis, containment, eradication and recovery, and post-incident activity.

Preparation

Incident response methodologies typically emphasize building an incident response capability and preventing incidents by ensuring that systems, networks, and applications are appropriately

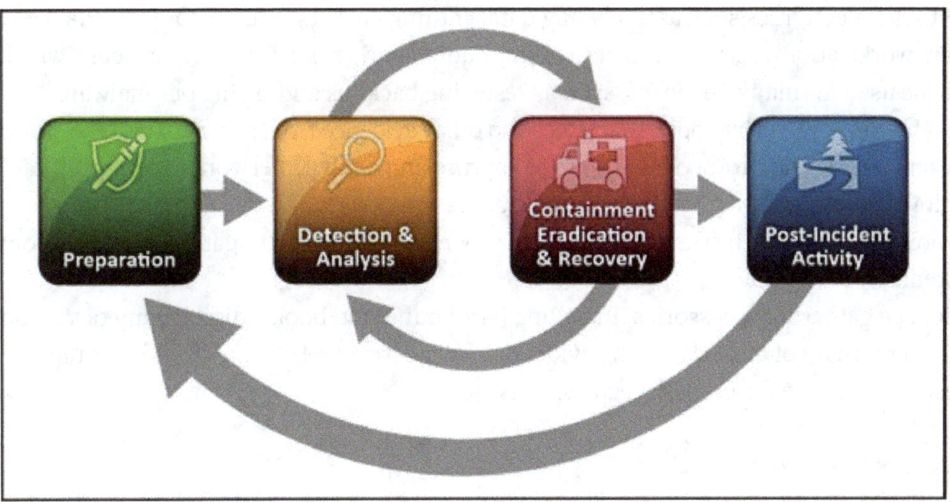

FIGURE 14.8 Incident response life cycle

and efficiently secure (The White House, 2016a). According to NIST Special Publication 800-61 Revision 2, the list below provides examples of tools and resources that may be of value during incident handling (Cichonski et al., 2012).

Incident Handler Communications and Facilities

- Contact information for team members and others within and outside the organization such as law enforcement and other incident response teams; information may include phone numbers, email addresses, public encryption keys (in accordance with the encryption software described below), and instructions for verifying the contact's identity
- On-call information for other teams within the organization, including escalation information
- Incident reporting mechanisms, such as phone numbers, email addresses, online forms, and secure instant messaging systems that allow users to report suspected incidents (at least one mechanism should permit people to report incidents anonymously)
- Issue tracking system for tracking incident information, status, etc.
- Smartphones to be carried by team members for off-hour support and onsite communications
- Encryption software to be used for communications among team members, within the organization, and with external parties; for Federal agencies, software must use a FIPS-validated encryption algorithm
- War room for central communication and coordination; if a permanent war room is not necessary or practical, the team should create a procedure for procuring a temporary war room when needed
- Secure storage facility for securing evidence and other sensitive materials

Incident Analysis Hardware and Software

- Digital forensic workstations and/or backup devices to create disk images, preserve log files, and save other relevant incident data

- Laptops for activities such as analyzing data, sniffing packets, and writing reports
- Spare workstations, servers, and networking equipment, or the virtual equivalents, which may be used for many purposes, such as restoring backups and trying out malware
- Portable printer to print copies of log files and other evidence from non-networked systems
- Packet sniffers and protocol analyzers to capture and analyze network traffic
- Digital forensic software to analyze disk images
- Removable media with trusted versions of programs to be used to gather evidence from systems
- Evidence gathering accessories, including hardbound notebooks, digital cameras, audio recorders, chain of custody forms, evidence storage bags and tags, and evidence tape, to preserve evidence for possible legal actions

Incident Analysis Resources

- Port lists, including commonly used ports and Trojan horse ports
- Documentation for OSs, applications, protocols, and intrusion-detection and antivirus products
- Network diagrams and lists of critical assets, such as database servers
- Current baselines of expected network, system, and application activity
- Cryptographic hashes of critical files to speed incident analysis, verification, and eradication (Cichonski et al., 2012, p. 22)

Exercises involving simulated incidents can be very significant for preparing staff for incident handling (see Cichonski et al., 2012). In addition, maintaining the number of cyber incidents affordably low is very important to protect the professional operations of the organization. If security controls are ineffective, higher volumes of cyber incidents may occur, overwhelming the incident response operations. As a result, the organization's responses will be slow and incomplete, leading to a negative impact on business and national security systems. In order to effectively deal with cyber incidents, it is essential to train incident response staff in risk assessment and network security configuration.

Detection and Analysis

The detection and analysis phase of incident response is one of the trickiest to commit to a routine process. Although cyber-intelligence officers and/or cybercrime investigators have many tools at their disposal that may assist in identifying that a security incident is taking place, many incidents are only detected because of the trained eye of an experienced analyst (Van der Kleij et al., 2017).

The incident response team should rapidly analyze and validate each incident; in particular, the team should quickly determine the incident's scope (e.g., targeted networks, systems, or applications); who or what originated the incident; what tools or attack methods are being employed; and what vulnerabilities are being exploited. The following are recommendations for facilitating incident analysis more effectively (Cichonski et al., 2012, pp. 29–30).

Profile Networks and Systems

Profiling is measuring the characteristics of expected activity so that changes to it can be more easily identified. Examples of profiling are running file integrity checking software on hosts to derive checksums for critical files and monitoring network bandwidth usage to determine what the average and peak usage levels are on various days and times. In practice, it is difficult to detect incidents accurately using most profiling techniques; organizations should use profiling as one of several detection and analysis techniques.

Understand Normal Behaviors

Incident response team members should study networks, systems, and applications to understand what their normal behavior is so that abnormal behavior can be recognized more easily. One way to gain this knowledge is through reviewing log entries and security alerts. Conducting frequent log reviews should keep the knowledge fresh, and the analyst should be able to notice trends and changes over time.

Create a Log Retention Policy

Information regarding an incident may be recorded in several places, such as firewall, IDPS, and application logs. Creating and implementing a log retention policy that specifies how long log data should be maintained may be extremely helpful in analysis because older log entries may show reconnaissance activity or previous instances of similar attacks. Another reason for retaining logs is that incidents may not be discovered until days, weeks, or even months later.

Perform Event Correlation

Evidence of an incident may be captured in several logs that each contain different types of data. A network IDPS may detect that an attack was launched against a particular host, but it may not know if the attack was successful.

Keep All Host Clocks Synchronized

Protocols such as the Network Time Protocol (NTP) synchronize clocks among hosts. Event correlation will be more complicated if the devices reporting events have inconsistent clock settings. From an evidentiary standpoint, it is preferable to have consistent timestamps in logs.

Maintain and Use a Knowledge Base of Information

The knowledge base should include information that handlers need for quick reference during incident analysis. Text documents, spreadsheets, and relatively simple databases provide effective, flexible, and searchable mechanisms for sharing data among team members.

Use Internet Search Engines for Research

Internet search engines can help analysts find information on unusual activity. For example, an analyst may see some unusual connection attempts targeting TCP port 22912. Performing a search on the terms "TCP," "port," and "22912" may return some hits that contain logs of similar activity or even an explanation of the significance of the port number.

Run Packet Sniffers to Collect Additional Data

Sometimes the indicators do not record enough details to permit the handler to understand what is occurring. If an incident is occurring over a network, the fastest way to collect the necessary data may be to have a packet sniffer capture network traffic.

Filter the Data

There is simply not enough time to review and analyze all the indicators; at minimum, the most suspicious activity should be investigated. One effective strategy is to filter out categories of indicators that tend to be insignificant. Another filtering strategy is to show only the categories of indicators that are of the highest significance; however, this approach carries substantial risk because new malicious activity may not fall into one of the chosen indicator categories.

Figure 14.9 illustrates retrieving a real-time list of running processes. The "top" command is used to provide CPU utilization under CPU stats, memory usage, and details about running processes.

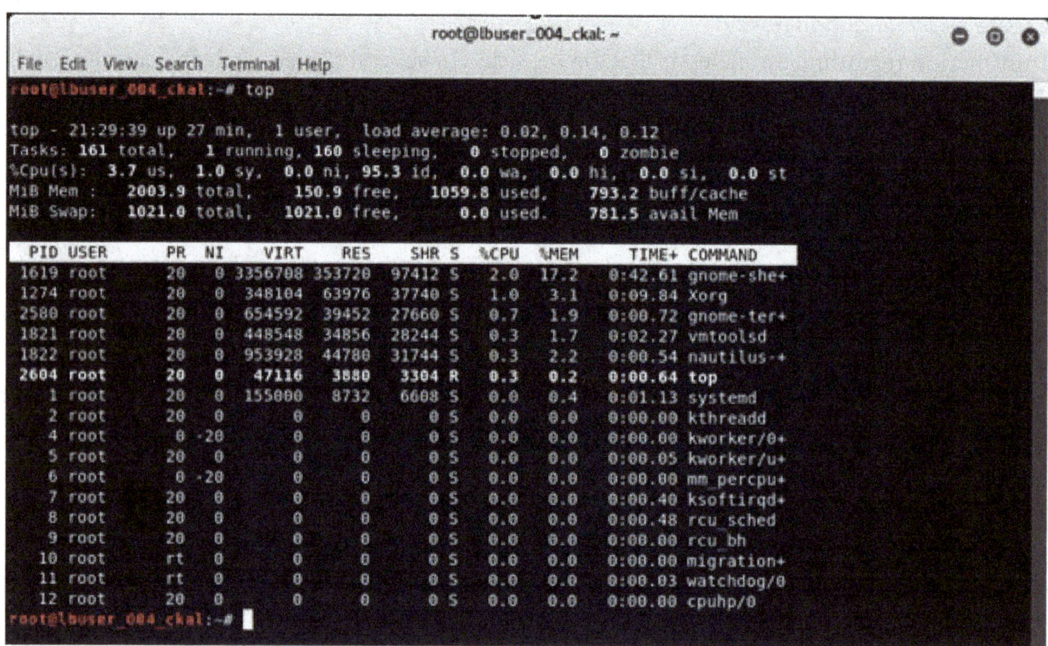

FIGURE 14.9 Retrieving a real-time list of running processes

Once an incident response team recognizes a cyber incident, they should immediately start recording all facts regarding the incident. Every document regarding the incident should be dated and signed by the incident handler. In fact, it can be utilized as evidence in a court of law if legal prosecution is pursued.

Containment Eradication and Recovery

Containment is significant before an incident overwhelms resources or increases damage. During the incident detection and analysis stage, the cyber incident response team is involved in mainly passive activities designed to point out and analyze the incident. After conducting this assessment, the incident response team needs to take active measures designed to contain the effects of the incident (e.g., shut down a system, disconnect it from a network), eradicate the incident from the network, and recover normal operations.

Evidence Gathering and Handling

It is crucial to properly document how all evidence, including compromised systems, has been preserved for legal proceedings so that any evidence can be admissible in court. A detailed log of the following information should be kept for all evidence (Cichonski et al., 2012):

- Identifying information (e.g., the location, serial number, model number, hostname, media access control (MAC) addresses, and IP addresses of a computer)
- Name, title, and phone number of each individual who collected or handled the evidence during the investigation
- Time and date (including time zone) of each occurrence of evidence handling
- Locations where the evidence was stored

Identifying the Attacking Hosts

In addition, the incident response team may desire to identify the attacking host(s). Based on NIST Special Publication 800-61 by Cichonski et al. (2012, p. 37), the following items describe the most pervasively implemented activities for the identification of attacking host(s).

Validating the Attacking Host's IP Address. New incident handlers often focus on the attacking host's IP address. The handler may attempt to validate that the address was not spoofed by verifying connectivity to it; however, this simply indicates that a host at that address does or does not respond to the requests. A failure to respond does not mean the address is not real. Also, the attacker may have received a dynamic address that has already been reassigned to someone else.

Researching the Attacking Host Through Search Engines. Performing an internet search using the apparent source IP address of an attack may lead to more information on the attack (e.g., a mailing list message regarding a similar attack).

Using Incident Databases. Several groups collect and consolidate incident data from various organizations into incident databases. This information sharing may take place in many forms, such as trackers and real-time blacklists. The organization can also check its own knowledge base or issue-tracking system for related activity.

Monitoring Possible Attacker Communication Channels. Incident handlers can monitor communication channels that may be used by an attacking host. For example, many bots use IRC as their primary means of communication. Also, attackers may congregate on certain IRC channels to brag about their compromises and share information. However, incident handlers should treat any such information that they acquire only as a potential lead, not as fact.

After an incident has been contained, eradication may be necessary to eliminate components of the incident (e.g., deleting malware and disabling breached user accounts, identifying and mitigating all vulnerabilities). For some incidents, eradication is either not essential or is implemented during recovery. In recovery, cybersecurity personnel restore systems to normal operation, confirm that the systems are properly functioning, and remediate vulnerabilities to prevent similar incidents.

Post-Incident Activity

One of the most significant parts of incident response is that the incident response team reflects new cyberthreats and improved technology on preventing future cyberthreats. Using collected incident data, we can point out systemic security weaknesses and threats as well as changes in incident trends along with risk assessment. Organizations must concentrate on collecting data that is actionable, rather than collecting data simply because it is available (The White House, 2016b). In accordance with NIST Special Publication 800-61 by Cichonski et al. (2012, p. 40), possible metrics for incident-related data include:

Number of Incidents Handled

The number of incidents handled is best taken as a measure of the relative amount of work that the incident response team had to perform, not as a measure of the quality of the team, unless it is considered in the context of other measures that collectively give an indication of work quality. Subcategories also can be used to provide more information. For example, a growing number of incidents performed by insiders could prompt stronger security controls on internal networks and stronger policy provisions concerning background investigations for personnel and misuse of computing resources.

Time Per Incident

For each incident, time can be measured. For example, you can measure total amount of labor spent working on the incident, elapsed time from the beginning to completing the incident assessment, and time for the incident response team to provide the initial report of the incident.

Objective Assessment of Each Incident

The response to an incident that has been resolved can be analyzed to evaluate how effective it was. For instance, you can review logs, forms, reports, and other documentation for adherence to established incident response policies and procedures, identifying which precursors and indicators of the incident were recorded to evaluate how effectively the incident was logged and identified.

Subjective Assessment of Each Incident

Incident response team members may be requested to evaluate their own operations. Another valuable source of input is to assess whether the owner of a resource that was compromised believes the incident was handled effectively and feels that the incident response was satisfactory.

In line with the incident handling process, there is the incident handling checklist, which can provide the major steps to be performed in incident handling. Table 14.1 outlines the incident handling checklist.

Summary

The objective of the cyber-intelligence operation should be to proactively respond to cyberthreats before cyber actors exploit and/or hit the networks and systems. In other words, the investigator must "run" faster than the cyber adversary by observing attacks and integrating

TABLE 14.1 Incident Handling Checklist

	Action	Completed
	Detection and Analysis	
1.	Determine whether an incident has occurred	
1.1	Analyze the precursors and indicators	
1.2	Look for correlating information	
1.3	Perform research (e.g., search engines, knowledge base)	
1.4	As soon as the handler believes an incident has occurred, begin documenting the investigation and gathering evidence	
2.	Prioritize handling the incident based on the relevant factors (functional impact, information impact, recoverability effort, etc.)	
3.	Report the incident to the appropriate internal personnel and external organizations	
	Containment, Eradication, and Recovery	
4.	Acquire, preserve, secure, and document evidence	
5.	Contain the incident	
6.	Eradicate the incident	
6.1	Identify and mitigate all vulnerabilities that were exploited	
6.2	Remove malware, inappropriate materials, and other components	
6.3	If more affected hosts are discovered (e.g., new malware infections), repeat the Detection and Analysis steps (1.1, 1.2) to identify all other affected hosts, then contain (5) and eradicate (6) the incident for them	
7.	Recover from the incident	
7.1	Return affected systems to an operationally ready state	
7.2	Confirm that the affected systems are functioning normally	
7.3	If necessary, implement additional monitoring to look for future related activity	
	Post-Incident Activity	
8.	Create a follow-up report	
9.	Hold a lessons learned meeting (mandatory for major incidents, optional otherwise)	

investigation tools. The cyber-intelligence cycle is a circular and repeated process to convert data into intelligence that can be disseminated to the intelligence customer. The cycle includes five steps: planning and direction; collection; processing and exploitation; analysis and production; dissemination and integration. Through active reconnaissance, port scanning and service discovery techniques, log and configuration analysis, and harvesting data from DNS and Whois, you will be able to gather information concerning vulnerabilities and abnormal behaviors on the network as well as detect new cyberthreats.

Incident response procedures and analyzing symptoms for incident response are based on US-CERT's NCIRP (National Cyber Incident Response Plan) and Presidential Policy Directive (PPD) 41, United States Cyber Incident Coordination. In that sense, the incident response process has an incident response life cycle: preparation, detection and analysis, containment, eradication and recovery, and post-incident activity. First, during preparation, in accordance with the results of risk assessments, you need to limit the number of incidents that will occur by selecting and conducting a set of controls. Detection of cybersecurity breaches is a precondition to alert the organization whenever a cyber incident occurs. The organization can reduce the impact of the incident by containing it and recovering from it. During the containment eradication and recovery phase, activity may cycle back to detection and analysis. After the incident is properly handled, the organization issues an incident response report that includes the cause and cost of the incident and the methods the organization should take to proactively mitigate future incidents.

Lab 14.1: Examining the DDOS_Attack.pcap File

Mr. Evil Noodle launched distributed denial-of-service (DDoS) attacks to damage protected computers and websites for the city of CIC and the CIC Police Department. DDoS attacks can be harder to detect due to the traffic coming from many places, and that also makes them much harder to stop.

In this project, as a cyber-intelligence agent or cybercrime investigator, you need to examine the DDOS_Attack.pcap file. To examine this file, you can follow the steps below:

1. From the desktop, open Wireshark.
2. In the Wireshark Network Analyzer window, from the menu bar, click File and select Open.
3. In the Wireshark, Open Capture File window, from the left pane, click This PC.
4. In the right pane, scroll down and navigate to Local Disk (C:) > 093009Data > Analyzing Attacks on Computing and Network Environments, select DDOS_Attack.pcap, and click Open.
5. In the DDOs_Attack.pcap window, from the menu bar, click Statistics and select Conversations.
6. In the Wireshark • Conversations • DDOS_Attack window, click the IPv4 · 4594 tab and observe the wide variety of IP addresses and the number of packets coming from each.
7. Click the TCP · 4594 tab and observe the details.

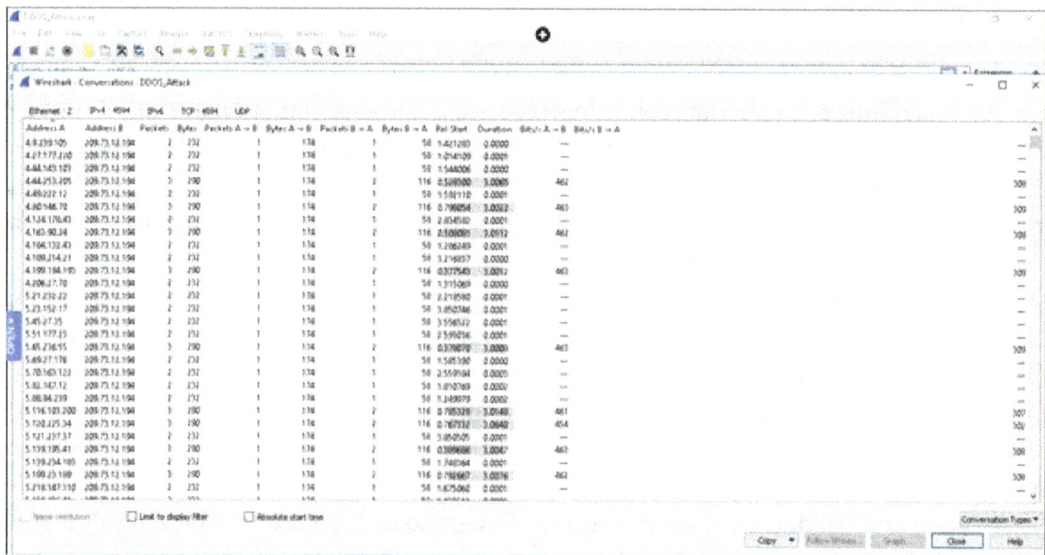

FIGURE 14.10 Examining the DDOS_Attack.pcap. File

Lab 14.2: Inspecting the Vulnerability in the Echo Server's Source Code

Mr. Evil Noodle is a citizen of China and resident of Ukraine. Mr. Evil Noodle researched and identified victims possessing information of interest, including trade secrets, confidential business information, and personal identifying information of victim employees and customers. Mr. Evil Noodle then gained unauthorized access to victims possessing the information sought by the conspiracy. As a cyberthreat-intelligence officer in the U.S. Navy, you detected abnormal activities and unauthorized access in their network; however, you are not sure about Mr. Evil Noodle. In line with this scenario, first, you need to inspect the vulnerability in the echo server's source code.

This vulnerability may be exploited by Mr. Evil Noodle to perform unauthorized actions within the U.S. Navy's computer system. To inspect the vulnerability in the echo server's source code, here is what you need to do:

1. Investigate the vulnerability in the echo server's source code:
   a. From the desktop, double-click **vulnerable-echo-server.c**. This application creates a TCP server that a client connects to. The client can send a string input to the server, and the server then echoes this string back to the client.
   b. In the **vulnerable-echo-server.c** text editor window, scroll down to view the **copy_data()** function.
   c. Close the **vulnerable-echo-server.c** text editor window.
2. Compile the source code to create the executable server, and then run the server:
   a. From the left sidebar, open the Terminal window.

```
root@lbuser001ptcs:~# gcc Desktop/vulnerable-echo-server.c -o echoserv
Desktop/vulnerable-echo-server.c: In function 'copy_data':
Desktop/vulnerable-echo-server.c:23:2: warning: implicit declaration of function 'memcpy' [-Wimplicit-f
unction-declaration]
  memcpy(vulnerable_buffer, buffer, buffer_len);
  ^~~~~~
Desktop/vulnerable-echo-server.c:23:2: warning: incompatible implicit declaration of built-in function
'memcpy'
Desktop/vulnerable-echo-server.c:23:2: note: include '<string.h>' or provide a declaration of 'memcpy'
Desktop/vulnerable-echo-server.c:13:1:
+#include <string.h>

Desktop/vulnerable-echo-server.c:23:2:
  memcpy(vulnerable_buffer, buffer, buffer_len);
  ^~~~~~
Desktop/vulnerable-echo-server.c: At top level:
Desktop/vulnerable-echo-server.c:27:1: warning: return type defaults to 'int' [-Wimplicit-int]
 main(int argc, char **argv)
 ^~~~
root@lbuser001ptcs:~# 
```

FIGURE 14.11 Compiling the source code

    b. In the Terminal window, execute the following command to compile the source code and output an executable named echoserv at the root folder:
   - gcc Desktop/vulnerable-echo-server.c -o echoserv

    c. In the Terminal window, execute the following command to run the server on localhost (127.0.0.1) and have it listen on port 9999:
   - ./ecoserv 9999

3. Verify that the server is operational:
    a. From the left sidebar, right-click the Terminal icon and select New Window.
    b. In the second Terminal window, execute the following command to connect to 127.0.0.1:
   - telnet 127.0.0.1 9999

    c. In the second Terminal window, execute the following command to connect to echo the message back to you:
   - hello

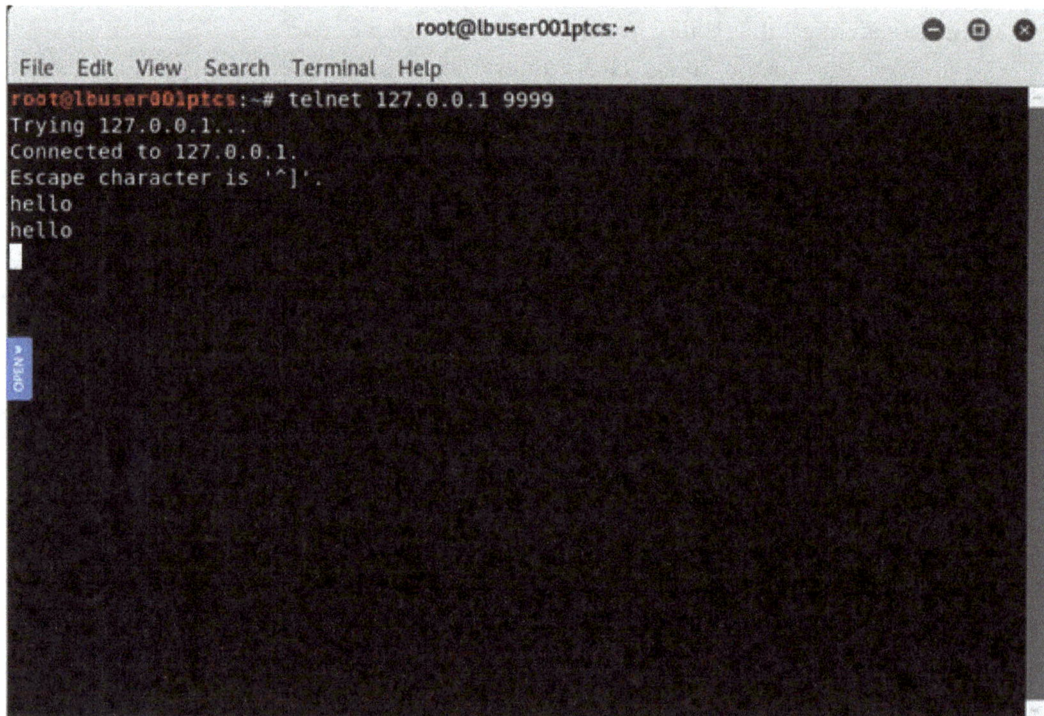

FIGURE 14.12 Command to connect to echo the message back

    d. Switch back to the first Terminal window and you will observe that the Terminal running the server also displays the hello message.
    e. Close the second Terminal window.

4. Investigate the configuration file and run Simple Fuzzer with it loaded:
    a. From the desktop, double-click **buff-fuzz.cfg**.
    b. In the **buff-fuzz.cfg** text editor window, verify that the **sequence** is **X** and the **maxseqlen** is **500**.

5. Run the fuzzer:
   a. From the left sidebar, right-click the Terminal icon and select **New Window**.
   b. In the second Terminal window, execute the given command to run Simple Fuzzer on the TCP echo server listening on port 9999 of localhost:
      - sfuzz -TO -f Desktop/buff-fuzz.cfg -S 127.0.0.1 -p 9999
   c. Switch to the first (server) Terminal window and verify that your message was sent and received by the server, which suffered a segmentation fault.

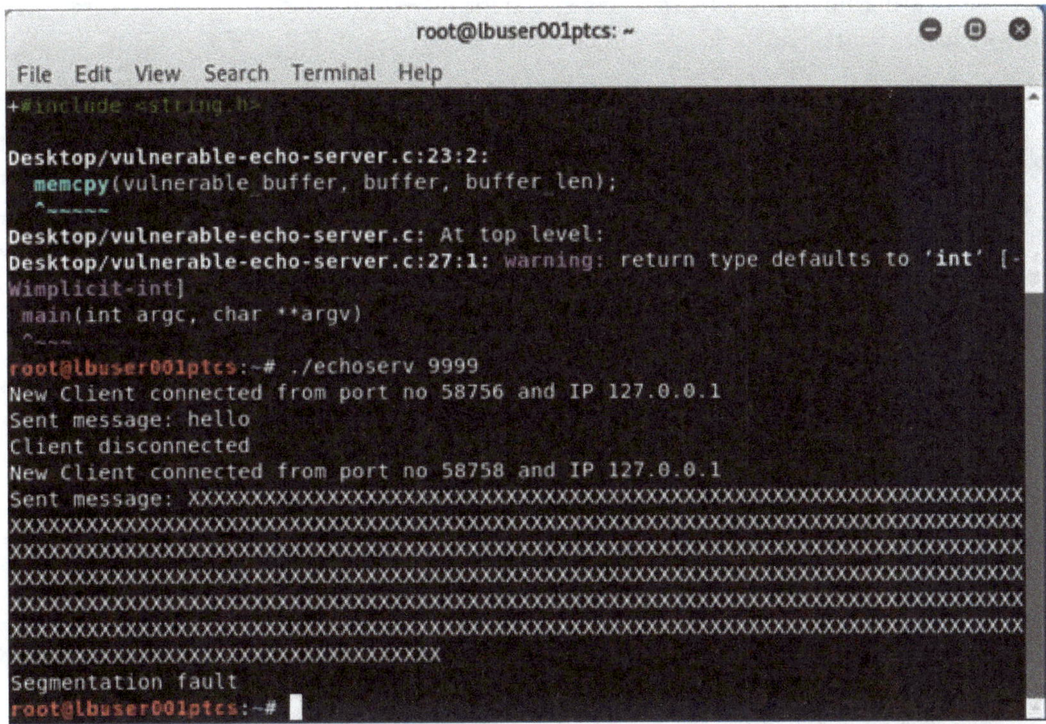

FIGURE 14.13 Segmentation fault

   d. In the first (server) Terminal window, execute the following command to try to connect to the echo server again using Telnet:
      - telnet 127.0.0.1 9999
6. Examine a more security-conscious version of the source code and compile it.
   a. From the desktop, double-click **fixed-echo-server.c**.
   b. In the fixed-echo-server.c text editor window, scroll down and look for the statement that calls the **copy_data()** function from before.
   c. Close the **fixed-echo-server.c** text editor window.
   d. Switch back to the first (server) Terminal window.
   e. In the first (server) Terminal window, execute the following command to compile the fixed source code:
      - gcc Desktop/fixed-echo-server.c -o fixed-echoserv

7. Run the fixed echo server and attempt to cause a buffer overflow:
   a. In the first (server) Terminal window, execute the following command to start the server:
      - ./fixed-echoserv 9999
   b. Switch back to the second Terminal window and execute the following command to run Simple Fuzzer on the TCP echo server listening on port 9999 of localhost:
      - Sfuzz -TO -f Desktop/buff-fuzz.cfg -S 127.0.0.1 -p 9999
   c. Switch back to the first (server) Terminal window and verify that your server is still running despite the fuzzing attempt.

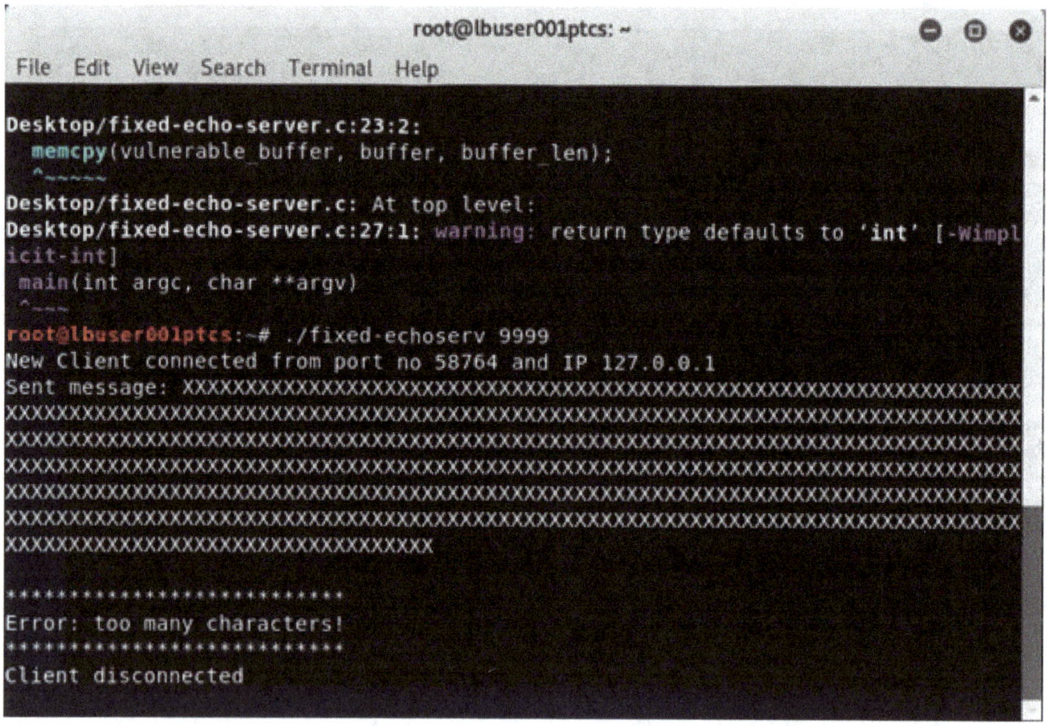

FIGURE 14.14 Verifying the server is running

Lab 14.3: Initiating an SSH Session from Your Windows 10 Client to Your Windows Server

In this project, as a member of the incident response team, you need to initiate an SSH (Secure Shell) session from your Windows 10 client to your Windows Server. SSH is a remote administration protocol that allows users to control and modify their remote servers over the internet.

1. Initiate an SSH session from your Windows 10 client to your Windows Server:
   a. From the taskbar, open File Explorer.
   b. In the File Explorer window, from the left pane, click Local Disk (C:) and then from the right pane, navigate to 093009Data > Performing Active Asset and Network Analysis and open putty.exe.

   c. At the Open File-Security Warning prompt, click Run.

   d. In the PuTTY Configuration dialog box, type Host Name (or IP address) as 10.0.0.19 and click Open.

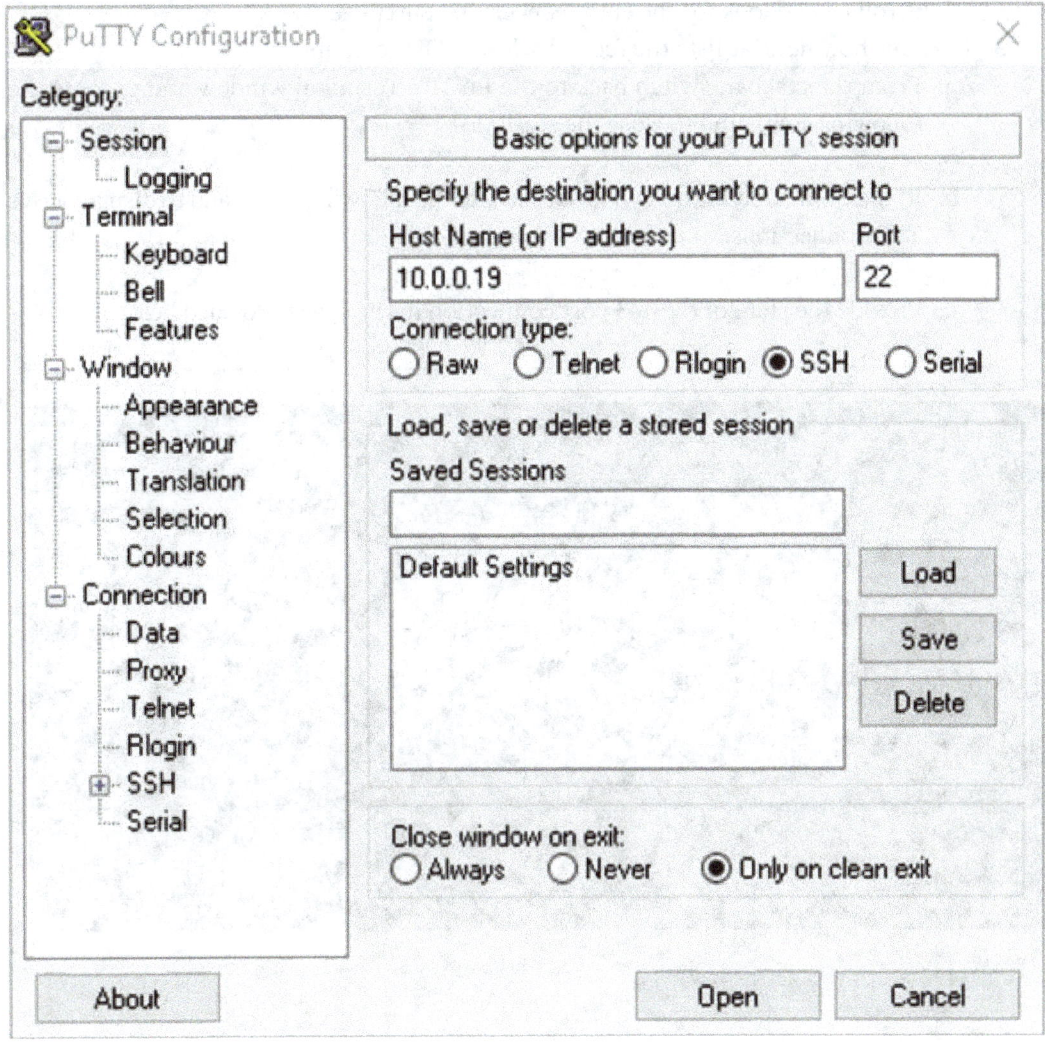

FIGURE 14.15 PuTTY configuration

   e. At the PuTTY Security Alert prompt, click Yes.

   f. In the PuTTY Terminal window, type the following details to log into the Terminal window in Virtual machine.

   g. In the PuTTY Terminal window, execute the following command to display the information and statistics about protocols in use and current TCP/IP network connections:

     • Netstat

   h. Minimize all windows.

2. Run **netstat** to view all open connections:

   a. Right-click the Start menu and select Search.

    b. In the Search box, type command prompt and click Command Prompt.

    c. In the Command Prompt window, execute the following commands:

- netstat /?
- netstat -ab

    d. Scroll up and look for the entry concerning putty.exe.

3. Examine how **netstat** lists the recently closed SSH session:

    a. From the taskbar, switch back to the PuTTY Terminal window and execute the following command to close the session:

- exit

    b. In the Command Prompt window, execute the following command to display active TCP connections:

- netstat -n

    c. Provide the state of the 443 port connection in the space provided below:

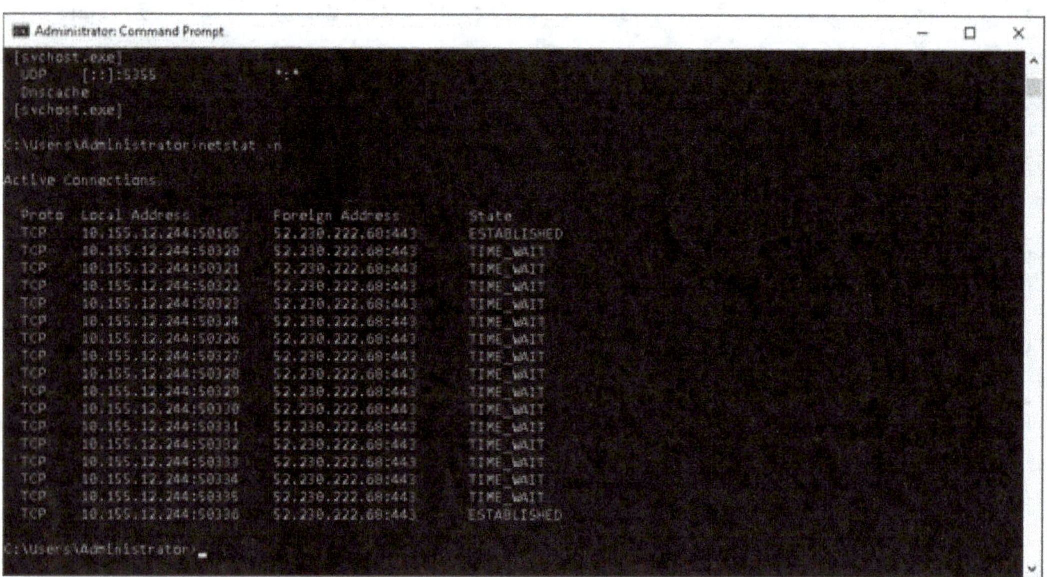

FIGURE 14.16 Displaying active TCP

References

Bartlett, G., Heidemann, J., & Papadopoulos, C. (2007, October). Understanding passive and active service discovery. In *Proceedings of the 7th ACM SIGCOMM Conference on Internet Measurement* (pp. 57–70).

Bul'ajoul, W., James, A., & Pannu, M. (2015). Improving network intrusion detection system performance through quality of service configuration and parallel technology. *Journal of Computer and System Sciences*, *81*(6), 981–999.

Bullock, J. A., Haddow, G. D., & Coppola, D. P. (2016). *Introduction to homeland security: Principles of all-hazards risk management*. Butterworth-Heinemann.

Castelluccia, C., Grumbach, S., & Olejnik, L. (2013, June). Data harvesting 2.0: from the visible to the invisible web. In *The Twelfth Workshop on the Economics of Information Security*.

Cichonski, P., Millar, T., Grance, T., & Scarfone, K. (2012). Computer security incident handling guide. *NIST Special Publication*, *800*(61), 1–147.

Department of Homeland Security. (2016, December). *National cyber incident response plan*. https://us-cert. cisa.gov/sites/default/files/ncirp/National_Cyber_Incident_Response_Plan.pdf

Duvenage, P., von Solms, S., & Corregedor, M. (2015, July). The cyber counterintelligence process: A conceptual overview and theoretical proposition. In *Proceedings of the 14th European Conference on Cyber Warfare and Security, ECCWS 2015* (pp. 42–51).

Lyytinen, K., & Damsgaard, J. (2011). Inter-organizational information systems adoption: A configuration analysis approach. *European Journal of Information Systems, 20*(5), 496–509.

Matsunaka, T., Yamada, A., & Kubota, A. (2013, March). Passive OS fingerprinting by DNS traffic analysis. In *2013 IEEE 27th International Conference on Advanced Information Networking and Applications* (AINA) (pp. 243–250). IEEE.

Okada, S., Fukuyama, N., Morinaga, M., & Miyazaki, H. (2013). *Packet capture system, packet capture method, information processing device, and storage medium* (U.S. Patent No. 8,605,617). U.S. Patent and Trademark Office.

Richelson, J. T. (2015). *The US intelligence community*. Hachette UK.

Scarfone, K., Souppaya, M., Cody, A., & Orebaugh, A. (2008). Technical guide to information security testing and assessment. *NIST Special Publication, 800*(115), 2–25.

Van der Kleij, R., Kleinhuis, G., & Young, H. (2017). Computer security incident response team effectiveness: A needs assessment. *Frontiers in Psychology, 8*, 2179.

Vardangalos, G. (2016). Cyber-intelligence and cyber counterintelligence (CCI): General definitions and principles. *International Strategic Analyses* (KEDISA), (1).

The White House. (2016a, July 26). *Presidential policy directive: United States cyber incident coordination*. https://obamawhitehouse.archives.gov/the-press-office/2016/07/26/presidential-policy-directive-united-states-cyber-incident

The White House. (2016b, July 26). *Annex presidential policy directive: United States cyber incident coordination*. https://obamawhitehouse.archives.gov/the-press-office/2016/07/26/annex-presidential-policy-directive-united-states-cyber-incident

Credits

Cyberterrorism: Criminal Patterns and Countermeasures

Kyung-Shick Choi

Introduction

As the accessibility and usage of cyberspace increase exponentially, terrorism has taken on a more convoluted and imperceptible form. In recent years, cyberterrorists leveraged sophisticated techniques to attack the nation's critical infrastructures for damaging targets' functionalities and stealing highly sensitive intellectual property, private information, and valuable assets. The scope of the recent hacking against the U.S. government, which extends "beyond nuclear laboratories and Pentagon, Treasury and Commerce Department systems, complicates the challenge for federal investigators as they try to assess the damage and understand what had been stolen" (Sanger & Perlroth, 2020). In addition to intricate cyberattacks, terrorists have used the internet as a technological facilitator to assist in proliferating attacks in the physical world through funding, communication, and propaganda efforts.

One of the most complex and sensitive cybercrime topics is cyberterrorism. How can we effectively prevent a potential cyber-disaster? In this chapter, we will learn important lessons from the past and try to come up with feasible preventive measures.

Cyberterrorism

Following the attacks on September 11, 2001, terrorism has transformed. This new form of terrorism has become more networked, diverse, and intricate. Due to the recent cyberattacks against European countries and the United States, cyberterrorism has globally received greater attention.

DDoS attacks and hacking of national government agencies initiated from China prove that these threats are real. Direct forms of cyberterrorism include DDoS attacks, the Stuxnet worm, hacking, and the use of botnets. Additional forms of cyberterrorism are utilization of the internet for logistics of traditional activity, recruitment, propaganda, training, education, financing, command and control, and procurement of supplies. The United Nations Office on Drugs and Crime (2012) defines this type of activity as "the use of the internet for terrorist purposes," and is concerned with international tensions derived from global cyberattacks.

Globally recognized, cyberterrorism is a fundamental threat to national and international security as well. Important questions to ask include:

- How is the internet being used by terrorist organizations?

  - Accessibility of prohibited information on the internet makes it a useful propaganda tool for terrorist organizations
  - Terrorists can contact potential recruits through websites, chat rooms, and social media pages
  - Easy to get away from law enforcement

- Why do terrorist organizations use the internet as one of their resources?
- How are law enforcement agencies responding against terrorist threats from the internet?

A violation of the safety of cyberspace, such as compromising computer systems and networks, can be seen as a threat as serious as an attack in the physical realm.

What Is Cyberterrorism?

There is controversy over the definition of cyberterrorism; however, it can generally be defined as an act of disrupting critical infrastructure such as energy, transportation, and public facilities by using cyber tools for the purpose of threatening the government or its people. Cyberterrorism utilizes software and networks in cyberspace to commit an act of terror.

Also included are infrastructure attacks whose aim is to disable or destroy a computer system that contains critical data. The USA PATRIOT Act of 2001 defines critical infrastructure as "systems and assets, whether physical or virtual, so vital to the United States that the incapacity or destruction of such systems and assets would have a debilitating impact on security, national economic security, national health or safety, or any combination of those matters." In order to be charged with an act of cyberterrorism, the person who actually committed the cyberterrorist act must be identified, which is nearly impossible if the attack originated from a different country. In addition to the activities listed above, the USA PATRIOT Act additionally requires serious financial loss, at least $5,000, physical injury to any person, and a threat to public health or safety or damage affecting a computer system used by or for a government entity in furtherance of the administration of justice, national defense, or national security (Moore, 2011).

Cyberterrorism is one type of cyberattack. Cyberattacks can either be cyberterrorism or cybercrime, depending on the circumstances surrounding national security or the social norms. As discussed in Chapter 1, cybercrime can be defined through criminal behaviors that use computing technology connected to the internet as a tool. Sharing or distributing bomb-making guides, warfare strategies, or information to assist an individual to learn hacking techniques that could later be used in a criminal activity may be considered forms of cyber-violence.

The results of cybercrime and cyberterrorism tend to overlap. Many crimes committed in cyberspace would constitute cyberterrorism if the crime is committed by a terrorist. Additionally, Moore (2011) addresses the importance of understanding proper classification of cyberterrorism, since many cybercrimes can be labeled as cyberterrorism.

Cyber-warfare is the act of aggression through the use of the internet by a state. When a state is the aggressor, the consequences are similar to cyberterrorism but can be much more serious (Schmid & Jongman, 2005).

Use of the internet by terrorists should be examined as a form of cyberterrorism. The Center for Strategic and International Studies (CSIS) defines cyberterrorism as:

> The use of computer network tools to shut down critical national infrastructures (e.g., energy, transportation, government operations) or to coerce or intimidate a government or civilian population.

The National (U.S.) Conference of State Legislatures has defined cyberterrorism as:

> The use of information technology by terrorist groups and individuals to further their agenda. This can include use of information technology to organize and execute attacks against networks, computer systems and telecommunications infrastructures, or for exchanging information or making threats electronically.

By examining these various definitions, we can see the term "cyberterrorism" can be used to refer to many types of cybercrimes which may or may not constitute cyberterrorism. Any type of criminal activity that transpires in cyberspace, including general forms of malicious computer hacking and/or cyberattacks, is included. Such acts of terrorism may be direct (e.g., attacks on infrastructure) or indirect through the generation of revenue via cybercriminal techniques (e.g., credit fraud and identify theft).

There is concern over a surprise cyberattack on the critical infrastructure that would, with a single event, be catastrophic in the United States. In fact, there may have already been cyber-attacks and attempts on the critical infrastructure in the United States, yet none have been directly linked to terrorist organizations. Some may argue that the fear of cyberterrorism on the critical infrastructure is largely due to a lack of education about the topic and media portrayals of terrible events.

However, several reports speculate that China has attempted to build a unit of computer hackers with the goal of launching cyberattacks against the United States. In recent years, China has already launched successful attacks against U.S. information systems.

Typology of Cyberterrorism

Aside from information attacks and infrastructure attacks, terrorists can utilize the internet for a plethora of other activities such as communication, planning and support functions, data mining, propaganda, and fundraising and financial activities (Ballard et al., 2002).

TABLE 15.1

CATEGORY	DEFINITION AND EXPLANATION
Information attacks	Cyberterrorist attacks focused on altering or destroying the content of electronic files, computer systems, or the various materials therein.
Infrastructure attacks	Cyberterrorist attacks designed to disrupt or destroy the actual hardware, operating platform, or programming in a computerized environment.
Technological facilitation	Use of cyber communications to send plans for terrorist attacks, incite attacks, or otherwise facilitate traditional terrorism or cyberterrorism.
Fundraising and Promotion	Use of the internet to raise funds for a violent political cause, to advance an organization supportive of violent political action, or to promote an alternative ideology that is violent in orientation.

Information and Infrastructure Attacks

As previously discussed, the three primary cyberterrorism attacks are Stuxnet worm, DDoS, and ransomware.

Stuxnet Worm

The Stuxnet worm is thought to have originally developed as a cyber-weapon, intended to attack the centrifuge in Iran's nuclear reprocessing plant in Natanz, Iran. The Stuxnet worm has been used as an exploit in the Microsoft Windows operating system to spread and attack Siemens industrial computers in the United States and Germany.

Because of the complexity of this computer virus, there have been suspicions that state intervention (United States and Israel) played a role in its creation. James Lewis, of the Center for Strategic and International Studies, outlined the Stuxnet worm incident as the first case of cyber-warfare.

Image 15.1

Distributed Denial-of-Service (DDoS)

Denial-of-service attacks are known as attacks on a computer system or a network. If this type of attack succeeds, the victimized computer system or network is then rendered inoperable. This method of attack involves overloading computer resources or ceasing a network by exhausting the bandwidth of the network.

Image 15.2

Ransomware

Kharraz, Robertson, Balzarotti, Bilge, and Kirda (2015), who studied a collection of ransomware samples between 2006 and 2014, divided 1,359 ransomware samples into 15 different ransomware families based on the distinct encryption, deletion, and communication techniques of each ransomware. Apart from locking or preventing access into the victims' computers and/or networks, some ransomware families encrypt the data and demand a decryption key in order to access computers or networks. Other ransomware families steal user information or change the browser settings while performing multiple malware installations. In addition, several ransomware families are even capable of deleting existing data (Kharraz et al., 2015). Generally, regardless of different forms and approaches, ransomware is designed to deny access to valuable information while holding the decryption key to extort millions or even billions of dollars from individuals, businesses, and governments.

The proliferation of ransomware can also be connected to terrorist activities. An example of terrorist activity and its connection to ransomware would be the Albanian Hacker case. According to Johnson (2016), in August 2015, a ransomware attack under the username of Albanian Hacker demanded payment of two Bitcoin, which had a value of $500 at the time, from an Illinois internet retailer in exchange for removing bugs from their computer. While seizing control of computers, Albanian Hacker generated a kill list used as one of the first kill lists issued by the Islamic State of Iraq and Syria (ISIS).

The WannaCry attack and its relationship with the Lazarus Group, sponsored by the North Korean intelligence agency, is an additional bona fide example of terrorist activities against another nation utilizing a new technology. The WannaCry outbreak shut down computers in more than 80 National Health Service organizations in England alone, resulting in 600 general practices having to return to pen and paper, and five hospitals simply diverting ambulances to handle more emergency cases (Hern, 2017). The WannaCry 2.0 global ransomware attack also

Image 15.3

resulted in damage to massive amounts of computer hardware and the extensive loss of data, money, and other resources (U.S. Department of Justice, 2018).

Additionally, the 2019 state and local government ransomware attacks in the United States may demonstrate a tie between ransomware and terrorist activities. All of these examples suggest that ransomware attacks are handy tools for meeting both the political and economic objectives of the government or terrorist groups.

Technological Facilitation

From technological facilitations to terrorist groups, cyber communication technology has dramatically shortened transmission time from one member to the next. Web forums, social networking sites, video conference chatting software, and Voice-over-Internet Protocol (VOIP) can all be instrumental in enhancing the efficiency of communication among terrorist organizations. IT largely facilitates terrorists' planning activities and support functions.

In April 2015, Ardit Ferizi—a citizen of Kosovo and the known leader of the Kosova Hacker's Security posted on his Twitter account, @Th3Dir3ctorY, "Getting to Know Kosovar Hacker's Security Crew plus an Exclusive Interview with Th3Dir3ctorY," complete with a link providing access to the information (*United States of America v. Ardit Ferizi*, 2015:6). On April 26 and 27, 2015, he exchanged tweets and confirmed that, "Yes brother/Im muslim al britani" (*United States of America v. Ardit Ferizi*, 2015:10). He continuously posted ISIS videos and provided an ISIS recruiter and leader with the personal information of 1,351 U.S. military members so that it could be used as a "hit list." Given these unlawful actions, he was charged with providing material support to ISIS. Ferizi was arrested in Malaysia, where he was accessing the internet to further his cyberterrorist agenda and was extradited to the United States to face prosecution. Ferizi pleaded guilty and was sentenced to 25 years in prison.

Terrorist organizations and individuals like Ferizi have developed extensive capacities for disseminating propaganda, recruiting members, and communicating support to terrorist organizational structures and operations. While documents such as *The Minimanual of the Urban Guerilla* (Marighella, 1969) and the 1977 edition of *The IRA Green Book* (Coogan, 2002) continue to offer practical guidance, the scope and sophistication of current activities could not have been predicted by their authors. Counterterrorism focuses on the capacity of cyberterrorists to employ sophisticated techniques to attack a nation's critical infrastructure. However, terrorists also routinely use advanced technology to facilitate and coordinate daily activities and conduct conventional attacks.

Communication

- Websites and online communication are a means of interacting with media and the public in general
- The internet is a communication tool between terrorists and terrorist groups
- Forums and chat rooms are the most popular means of communication
- Global recruitment of new members
- Networking
- Communication relies on techniques that are not easy to track or crack (e.g., encryption)

Online communication is an area not strictly regulated by the authorities and hence remains one of the most common ways that terrorists make use of the internet. Websites, forums, blogs, emails, instant messaging, voice-over-IP, etc. are as frequently used by terrorists as by everybody else.

Online communication facilitates the exchange of information among terrorists and terrorist organizations. In general, this exchange plays an important role in communicating with the media and the public.

In recent years, digital cryptography has become very easy to learn. Now, it is even easier for a person to hide information in a digital file (e.g., an image), giving the appearance it is harmless. There are several types of software programs that can easily encrypt messages and information, which can be accessed via the internet, decrypted, and utilized for the purposes of a terrorist attack.

Reports of data encryption used by terrorist organizations were speculated in connection with the September 11, 2001, attacks on the United States. One report shares the belief that Al Qaeda used intelligence hidden via encryption on pornographic websites and sports chat rooms. This information could have been accessed by any person at any time; however, the massiveness of the files stored digitally made it nearly impossible to search for terrorist-related information. Islamic state terrorists used online gaming tools (e.g., PlayStation) and encrypted apps to hide plotting for the 2015 Paris attack (Sanger & Perlroth, 2015).

In addition to the more general communications, the internet provides an array of options for global networking and recruitment. Because of the widely available internet technology, the internet recruitment for new terrorist members has become a fully organized activity and is increasingly characterized by individuality and self-recruitment.

Planning Activities and Supporting Functions

- Instructions for attacks and activities can be downloadable, open to the public in forums, and often very easy to follow
- Training through guidebooks or handbooks (e.g., "Terrorist's Handbook," "Mujahideen Poisons Handbook," "How to Make Bombs") are freely available on the internet
- Supporting functions (e.g., communication, propaganda, coordination, administration) during attacks between individuals and organizations
- Involving voluntary assistance from "outsiders" as a reaction to the attacks and/or political situation

Instructions for less sophisticated attacks and activities, such as how to launch a DoS attack or how to make simple bombs, are freely available on the internet. The instructions are easily accessible and easy to follow. Moreover, interactive tutorials, guidebooks, and various online training opportunities allow for a perfect virtual training camp or open university to all interested participants.

Additionally, the internet plays a role in supporting activities such as communication, propaganda, coordination, and administration during attacks between individuals and organizations.

As mentioned above, the online environment makes it relatively easy to involve voluntary assistance from outsiders, thereby magnifying the effects and consequences of these attacks.

Data Mining

- Researching information, contacts, and backgrounds for terrorist activities and targeting analyses
- Often using open sources for information (e.g., Google Earth, Google Maps, Microsoft Virtual Earth, NASA WorldWind)
- Combination of all unclassified information available on the internet adds up to something that must be classified (Brunst, 2010)

Data mining is strongly interlinked with planning and supporting terrorist activities. The vast amount of information freely available on the internet is surprising. Open sources— including Google Earth and NASA WorldWind—provide more information than traditional maps. Moreover, researching the background of persons, targets, and security measures now only involves a few clicks.

As concluded by a quote in an article by P. W. Brunst (2010), the "combination of all unclassified information available in the internet adds up to something that ought to be classified."

Fundraising and Promotion

Terrorist groups use digital technology for the purpose of disseminating propaganda intended to garner support from individuals. Promotion involves any use of cyber communications technology to boost the cause of a terrorist group, to recruit members, to solicit information, or to generate revenue (e.g., fundraising). Many major terrorist organizations have an internet presence and ability to use digital technology as a tool, just like any legitimate business would.

Choi, Lee, and Cadigan (2018) analyzed how Al Qaeda and ISIS use cyber-resources in their propaganda processes using a global cyberterror data set by combining both cyber-resources (Facebook, online forums, Twitter mentions, websites, and YouTube videos) and legal documents to individual cases that were mentioned in reports.

The study indicates that cyberattack strategies of Al Qaeda and ISIS have a different configuration. YouTube videos were predominantly used for propagating certain ideologies and for recruiting members for both Al Qaeda and ISIS. However, there was an isomorphic pattern in the number of suspects who were subscribed to Al Qaeda and ISIS ideologies, respectively. Al Qaeda–affiliated cyberterrorists used YouTube videos as both individual sources and embedded sources through Facebook and Twitter, whereas ISIS used Twitter accounts to show their successes and victories by disseminating messages and videos of beheadings and bombings.

Online Propaganda

The internet is a low-cost, fast, and global medium that can disseminate the following:

- Terrorist beliefs and information about specific organizations and their history, and their leaders' profiles, views, beliefs, aims, and ambitions
- Publicity (newsletters, videos, photos, etc.)
- Threats, usually via terrorist websites

The internet is an attractive, low-cost tool; it is fast and provides a global platform to present terrorist beliefs. The internet makes it possible to spread information about specific organizations' history, leaders' profiles, their views, beliefs, aims, and ambitions. Successful and widespread propaganda not only advances terrorism as such, but it may also include threats to perform new acts and create fear among the population.

The main tools for propaganda are terrorist websites and various freely disseminated newsletters, videos, photos, and more. According to the U.S. State Department, 4,500 terrorist websites were known to exist in 2005.

Fundraising and Financial Activities

- To raise funds for their activities directly via websites (e.g., donations of money and other items, profits from e-stores)
- To engage in resource mobilization through illegal means (e.g., credit card fraud, the use of electronic communication and encryption in drug trafficking, scam emails)
- Online environment offers anonymity for supporters

The most worrisome part of terrorists using the internet has been financing. Through websites, funds for terrorist activities can be raised directly—for example, donations of money and other items, profits from e-stores, and/or engaging internet infrastructure in resource mobilization via illegal means, credit card fraud, using electronic communication and encryption in drug trafficking, and scam emails.

Use of Bitcoin and other cryptocurrency by terrorist organizations and militant groups has also been identified. According to Encila (2019), Hamas, a Palestinian militant fundamentalist organization, had financed its operation using Bitcoin. The recent seizure of cryptocurrency from Al-Qaeda, ISIS, and the al-Qassam Brigades, Hamas's military wing, by the U.S. Department of Justice (DOJ) also demonstrates how terrorist groups adopt the cyber age to finance their activities (U.S. Department of Justice, 2020).

Cyberterrorism Case Studies

The number of threats posed by cyber incidents and the growing vulnerability of modern information societies to such offenses is alarming.

In recent past years, international communities have seen a new trend of politically motivated cyberattacks. In turn, this has underlined the need to tackle cybercrime in national legislation. Moreover, these incidents have raised the issue of the necessity of a comprehensive approach to cybersecurity in general.

The following lesson introduces four major cyberattacks: Estonia (2007), Lithuania (2008), Georgia (2008), and Sony (2014). Each describes the background, targets, means, and effects of the attacks.

Case Study 1: The Cyberattacks against Estonia in 2007

Estonia is a small but highly IT-developed country in northern Europe. The state offers the citizens wide availability of more than one thousand different electronic services, and the internet is actively accessible and utilized both in public and private spheres. Signing documents by electronic signature and voting in online parliamentary elections is the norm for Estonians, and 95% of all banking operations are strictly performed electronically. These are among a few examples that illustrate the extensive use of and at the same time dependence on information and communication technologies (ICT) (Davis, 2017).

Political tensions were created because of the relocation of the Soviet Bronze Soldier of Tallinn in 2007. This tension resulted in riots in the streets that soon turned into riots in cyberspace.

Estonia 2007: Targets

- Public institutions: President, government, parliament, ministries, state audit office websites, etc.
- Public and private e-service: Banks and media, accessible to public information
- Network infrastructure: Internet service providers (ISPs)
- Other targets: Political parties, etc.

The attacks mainly targeted information distribution channels of both public and private sectors. The websites of public institutions, such as the president, government, parliament, ministries, and state audit office, as well as public and private e-services like banks and media were under assault. Additionally, ISPs and the websites of some of the political parties were attacked.

An emotionally charged phase hit off the cyberattacks, characterized by relatively simple and ad hoc coordinated attacks. This included openly distributed simple instructions in various internet forums as well as the launch of DoS attacks and website defacements.

The next phase of the attacks was much more sophisticated, including the use of large botnets and professional coordination. The effects of the DDoS attacks were more severely noticed by users outside Estonia, as many foreign connections were cut off in order to cope with the excessive traffic. At the most active moment, 58 websites were shut down at once. Additionally, heightened use of mass unsolicited email was noted against government email servers and individual email accounts (Davis, 2017).

Estonia 2007: Cyber War I

- Argued to be the first cyberattack "against a nation state"
- Set the standard of talking publicly about cyber incidents and raised the issue of "cyberterrorism" as a part of the wider spectrum of cyber incidents
- Proved that a country can react ad-hoc, based on a regulatory framework and practice
- Demonstrated the importance of international cooperation

The origin of the attacks remains unclear.

International attention was quickly gained by the Estonian cyberattacks and started an open debate on the spectrum of cyber incidents, national responses, relevant legal framework, and the growing vulnerability of modern information societies.

Once again, the attacks proved that cyberthreats are a cross-border issue and concern various jurisdictions. The attacks showed that a successful investigation requires international cooperation. At this time, the origin of the attacks remains unclear as the involvement of computers from 178 countries was reported.

Case Study 2: Cyberattacks Against Lithuania in 2008

Lithuania, a country situated in eastern Europe, is a modern information society with ICT deployment statistics steadily growing. In 2007, the country had 48% of the population using the internet and various online public services. A comprehensive IT regulatory framework has been in place since 2002 (McLaughlin, 2008).

Image 15.5

In 2008, numerous cyberattacks were faced by Lithuania after the parliament signed into law a bill to ban public use of Soviet and communist symbols and adopted an amendment concerning the freedom of speech and freedom of assembly (McLaughlin, 2008).

Lithuania 2008: Targets

- Defaced the websites of about 300 public and private institutions (Chief Institutional Ethics Commission, Lithuanian Social Democratic Party, etc.)
- Email spam
- Exploited the vulnerabilities of an ISP

Defacement of websites was the main reported attack. More than 300 websites had original content replaced with communist images related to the Soviet Union. Both public and private institutions suffered website defacement, including the Chief Institutional Ethics Commission and the Lithuanian Social Democratic Party. The attackers also used internet forums and blasted email spam in efforts to highlight their manifesto and exploit the vulnerabilities of one ISP (McLaughlin, 2008).

The Lithuania case is significant because the public institutions were warned about the possible attacks. Despite warning the institutions, they did not manage to inform the private sector of the threat in time. The attacks also indicated the connection between political decisions and cyberattacks.

Defacement is considered to be a criminal offense under the Lithuanian penal code. However, since the origin of the attacks was unclear, the investigation led to a dead end.

Moreover, the attacks raised the issue of service-level agreements for e-services. Therefore, governments in countries that have a high degree of cyber vulnerability must consider additional guarantees for the smooth operation of their information infrastructure.

Case Study 3: Cyberattacks Against Georgia in 2008

Georgia is situated in the Caucasus region south of Russia. Less than 10% of the population are active internet users and Georgia lacks nationwide information society services. The country has a relatively weak IT regulatory framework. Additionally, Georgia's information infrastructure relies heavily on its neighboring countries. Notwithstanding the low dependence on ICT, the country encountered cyberattacks in 2008. Interestingly, the cyberattacks fell within the context of a broader armed conflict between the Russian federation and Georgia (Markoff, 2008).

The targets of the cyberattacks are similar to the previous two cases. The websites of public institutions, government sites, as well as newspapers, media agencies, and banks were under attack.

Georgia 2008: Targets

- Public institutions and government sites: Georgian president's website, website of the Ministry of Foreign Affairs, the parliament, etc.
- Public and private targets: (newspapers, media agencies, banking, etc.)
- Other targets

The main means of attack were defacement and distributed denial-of-service attacks. Several websites were defaced with pictures of twentieth-century dictators such as Adolf Hitler. During this attack, instructions to launch the attacks, malicious software, and email addresses for spamming were distributed in blogs and forums (Markoff, 2008).

Image 15.7

The Georgian cyberattacks took place at the time of a nationally declared state of war. Because of this, the attacks were not the priority of national security. Nevertheless, the well-coordinated information blockade changed the typical belief that only highly developed IT societies were vulnerable to cyberthreats.

Looking at this through a legal lens, it can be concluded that the case covers the interrelation of three different fields of law: law of armed conflict, criminal law, and the legal relation of ICT infrastructure.

Case Study 4: The Sony Hack Case in 2014

On November 24, 2014, Sony's computer system fell victim to hacking. Skulls appeared on employees' screens along with a threatening message to expose confidential Sony data. An unknown hacker group, Guardians of Peace, claimed to be the masterminds behind the cyberattack. Sony's private company information (including confidential emails, employees' social security information, and bosses' salaries) had been exposed to the public. Sony's new and unreleased movies were also stolen and shared through P2P websites (Nakashima et al., 2014).

Speculation of North Korea's cyberattack was derived from their initial threat in June to Sony for the upcoming release of the comedy movie *The Interview*. This movie featured two journalists who were granted an audience with the North Korean leader, Kim Jong-un. The plot of the movie then focuses on CIA efforts to recruit the journalists to assassinate Kim Jong-un. The consequences from the hack intensified after the Guardians of Peace claimed that they would perform a 9/11-type attack on movie theaters showing the Sony film (Nakashima et al., 2014).

These threats led to cancellations of the film's New York premiere; Sony announced they would not release the film as planned December 25, 2014. Later, Sony allowed a theatrical and online release of the movie at Christmas, which earned $15M and was downloaded more than two million times in its first few days (Nakashima et al., 2014).

- 22 November: Sony computer systems hacked, exposing embarrassing emails and personal details about stars
- 7 December: North Korea denies accusations that it is behind the cyberattack, but praises it as a "righteous deed"
- 16 December: Guardians of Peace hacker group threatens 9/11-type attack on cinemas showing film; New York premiere cancelled
- 17 December: Leading U.S. cinema groups say they will not screen film; Sony cancels Christmas Day release
- 19 December: FBI concludes North Korea orchestrated hack; President Obama calls Sony cancellation "a mistake"
- 20 December: North Korea proposes joint inquiry with United States into hacks, rejected by the United States
- 22 December: North Korea suffers a severe internet outage
- 23 December: Sony confirms a limited Christmas Day release for *The Interview*
- 2 January: The United States imposes sanctions on North Korea in response to the cyberattack

Similarities

Cyberattacks are fast, cheap, and moderately destructive, but no one would plan to fight using only cyber weapons. Cyberattacks are not destructive enough to damage an opponent's will and capacity to resist. Cyberattacks will not defeat a large and powerful opponent. In most scenarios involving state actors, these attacks will not be "cyber only" incidents unless an opponent makes major miscalculations or their tolerance for risk is high, and risk tolerance may be a function of their belief that the action is covert (and thus less likely to trigger a damaging response) (Lewis, 2013, p. 14).

The four cases discussed share a number of similarities.

- Each attack was launched due to political and social motivation.
- The attacks were carried out mostly via the same means: defacement, DDoS attacks, online propaganda, etc.
- The targets remain more or less the same covering both public and private spheres.
- All of the attacks appear to be coordinated and instructed, and the attackers were never identified.

Notably, the responses to the attacks required the cooperation of public and private institutions as well as international organizations.

Strategic Plan Against Cyberterrorism

Recently, the trend of cyberattacks indicate an alarming rise in political motivation, and it is almost impossible to identify the attacker(s). A successful defense would be the sum of political, legal, and technical measures, which can be facilitated through establishing international information exchange and international cybersecurity policies.

Arms Control Policy

In an effort to have the costs outweigh the benefits of cyberterrorism, governments around the world should implement an arms control policy within cyberspace. Though a policy like this can be beneficial, it can often be extremely complicated to establish within the cyber-world.

According to Borghard & Lonergan (2018), who touched upon four main reasons traditional arms control regimes are not appropriately pertinent to the cyber-world, implementing a policy like this into cyberspace can be difficult for several reasons. First, it can be challenging to measure how strong a country is, particularly the strength of its cyber capabilities. Second, due to the rapid development of technological innovations, it can be challenging to predict cyber capabilities, which complicates the process of establishing policies. Third, developing compliance among nations would require them to share private government information, which can exploit vulnerabilities and allow other countries to capitalize on said vulnerabilities. Lastly, it can be difficult to enforce as well as punish the violation of established policies. Even if it may seem to be too challenging to construct a universal legal framework, establishing a cyber-arms control regime can help to enforce, monitor, and mitigate acts of cyberterrorism. This requires international cooperation and constant efforts from government agencies, private sectors, and educational institutions.

As Nye (2015) indicated, implementing a policy like this can help to establish an agreement between nations to diminish criminal acts within the cyber-world. For example, countries could agree not to target or attack certain sectors of civilian infrastructure. According to the UN (as cited in Nye, 2015), each nation's government should accept responsibility when another nation seeks help due to an attack, nations should agree to not interfere with emergency response teams that are assisting other nations who fell victim to a cyberattack, and lastly, nations should increase the amount of transparency in regards to their cyber policies.

Incident Response Plan

According to the FireEye report (2020), the global median dwell time of external and internal cyberattacks is 141 days and 56 days, respectively. In the time between the initial breach and detection, the computer systems and data are compromised or encrypted. The report predicts that the significant threats of ransomware from established cybercrime groups will continue to monetize access to victim environments as a primary source of generating revenue.

System and data backup are crucial; however, it may be a short-term strategy considering extensive cleansing, recovery, and investigation time. Thus, organizations in both the public and private sectors need to rapidly respond to the impact of a cyberattack. Constructing a robust incident response plan is vital to proactively prevent a catastrophic cyberattack.

Witty, Hoeck, and Gregory (2019) suggest that there are three phases for best-practice implementation, which consist of the following: Phase 1) planning for pre-attack; Phase 2) response and recovery for post-attack; and Phase 3) stand down for post-attack. Phase 1 includes planning activities, which means organizations need to understand the characteristics of a cyberattack and its potential organizational impact. Forming organizational cyberattack response teams and establishing strong crisis management and communications processes are essential to boost the organization's capability to control the damage of a potential cyberattack and ensure collaboration throughout all phases of the incident cycle. Phase 2 contains response and recovery activities, which requires the incident response team to advise the crisis management team lead, who then designates an incident commander for the organization's overall response to the incident. Monitoring the timeline of the response and assessing the integrity of applications and data backups for recovery should be followed by a forensics check on all backups. The stand-down phase is the final stage of performing checkups to ensure the safe business and IT operations environment and conduct a review of the response and recovery processes for future improvement.

Implementing the foundation of an effective incident response plan in organizations is one of the most viable action plans to keep the system and data safe, secure, and to remove the potential for future cyber-terrorist attacks.

As you construct a strategic plan, it is important to consider two crucial alternatives:

- There are positive changes in law enforcement agencies such as establishing information reception capabilities and improving information sharing across all agencies; however, there is still room for improvement. This requires international cooperation between government agencies.
- Even if it is difficult to focus on a specific target due to complexities of interactions in cyberspace, law enforcement agencies must expand their technological capability to monitor targeted terrorist organizations (Yoon et al., 2012).

International Cooperation

The Council of Europe has made it compulsory to:

- Establish measures for swift and smooth international cooperation on criminal cases for effective punishment regarding cyberterrorism
- Generate international and domestic support for the investigation, prosecution and detection of cyberterrorist acts
- Promote fair international cooperation

The United Nations Convention Against Transnational Organized Crime serves the purpose of regulating serious crimes, such as hacking, DDoS attacks, and use of threat and extortion for economic profit. At least 147 countries have signed and completed the ratification process. In a case of cyberterrorism, a close cooperation system is necessary among the involved nations to track down the suspect of hacking or DDoS crimes. The perpetrator of a hacking or DDoS attack is likely to be an intelligent criminal. Such a suspect may have the ability to delete all traces of involvement, and there is also the possibility that various system access logs and service user logs can be deleted after a certain duration of time. Thus, immediate tracking of the suspect is crucial (Yoon et al., 2012).

The G8 High-Tech Crime 24/7 Network is a multinational network of law enforcement agencies from more than 70 countries, established for the purpose of persistent cooperation for investigation of transnational cybercrimes. The formal system of international cooperation through official diplomatic channels takes too much time and is inappropriate for investigations in transnational cybercrimes. It is important to recognize this system adopts a cooperation process that uses informal contacts (phone, email, fax) to request judicial assistance first and then go through the official channels later, which gives it the unique advantage of swiftness. For example, the CLOUD Act, signed into law in March 2018, promotes the possibility of agreements between the U.S. and EU governments on law enforcement access to digital evidence. In addition, the EU's proposed and discussed "e-evidence" regulation would streamline law enforcement access to data among its 28 member states. These two actions will greatly change the way the 24/7 Network operates.

The common element between the perpetrators and the protectors is that they use the same tools. No matter where the perpetrators are, the tools that they use to commit cyberterrorism are the same tools that security experts employ and software developers generate.

In light of this fact, international cooperation on sharing known flaws and exploits in hardware and software with the appropriate officials and security experts has become one of the most essential safeguards against cyberterrorism. Security experts and IT developers study the means to hack their own computer systems and make efforts to protect their systems before any cyberterrorists or cybercriminals are able to exploit them.

Utilizing available venues where IT developers, security experts, and government officials can convene and share flaws and exploits, and prevent their use, would be a very important step against cyberterrorism.

Law Enforcement Training

Crimes, specifically cyberterrorism, that occur in cyberspace differ from traditional crimes. Beyond the absence of physical evidence and eyewitnesses, the concept of a crime scene can be largely ambiguous for cyberterrorism events. Similar to cybercrime, cyberterrorism creates a "fractured crime scene." Meaning the attacker may be physically located in one country, route the attack through a server in a different country, and finally produce harm in a target country. In order to effectively combat cyberterrorism, a fresh response from law enforcement officers trained in both terrorist behavior and information technologies is required; however, this type of training is quite uncommon in the United States.

It has been suggested that the federal government should provide support to local departments. Primarily, most departments are in need of funding for cyberterrorism initiatives, specialized training programs, and equipment. Supplying local law enforcement with the financial support they need, developing training on effective counter-cyberterrorism procedures, and providing the technology to detect and investigate cyberattacks would present invaluable resources for a local department's preparedness.

Additionally, the federal government could facilitate investigating cyberterrorism by promoting alliance between local law enforcement and internet service providers (ISPs). Permitting ISPs to use the latest technologics to record connection histories would provide vital information to law enforcement investigations of cyberterrorist incidents. Encouraging ISPs to share this information with local departments—specifically following such an attack—represents an important opportunity available to the federal government in combating cyberterrorism. The Cyber Security Enhancement Act of 2002, a subsection of the Homeland Security Act, offers legal protection for ISPs that assist and cooperate with investigations.

Federal law enforcement also needs to collaborate with state and local law enforcement in the investigation of cyberterrorism and apprehension of cyberterrorists. Cybercrime training programs are designed to instruct law enforcement officers on how to detect and respond to digital evidence. Primarily, these training initiatives are located among the FBI's 16 Regional Computer Forensic Laboratories (RCFLs) led by the RCFL National Program Office (NPO).

Established in 2002, the NPO manages the creation of a nationwide RCFL program, whose responsibilities include digital evidence examination related to investigations of cyberterrorism, online child pornography, white collar crimes, and a number of other cybercrimes (Regional Computer Forensics Laboratory, 2012). According to the Department of Justice's 2011 Annual Report for the FBI's Regional Computer Forensic Laboratories program, over 32,000 individuals were trained by the RCFL between the years 2004 and 2011, resulting in approximately one officer who has attended RCFL-sponsored cyber training in every three (1:3) U.S. law enforcement agencies.

Another major agency dedicated to help fight cyberterrorism is The National Cyber Forensics and Training Alliance (NCFTA). The NCFTA was created in 1997 to combine expertise, communication, and resources by facilitating the assistance of law enforcement agencies, private industry, and academia. The initiatives enact international cooperation via internship programs to foster a greater knowledge of cybercrime investigation. In 2010, investigators from Ukraine, Lithuania, the Netherlands, Australia, Great Britain, and Germany were invited to a 90-day internship program to share knowledge and collaborate with investigations. To combat the fundamental global source of cyberterrorism, training on an international framework is necessary. The NCFTA has been addressed by Barak Obama's administration as an "effective model" in national cybersecurity. The cooperation with private enterprises should not be overlooked; it can offer not only the best subject-matter expertise but also some of the most significant intelligence.

Global Cooperation in Law Enforcement

The world believes that U.S. federal law enforcement should strive to organize international alliances to combat cyberterrorism. One such organization is the United Nations International Multilateral Partnership Against Cyber Threats (IMPACT). Since 2011, IMPACT has acted as the cybersecurity executing arm of the International Telecommunication Union (ITU). While IMPACT is not aligned specifically to fight cyberterrorism, the membership of 193 countries provides a network of international agencies, corporations, academics, and field experts that serves to assist in recognizing and responding to cyberthreats. The information shared among these reliable global resources can assist in increasing the identification of cyberterrorist events and provide "best practices" for such instances. Furthermore, the FBI reports partnerships with various nation-states to provide international cooperation in the investigation of cyberattacks; in the event of cyberterrorism, this resource could easily be utilized.

Prevention Efforts from State/Local and Federal Law Enforcement Agencies

The response and investigation of cyberterrorist attacks is one aspect of law enforcement's role in combating cyberterrorism. Prevention of cyberterrorist attacks represents a proactive approach. This type of approach can greatly minimize terrorist threats. Increasing the rate of correct identification of cyberterrorists via information sharing and publicizing punishment of cyberterrorist activity may provide some deterrent effect. However, responses that focus on the prevention and early detection of cyberterrorism will potentially have a larger impact on cyberterrorism threats.

Prevention at the Federal Level

Some experts argue that the focused preventive role of the federal government should be increasing international communication on cyberterrorism. IMPACT, the multinational organization mentioned previously, is an UN-sponsored organization working with national governments, private corporations, and field experts to assist these entities with sharing information. They intend to lead the development of strong cyber protection on a global scale. The U.S. federal government representatives of this coalition can transfer the lessons learned from other countries to agencies and corporations within its borders. Likewise, threats encountered within the United States can serve to educate the global IMPACT partners and promote greater global cybersecurity, ultimately impeding the cyberterrorism agenda.

Prevention at the State/Local Level

The greatest effect of minimizing the cyberterrorism threat at the state and local level would be active community engagement. It is critical for all community members to be aware of potential cyberterrorist activity on the internet. Local and state law enforcement should inform their citizens to be alert and watch for suspicious online activity and then to report such behavior to a dedicated unit. Ideally, law enforcement would have educated a community network on how to recognize potential cyberterrorist groups and how to bring them to law enforcement's attention. This preventative step would increase the reporting concerning potential terrorist activities online.

Educating the Public and Corporate Entities on the Cyberterrorism Threat

To minimize the potential risks, educating the general public can be an effective strategy. Another aspect of civilian education is providing a general overview of cybersecurity. By informing the public on the importance of securing home wireless networks via simply employing strong password protection for internet access, the potential to minimize the risk of cyberterrorist attacks is great. Furthermore, the national government should mandate ISPs to require passwords for every wireless router created and encourage them to disseminate the most recent technologies in cybersecurity.

As discussed, cyberterrorism uses social media to target civilians who have access to the internet; therefore, it is imperative that citizens are trained to recognize potential risks and know the best way to report and detect potential cyberterrorism. While public awareness programs and campaigns for citizens are important and require effective tools to recognize the different types, characteristics, and issues within cyberterrorism, major online platforms, such as Facebook, Twitter, and YouTube, must consider strengthening their ability to identify and report posts and videos with cyberterror-related content.

Historically, government agencies have encountered a problem of partnering on cyberterrorism initiatives with private companies due to their fear of sharing valuable information with competitors, the negative press associated with cyberattacks, and the discovery of vulnerable infrastructure. The suggested collaboration represents a vital endeavor since private companies run nearly the entire U.S. critical infrastructure.

Corporate entities should be required to adopt the newest authentication technologies to strengthen control over their own networks. Corporations need to systematically locate vulnerabilities within their business networks and find ways to strengthen their online presence against the threat of attacks. InfraGard provides such a service to over 50,000 private and public entities within the United States. This FBI program offers private corporations and individuals information about the techniques of cyberterrorist activity and advice on how to reduce their victimization risk. Through local chapters led by an FBI Special Agent Coordinator, interested individuals in both private and public sectors can partner with fellow community members on how to react in cyberterrorist attack events within their community. This collaboration serves to reduce the threat of cyberterrorism while also increasing information sharing on cyberterrorist events and minimizing the harm caused by cyberterrorists.

Ideally, corporations should be proactive by building a safe network with security being the principal concern from the start. Additionally, corporations should take ownership over training their employees on the threat of cyberterrorism by teaching best practices and mandating appropriate online behaviors at their workspace. Furthermore, corporations should report cyberterrorism events and seek prosecution.

Summary

Although terrorists have yet to take responsibility for any internet attacks that have caused great damage, they are increasingly using the internet for propaganda, financing, communication, data mining, planning, and support to fulfill their purposes.

With cyberterrorism becoming a progressively virtual movement using cyberspace as a strategic battle ground, awareness of threats posed by the use of the internet for terrorist purposes needs to be raised since today's global societies are highly dependent on the internet and various electronic services. Special attention should be paid to the possible attacks against critical infrastructure as they can have the most alarming outcomes.

The case studies within this unit illustrate the shocking rise in politically motivated cyber-attacks involving cross-border investigations. The global community should be more alert to cybersecurity policies and prepare their national legislation for politically motivated cyberattacks in order to effectively oversee offenses. Defense of such attacks demands national efforts from both the private and public sectors and international cooperation with political, legal, and technical measures.

Lab 15: Defacement

In early January 2021, Mr. CIC, a special agent in the FBI, learned that a man known as Mr. Evil Noodle, who is allegedly an X county computer programmer and part of a state-sponsored hacking organization, is responsible for several of the most expensive computer intrusions of all time. This includes intrusions such as the cyberattack on American Pictures Entertainment, a series of attacks targeting banks across the world that collectively attempted to steal more than one billion dollars, and ransomware attacks that affected tens of thousands of computer systems around the world.

According to allegations, Mr. Evil Noodle was an active participant in a wide-ranging criminal conspiracy undertaken by a group of hackers who were employed at a company operated by the X government. The front company—X Joint Venture—was affiliated with Lab 666, one of X government's hacking organizations. The conspiracy involved this hacking group, which various cybersecurity researchers have labeled as the Lizard Group. Recently, a federal arrest warrant was issued for Mr. Evil Noodle in the United States after he was charged with one count of conspiracy to commit wire fraud and one count of conspiracy to commit computer-related fraud and computer intrusion.

It was reported that Mr. Evil Noodle and his colleagues wanted to express their anti-American political agenda by defacing numerous American government and corporate websites.

Zone-h.com is a useful resource that can be utilized to track defacement incidents posted by cyber groups and/or individuals. Please note that users may find some content offensive.

Please answer the following questions (2–3 pages):

1. For the purposes of this exercise, how can Mr. Evil Noodle deface the government and corporate websites?
2. What potential vulnerabilities were present that increased these websites' risk of becoming victimized?
3. If Mr. CIC and Mr. Hero were to investigate this case, how would they go about doing so?
4. Prevention: How would one stop Mr. Evil Noodle from committing potential defacement attacks?
5. Discussion: What are your thoughts on this type of case? What would your reaction be as a victim?

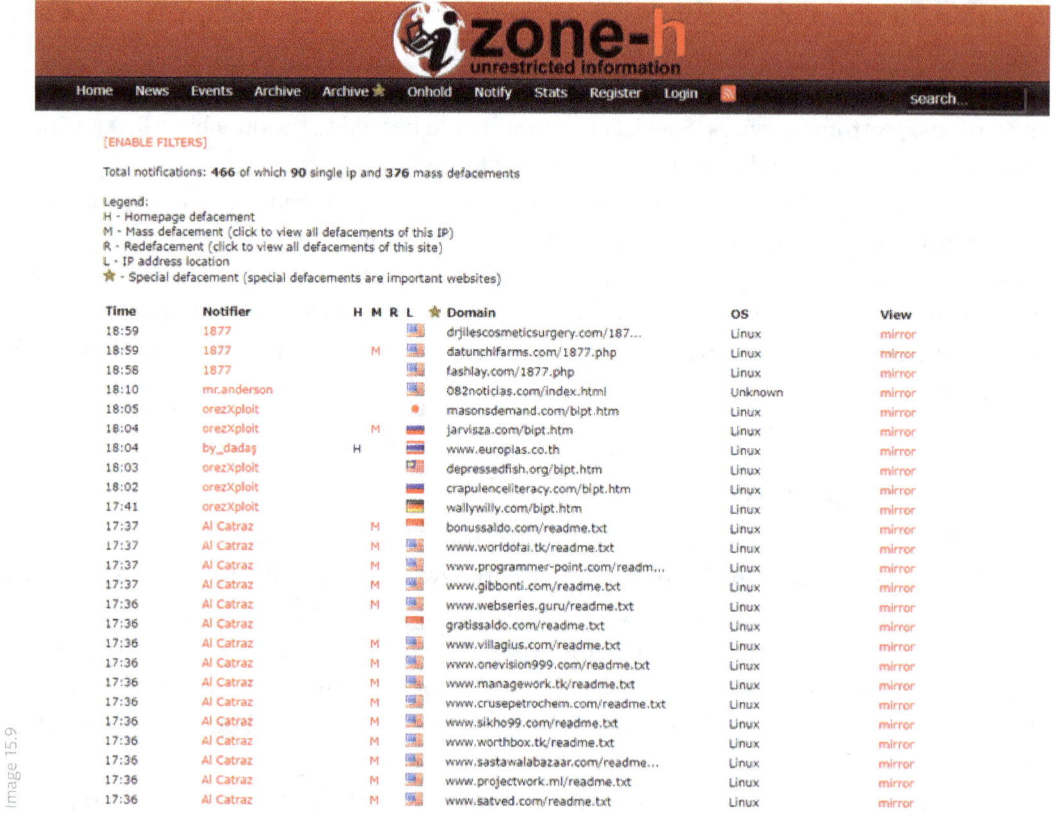

Image 15.9

Check http://www.zone-h.org/ and search recent defacement activities against government and corporate websites.

References

Ballard, J. D., Hornik, J. G., & McKenzie, D. (2002). Technological facilitation of terrorism: Definitional, legal, and policy issues. *American Behavioral Scientist*, 45(6), 989–1016.

Borghard, E. D., & Lonergan, S. W. (2018, January 16). *Why are there no cyber arms control agreements? Council on Foreign Relations*.

Brunst, P. W. (2010). Terrorism and the internet: New threats posed by cyberterrorism and terrorist use of the internet. In *A War on Terror?* (pp. 51–78). Springer, New York, NY.

Choi, K. S., Lee, C. S., & Cadigan, R. (2018). Spreading propaganda in cyberspace: Comparing cyber-resource usage of Al Qaeda and ISIS. *International Journal of Cybersecurity Intelligence & Cybercrime*, 1(1), 21–39.

Coogan, T. P. (2002). *The IRA*. St. Martin's Press.

Davis, J. (2017, June 05). Hackers take down the most wired country in Europe. *Wired*. https://www.wired.com/2007/08/ff-estonia/

Encila, J. (Aug 19, 2019). Terror groups use Bitcoin, instead of traditional cash, to fund their operations. *Business Times*. https://en.businesstimes.cn/articles/117276/20190819/hamas-terror-groups-Bitcoin-crypto.htmFBI_RCFL 2003-2012, *FBI Regional Computer Forensic Laboratory Annual Reports 2003–2012*, FBI, Quantico.

FireEye. (2020, February). *FireEye mandiant m-trends 2020 report reveals cyber criminals are increasingly turning to ransomware as a secondary source of income*. https://investors.fireeye.com/newsreleases/news-release-details/fireeye-mandiant-m-trends-2020-report-reveals-cyber-criminals

Hern, A. (2017, December 30). WannaCry, Petya, NotPetya: How ransomware hit the big time in 2017. *The Guardian*. https://www.theguardian.com/technology/2017/dec/30/wannacry-petya-notpetya-ransomware

Johnson, T. (July 20, 2016). Computer hack helped feed an Islamic State death list. *McClatchy DC Bureau*. http://www.mcclatchydc.com/news/nation-world/national/article 90782637.html.

Kharraz, A., Robertson, W., Balzarotti, D., Bilge, L., & Kirda, E. (2015, July). Cutting the Gordian knot: A look under the hood of ransomware attacks. In *International Conference on Detection of Intrusions and Malware, and Vulnerability Assessment* (pp. 3–24). Springer Berlin Heidelberg.

Lewis, J. A. (2013). *Conflict and negotiation in cyberspace*. Center for Strategic and International Studies.

Marighella. (1969). *The minimanual of the urban guerilla*. https://www.marxists.org/archive/marighella-carlos/1969/06/minimanual-urban-guerrilla/.

Markoff, J. (2008, August 12). Before the gunfire, cyberattacks. *New York Times*. https://www.nytimes.com/2008/08/13/technology/13cyber.html?_r=0

McLaughlin, D. (2008, July 02). Lithuania accuses Russian hackers of cyber assault after collapse of over 300 websites. *The Irish Times*. https://www.irishtimes.com/news/lithuania-accuses-russian-hackers-of-cyber-assault-after-collapse-of-over-300-websites-1.942155

Moore, R. (2011). *Cybercrime: Investigating high-technology computer crime*. Anderson.

Nakashima, E., Timberg, C., & Peterson, A. (2014, December 03). Sony Pictures hack appears to be linked to North Korea, investigators say. *The Washington Post*. https://www.washingtonpost.com/world/national-security/hack-at-sony-pictures-appears-linked-to-north-korea/2014/12/03/6c3c7e3e-7b25-11e4-b821-503cc7efed9e_story.html

Nye, J. S. (October 1, 2015). The world needs an arms-control treaty for cybersecurity. *Harvard Kennedy School Belfer Center*. https://www.belfercenter.org/publication/world-needs-arms-control-treaty-cybersecurity

Sanger, D., & Perlroth, N. (2015, November 17). Encrypted messaging apps face new scrutiny over possible role in Paris attacks. *The New York Times*. https://www.nytimes.com/2015/11/17/world/europe/encrypted-messaging-apps-face-new-scrutiny-over-possible-role-in-paris-attacks.html

Sanger, D., & Perlroth, N. (2020, December 17). More hacking attacks found as officials warn of 'grave risk' to U.S. government. *The New York Times* https://www.nytimes.com/2020/12/17/us/politics/russia-cyber-hack-trump.html

Schmid, A. P., & Jongman, A. J. (2005). *Political terrorism: A new guide to actors, authors, concepts, data bases, theories, & literature*. Transaction Publishers.

United Nations Office on Drugs and Crime. (2012). *The use of the internet for terrorist purposes*. United Nations Office on Drugs and Crime in collaboration with the United Nations Counter-Terrorism Implementation Task Force.

U.S. Department of Justice. (September, 2018). North Korean regime–backed programmer charged with conspiracy to conduct multiple cyber-attacks and intrusions. https://www.justice.gov/opa/pr/north-korean-regime-backed-programmer-charged-conspiracy-conduct-multiple-cyber-attacks-and

U.S. Department of Justice. (August, 2020). Global disruption of three terror finance cyber-enabled campaigns: Largest ever seizure of terrorist organizations' cryptocurrency accounts. https://www.justice.gov/opa/pr/global-disruption-three-terror-finance-cyber-enabled-campaigns

United States of America v. Ardit Ferizi a/k/a Th3Dir3ctorY (2015, October 6). No. 1: 15-MJ-515. United States District Court for the Eastern District of Virginia.

Witty, R., Hoeck, M., and Gregory, D. (2019, August 14). *How to prepare for and respond to business disruptions after aggressive cyberattacks*. Gartner. https://www.gartner.com/en/documents/3956262/how-to-prepare-for-and-respond-to-business-disruptions-a

Yoon, H. S., Yun, M., Freilich, J., Chermak, S., & Morris, R. G. (2012). Cyberterrorism: Trends and responses. Korean Institute of Criminology, 3–303.

Credits

Printed in the USA
CPSIA information can be obtained
at www.ICGtesting.com
LVHW081735200724
786001LV00008B/47

9 781516 536368